Modeling Physiology of Crop Development, Growth and Yield

Modeling Physiology of Crop Development, Growth and Yield

Afshin Soltani

Gorgan University of Agricultural Sciences and Natural Resources, Gorgan, Iran

and

Thomas R. Sinclair

Crop Science Department, North Carolina State University, North Carolina, USA

Models/programs described in this book can be requested from the authors or can be downloaded from https://sites.google.com/site/CropModeling. The authors will update this website with extra exercises, revised crop parameters estimates, specific crop models derived from general models presented in the book and any other helpful information.

www.cabi.org

CABI is a trading name of CAB International

CABI
Nosworthy Way
Wallingford
Oxfordshire OX10 8DE
UK

Tel: +44 (0)1491 832111
Fax: +44 (0)1491 833508
E-mail: cabi@cabi.org
Website: www.cabi.org

CABI
875 Massachusetts Avenue
7th Floor
Cambridge, MA 02139
USA

Tel: +1 617 395 4056
Fax: +1 617 354 6875
E-mail: cabi-nao@cabi.org

A catalogue record for this book is available from the British Library, London, UK.

Library of Congress Cataloging-in-Publication Data

Soltani, Afshin.
Modeling physiology of crop development, growth and yield / Afshin Soltani & Thomas R. Sinclair.
p. cm.
Includes bibliographical references and index.
ISBN 978-1-84593-970-0 (alk. paper)
1. Crops--Physiology--Mathematical models. 2. Crops--Development.--Mathematical models. 3. Crops--Growth--Mathematical models. 4. Crop yields--Mathematical models. I. Sinclair, Thomas R., 1944- II. Title.

SB112.5.S63 2012
631--dc23

2011037995

ISBN-13: 978 1 84593 970 0

Commissioning editor: Rachel Cutts
Editorial assistant: Alexandra Lainsbury
Production editor: Fiona Chippendale

Typeset by SPi Publisher Services, Pondicherry, India.
Printed and bound by CPI Group (UK) Ltd, Croydon, CR0 4YY.

Contents

Preface

Intuition has always been a major "tool" for physiologists, breeders, and agronomists in proposing traits or management practices to increase crop performance and yield. Often the intuitive ideas that triggered productive experiments led to important advances for increasing crop yield. However, many of the obvious alterations of plant traits and management regimes have been exploited. For example, the next generation of crop improvements will require modifications of complex interactions in the plant or the cropping system. The outcomes of such modifications over a range of environmental conditions are less clear since many of these modifications have both a positive and negative impact on the plant and crop system.

The intuitive "model" of how a crop works in the head of the experimentalist used to gain past advances in crop performance will not be adequate to grasp the full complexity of the cropping system. Future advances by physiologists, breeders, and agronomists will require a new tool to consider the many interactions between plants and the environment, and the interactions within the plant. The hypotheses of how crops develop, grow, and form yield can be incorporated into quantitative simulation models to examine a range of ideas in many environments. The new tools of simulation models provide quantitative output about the probabilities of yield gain across growing season, and the magnitude of the yield changes. The intuitive ideas of the experimentalist can be evaluated within the quantitative format of a model before initiating a large experimental effort to understand the complex response to a trait or management modification.

A model is an expression of hypotheses about crop performance within a quantitative framework. That is, hypotheses in the model demand that a hypothetical behavior be expressed as a mathematical function. These functions describe the expected response in a process based on specific variables defined in a mathematical function. The necessity of expressing hypotheses as mathematical functions requires a high level of experimental study and rigorous logic. The effort to gain the insight and understanding to express quantitative functions is in itself one of the most important consequences in efforts to model a system. That is, the demands of developing and exploring quantitative hypotheses leading to a greater knowledge of system components, i.e. heuristic benefits, may well be the greatest benefit of the modeling effort.

Since the model is a collection of hypotheses, any individual model is a simplified expression of reality. The complexity of the biological–environmental world is so great that the full detail of reality can never be included in a model. Hence, one of the largest responsibilities in constructing a model is to understand explicitly the assumptions and approximations that are being invoked in the representations of reality. That is, under what conditions will the representations of reality in the model be satisfactory and unsatisfactory for the objectives for which the model was constructed. Defining these conditions is crucial because no model can be satisfactory for all conditions, and in fact, individual crop models will be appropriate only for a limited range of circumstances and objectives.

It is essential to remember that the study of hypotheses in a model is not a substitute for experimental laboratory, greenhouse, and field studies. The ultimate resolution of a problem will be the experimental results. However, study of the problem before initiating the experiments can focus the key features of the experiments and the outputs that need to be measured. The models can make the experiments more efficient and effective, and may allow extrapolation of results to different environments and situations.

Since heuristic discovery is a key feature of constructing a model, a high priority should be given to developing models that are transparent. Models should not be applied to new uses unless the assumptions and interactions are readily apparent to the model builder and, importantly, to anyone else who chooses to use the model. The model has to be organized in a logical fashion and the interactions among components are accessible to those using the model. Further, the quantitative hypotheses describing individual processes needs to be readily apparent along with a clear understanding of parameter values used in the functions.

Transparency is enhanced by using the simplest expression of a hypothesis allowed by the objective in constructing the model. In formulating hypotheses, transparency should be a much higher goal than introducing reductionist complexity in the hopes that such detail might somehow improve the model. Remember Albert Einstein said "Things should be made as simple as possible, but not any simpler".

Another important aspect of transparency is that the parameters used to quantify each hypothesis should be directly measurable and represent specific, quantified expressions of a process. Therefore, the parameters in the model should be independent of any particular model exercise. Unfortunately, some models require that coefficients in the model be adjusted so that the overall output of the model matches the observation. This approach, euphemistically identified as "calibration", reduces the modeling to an empirical exercise in which coefficients are adjusted to statistically improve overall output. The father of crop modeling, C.T. de Wit, described this approach as "going in circles chasing your own tail".

In this book, we develop and present quantitative hypotheses for the key processes in crop development, growth, and yield. Our emphasis is on functions that offer comparatively simple hypotheses, yet are appropriate over a range of conditions. However, we attempt to note alternative approaches that have been explored in models.

The book is organized into four parts. The first part (Chapters 1–5) presents fundamental perspectives required in developing crop models. This section also presents examples of how the crop models can be applied to explore important issues in crop improvement and production. The second part (Chapters 6–12) develops expressions for crop development (i.e. phenology), growth, and partitioning into plant components including grains. The functions developed in this section are used to assess the "potential" growth and yield of a crop when it is not limited by availability of water or nutrients, and there is no limitation as a result of insects, diseases, and weeds. The third part (Chapters 13–16) accounts for water limitation of specific processes in crop growth. Finally, the last part (Chapters 17–19) accounts for nitrogen limitation of plant growth processes.

Each chapter ends with exercise(s) for the reader to better understand how the functions operate and the implications of adjustments in the parameters of the functions. Three levels of exercises are generally suggested at the end of each chapter. One level is a guided exercise using the submodel/model to explore modifications, usually adjustments in individual parameters. A second level of exercise is a suggestion for further exploration with the submodel/model to examine the consequences of modifications of parameters in the model or changes in the input environmental conditions. A third level of exercises is independent explorations of crop or environmental situations that are particularly relevant to the reader.

We hope this book is informative for all of those who have an interest in quantitative understanding of crop development, growth, and yield, whether or not his/her primary interest is in constructing a model. We believe the challenge of expressing quantitative hypotheses is the important heuristic activity of modeling, whether the model is ever applied to a practical problem. Of course, we feel the application of models to the major questions of crop and management improvement can be a major benefit in using models. Model studies can focus the experimental investigations that will be needed to improve understanding and performance of crop systems.

1 What is a Crop Model?

The essence of modeling is to formulate a "picture" about how a system operates. What is the system? How does it work? How is it used? A model of a bicycle rider pedaling the bicycle to make it move is an example of such a simple model. This system moves in response to the energy supplied by the rider to the pedals. The bicycle-rider system offers a simple means of transportation. Such simple models of life experiences are used by everyone every day as we imagine how things operate.

The models of systems can generally be classified into one of four groups (Haefner, 2005):

1. Conceptual or verbal models depend on common language to describe a system and how it works. For example, chapters on photosynthesis or water relations in physiology or ecology textbooks can be considered conceptual models.
2. Diagrammatic models are pictorial representations of a system; they are sets of figures and charts to describe a system and its components and relationships between them. Ecological "box-and-arrow" diagrams of energy flows, water and nutrient cycles and physiological diagrams of metabolic pathways such as the Krebs cycle are examples of these models.
3. Physical models are physical mock-ups of a system. Depending on the system of interest, a physical model may be larger (e.g. an atom or a DNA strand) or smaller (e.g. an airplane or the solar system) than the actual system.
4. Mathematical models are descriptions of systems in mathematical language; the sets of mathematical (usually algebraic or differential) equations that describe a system and how it works.

In this book we are interested in developing mathematical models that represent development, growth, and yield of crops under a range of conditions. These mathematical models take model development to a level of quantifying interactions of the components in the model. That is, in the bicycle-rider system functions are developed to relate the energy supplied by the rider to the energy required to move the bicycle. Of course, a more complete model will include quantification of environmental variables that influence the system. In the bicycle-rider example, it may be necessary to consider such environmental factors as the angle of the ground on which the bicycle is to move, and direction and velocity of the wind. Mathematical functions can be developed for each of these relationships and brought together to allow simulation of the distance and speed of a bicycle ride through the countryside, or in the Tour de France. Since a series of calculations is required, even though the model is relatively simple, a computer is used to make calculations. Hence, in the computer the quantified model allows a simulation of the performance of the bicycle rider in terms of speed and distance.

Like the mechanical system of the bicycle rider, crop systems can be developed to do quantified simulations of crop growth. In this way, the key mechanisms that influence the yield of the crop are identified and estimates of yield under different environment conditions can be simulated. A simple crop model for simulating sugarcane growth is developed below.

Sugarcane Growth Model

Sugarcane is an important crop worldwide for the production of sucrose for direct consumption by humans, and for providing the raw resource in the fermentation for the production of ethanol as a biofuel. Brazil depends on sugarcane ethanol as a major source of automobile fuel. A key question in making decisions about the construction of a sugar mill or a fermentation factory is the yield that can be produced by the sugarcane fields in the vicinity of the proposed new mill or factory. A simple sugarcane growth model can address this question.

Like the bicycle-rider system, the critical currency in the sugarcane system is the energy input to the system and the output, in this case the growth of the crop. The energy input, of course, is the amount of light or more properly solar radiation that is intercepted by the sugarcane leaves. Assuming we are only interested in the period when the leaf canopy is closed and intercepting nearly all the incident solar radiation, the critical input information is the solar radiation (MJ m^{-2} day^{-1}, megajoules per square meter per day) obtained from weather observations. Daily solar radiation is sometimes reported by weather stations, or can be derived by information about the temperature regime at that location.

The next task is to define the relevant timeframe for sugarcane growth. In Florida, for example, seasonal temperatures limit the vast majority of sugarcane growth to the period of 15 May to 15 November. The daily solar radiation can be entered into a Microsoft Excel file to expedite the daily simulations of growth.

A critical variable required by the model is the amount of accumulated crop mass that results from interception of the photosynthetically active radiation (PAR) component of incident solar radiation, which is identified as radiation use efficiency (RUE, g MJ^{-1} PAR). As it turns out, RUE for an individual crop species remains fairly constant under non-stressed conditions (i.e. free from pests, diseases, and weeds, and experiencing non-limiting conditions with respect to temperature, water, and nutrients). Use of a RUE value means that the individual processes of photosynthesis are combined in the RUE term and there is no need for detailed modeling of these complex processes. Table 1.1 presents non-stressed values measured for major crop species. For sugarcane, the constant value of RUE is about 3.6 g crop mass produced for every megajoule of intercepted photosynthetic active radiation (Sinclair and Muchow, 1999).

The sugarcane model operates each day by inputting solar radiation data from an Excel file. Since solar radiation is available as total radiation, the first step in the model on each day is to convert the solar radiation data to a PAR value. This conversion is readily done since PAR is about half of total solar radiation so the input radiation values are multiplied by 0.5. The daily amount of sugarcane growth is then simply calculated by multiplying the PAR value by sugarcane RUE. The daily growth can be outputted to an Excel file. Adding the daily growth of the crop to the existing mass of crop gives estimates of the increase in accumulated crop mass through the growing season. At the end of the growth period, the model has generated an estimate of sugarcane yield for any particular season.

A summary of the sugarcane model can be visualized in a simple pictorial flow diagram (Fig. 1.1), where the flows of energy and mass are illustrated connecting the boxes that define the processes acting upon the energy or mass.

The implementation of this simple sugarcane model is exemplified for sugarcane growth in south Florida in the year 2010. The daily solar radiation (SRAD) data stored in the Excel file for the period from 15 May to 15 November is plotted in Fig. 1.2. These are the basic input data to the model, showing the daily variation in weather conditions with solar radiation trending downward following the summer solstice.

Table 1.1. Radiation use efficiency (g MJ^{-1} PAR) in major crops (Sinclair and Muchow, 1999; Keating *et al.*, 2003).

Crop species	Radiation use efficiency (g MJ^{-1})
Cotton	1.8
Chickpea	1.8
Pea	1.8
Soybean	2.0
Peanut	2.0
Canola	2.0
Sunflower	2.1
Barley	2.1
Wheat	2.2
Rice	2.2
Maize	3.5
Sorghum	3.5
Sugarcane	3.6

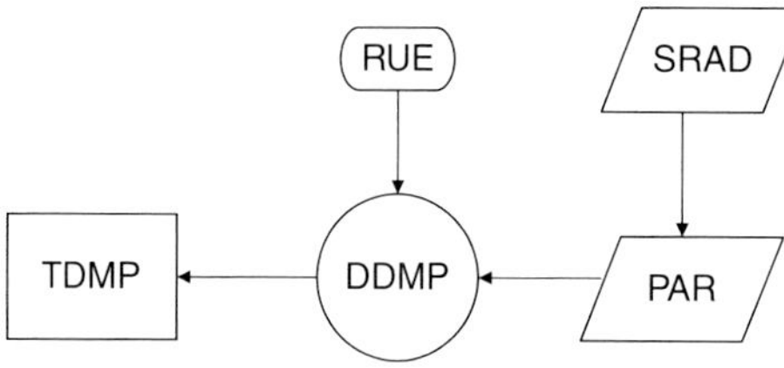

Fig. 1.1. Flow diagram for sugarcane model. SRAD is the daily solar radiation (MJ m^{-2} day^{-1}), PAR the daily photosynthetically active radiation (MJ m^{-2} day^{-1}), RUE the radiation use efficiency (g MJ^{-1}), DDMP the daily dry matter production (g m^{-2} day^{-1}), and TDMP the total cumulative dry matter production (g m^{-2}). Shapes are defined in Table 1.2.

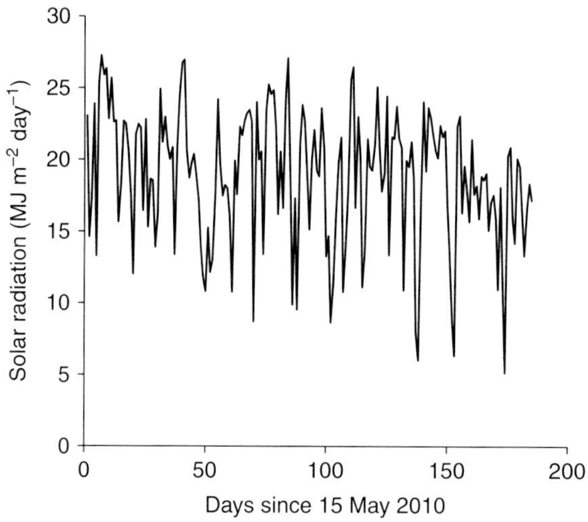

Fig. 1.2. Daily values of solar radiation from 15 May 2010 to 15 November 2010 in Belle Glade, Florida, USA.

The first calculation in the sugarcane model is to estimate PAR by multiplying solar radiation by 0.5.

$$PAR = 0.5 \times SRAD \tag{1.1}$$

To estimate daily growth or dry matter production (DDMP, g m^{-2} day^{-1}) on each day, PAR is multiplied by RUE.

$$DDMP = PAR \times RUE \tag{1.2}$$

The resultant DDMPs for this example are shown in Fig. 1.3. By summing the daily values of dry matter production, total crop cumulative dry matter on each day ($TDMP_i$, g m^{-2}) is calculated (Fig. 1.4):

$$TDMP_i = TDMP_{i-1} + DDMP \tag{1.3}$$

The results in Fig. 1.4 show that the total dry matter production during the major growth period for this sugarcane crop during this specific year is predicted to be

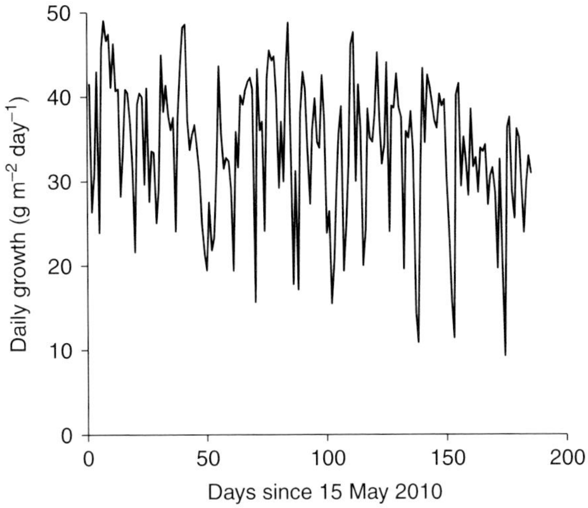

Fig. 1.3. Calculated daily values of dry matter production by a sugarcane crop during full-cover period in 2010 at Belle Glade, Florida, USA.

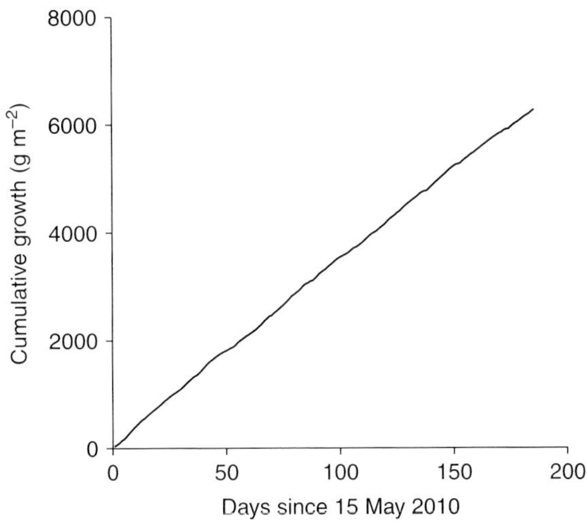

Fig. 1.4. Cumulative dry matter production by a sugarcane crop after 15 May 2010 at Belle Glade, Florida, USA.

6267 g m^{-2}. Since the fraction of the sugar in the final mass of the crop is about 0.5, this model gives an estimated sugar yield under non-stressed conditions of about 3134 g m^{-2} (or about 31 t ha^{-1}). An estimate of the fresh weight of the sugarcane crop is also possible since the fresh crop must be transported to the sugar mill or ethanol factory. Commonly, a fresh sugarcane crop may be about 70% moisture content of harvested plants, so the transportation system will need to be able to accommodate roughly 10,654 g m^{-2} or 106.5 t ha^{-1}.

Terms Used to Describe Models

While the features of the simple sugarcane model are obvious, discussion of more complex models requires a common understanding of some of the terminology in discussing models. Some of these critical terms are described below:

- System: System is any part of the universe in which we are interested. A system consists of components (objects) that have interactions with each other and the environment. A leaf, a plant, a field, the globe itself, an airplane are each systems. Some systems are alive and others are not. The sugarcane crop during its major growth period in Florida was the system of interest in the above example.
- Model: As discussed previously, a model is a description of the system that allows examination of the components of the system. It is important to remember that a model is not reality but a simplified representation of reality. The representation of reality is done using functions as in Eqns 1.1 to 1.3 above.
- Modeling: Modeling is the process of building or developing a model to bring together the functions that describe the components of the resultant model.
- Simulation: Simulation is the act of executing a model and obtaining output about the variables of interest. Simulation is application of the model to the topic of interest.
- Systems analysis: System analysis is the act of taking the results of the simulation and synthesizing the conclusions that can be drawn from the simulation results.

Variables in Models

In crop models, a variable is a symbolic name given to some known or unknown quantity or value, for the purpose of allowing the name to be used independently of the value it represents. In the above example, the variables included SRAD, PAR, RUE, and DDMP. Again, to clarify discussions about models, variables used in modeling can be classified into four groups (Table 1.2; Goudriaan and van Laar, 1994):

1. *State variables*: State variables indicate the current status of the system. For example, days since 15 May and cumulative dry matter production in the sugarcane model are state variables.
2. *Rate variables*: These variables indicate the rate or speed of change in a state variable. Daily rate of dry matter production (DDMP) in the sugarcane model is a rate variable.
3. *Parameters*: These variables are characteristics of the system and their values usually remain constant during a simulation. RUE in the sugarcane example is a parameter because it remained constant throughout the simulation.
4. *Driving variables*: These variables are commonly environmental variables that affect the rate variables. Daily solar radiation in the sugarcane example is a driving variable.

Table 1.2. Symbols that are used in this book to identify different variables and crop management and soil inputs in relational diagrams.

Variable group	Name	Symbol
Driving variable	Parallelogram	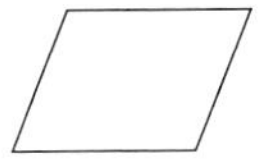
Crop parameter	Rounded rectangle	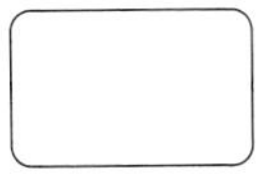
Rate variable	Circle	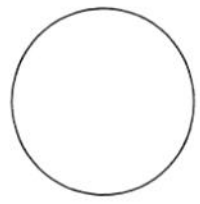
State variable	Rectangle	
Management and soil inputs	Folded corner rectangle	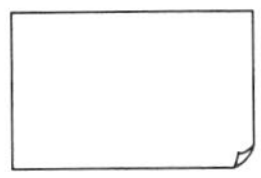

Classification of Mathematical Models

Mathematical models, including cropping systems, can be classified into different categories based on criteria dealing with several aspects of the model (Gershenfeld, 1998; Haefner, 2005):

1. *Representation of processes*: Mathematical models can be divided between the way processes are represented (Monteith, 1996). Process-oriented models are explicit representations of mechanistic processes in the system. Cause and effect are described in equations and these relationships are used to create the model. Empirical models rely on statistical equations that do not represent the mechanistic processes. A regression model that relates crop yield at a given environment to average temperature and total precipitation during the growing season is an empirical model. It should be noted that all process-oriented models become empirical at one or more lower organization levels.
2. *Static versus dynamic*: A static model does not account for the element of time, while a dynamic model does. In other words, in a static model, the development of the future status of the system is not predicted, while a dynamic model has the capability to predict future conditions of the system. Dynamic models typically are represented with difference or differential equation (Haefner, 2005).
3. *Continuous versus discrete*: In continuous models the mathematics represent time continuously, so time can take any value (e.g. 2.3 days), but in discrete models time is an integer only (e.g. 2 days).

4. *Deterministic versus stochastic*: In a deterministic model, every set of variable states is uniquely determined by parameters in the model and by sets of previous states of these variables. Therefore, deterministic models perform the same way for a given set of initial conditions. Conversely, in a stochastic model, randomness is present, and variable states are not described by unique values, but rather by probability distributions. Stochastic models are used in population ecology and weather data generation in meteorology (Haefner, 2005).

Considering the above mentioned classification, the sugarcane model described previously is a deterministic, discrete, dynamic, and process-oriented mathematical model.

Models Are Not Reality!

It is important to always remember that models are not reality. The ultimate arbiters in describing the real world are results from observations or experiments. Models offer a simplified concept of the real world that can offer powerful tools to understand the major factors influencing the cropping system. Models offer a very useful method for interpreting and understanding laboratory and field experiments.

Vast progress in all scientific disciplines since the 16th century has been stimulated by the evolution and testing of hypotheses instead of relying on imagination (Monteith, 1996). As pointed out by Monteith (1996), there is a danger that, because modeling in the form of computer simulation is such a powerful analytical tool, it will weaken the link between hypothesis and real-world (as opposed to imaginary) experiments. Even on a relatively modest microcomputer it is possible to conduct hundreds of imaginary experiments in the course of a morning.

By adding to a model or making changes in its structure or to parameter values, the modeler can often match information from a real experiment (Monteith, 1996; Sinclair and Seligman, 1996). Crop models cannot be built without invoking a set of hypotheses and the simulation results cannot be rigorously tested without measurements that describe the performance of the crop over a wide range of environments. Such information is rarely available. There is therefore a sense in which the information provided by crop models today resembles the speculation that surrounded attempts to describe the natural world 370 years ago (Monteith, 1996).

Exercise

Evaluate the growth potential of various crops in your environments of interest by adapting the sugarcane model to use the appropriate RUE value (Table 1.1) and local solar radiation data.

2 Fundamental Guides in Constructing Crop Models

There are distinct steps required in nearly all efforts in constructing models, including crop models. While the developers of crop models do not always explicitly proceed through each step, it is valuable to be clear about these steps. If each step is not fully developed, the resultant model may be seriously compromised if its use is attempted outside the context in which the developer visualized it. It is likely a more robust model will result from the modeling process that recognizes each of these stages (Sinclair and Seligman, 2000; Haefner, 2005). The linkages among the modeling process stages are illustrated in Fig. 2.1.

- Definition of goals and objectives.
- Preparation of assumptions (hypotheses).
- Mathematical formulation.
- Programming.
- Parameter estimation (parameterization).
- Model evaluation.

In this chapter, each of these stages in model construction is discussed. Further, we will illustrate each stage in model construction by developing a model of leaf production on sugarcane plants through the vegetative development of the crop.

Definition of Objectives

At the beginning of the modeling process, the objectives of the modeling effort should be explicitly and fully defined. A clear statement of specific objectives is essential to define the needs and nature of a crop model (Sinclair and

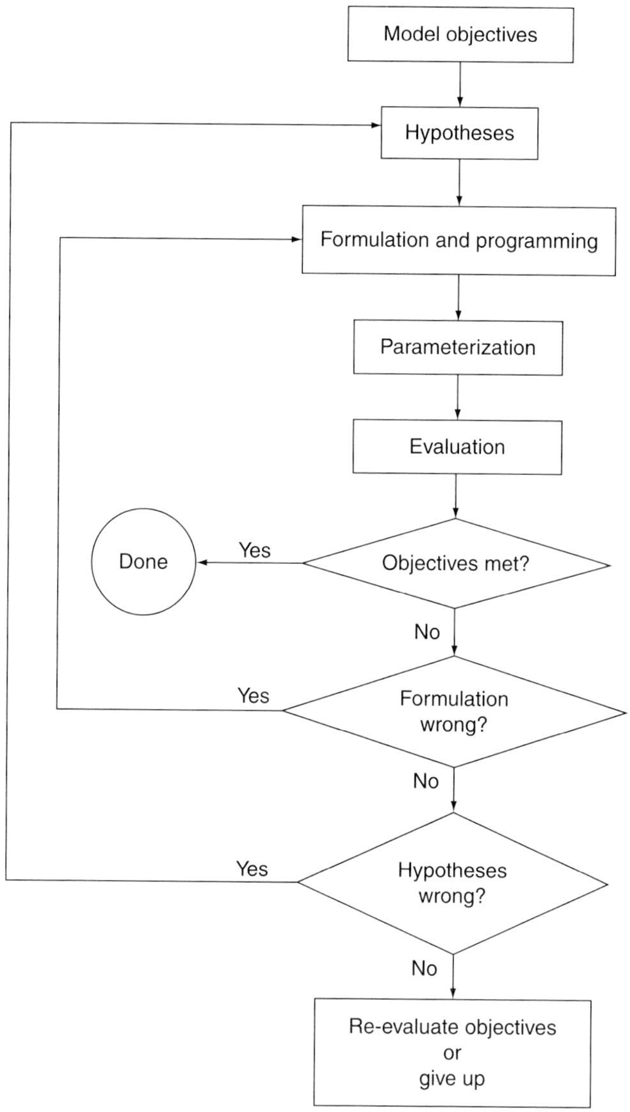

Fig. 2.1. The modeling process and its stages.

Seligman, 1996). To resolve the objective, the following questions need to be addressed (Sinclair and Seligman, 1996; Haefner, 2005):

1. What is the system to be modeled? What specific situation is to be investigated?
2. What are the major questions to be answered by the model? How will the model be used? What problem is to be studied with the model?
3. When does the modeling process end? What is the expected accuracy of the model? How good must the model be? To what will it be compared?
4. How will the model output be analyzed, summarized, and used?

It is more likely that success will be achieved when the objectives are clear, modest, and tractable (Sinclair and Seligman, 1996). Criteria for judging the acceptability of a model should be defined in relation to the model's objectives (Sinclair and Seligman, 2000). It is possible to quantitatively define stopping rules (question 3), in terms of statistical criteria concerning model predictions relative to a sample of observations. For example, a coefficient of variation less than 15% and/or correlation coefficient greater than 0.7 between the predicted and observed values might be acceptable criteria for a model such as a wheat phenology model that is used to predict days to flowering.

In Chapter 1, the potential growth of sugarcane was modeled for crops with full canopies that completely intercepted all incident radiation. Of course, this assumption is not valid when considering the whole growing season because the crop begins with no leaf area and leaves must develop to allow complete radiation interception. Therefore, for a more sophisticated estimate of crop growth, there needs to be a submodel to develop crop leaf area. Therefore, the likely objective for such a submodel is:

To model the development of crop leaf area under non-stressed conditions.

Itemize Critical Assumptions

Based on the objective, a list of specific hypotheses to be included in the model is prepared. Initially, it may be useful to list the hypotheses in the form of words and sentences. Some of the following points need to be considered in identifying hypotheses to be used to construct the model.

1. Models need to have generality. That is, model structure and application should not be limited to specific circumstances or given locations. At the same time, it must be remembered that development of universal crop models is impossible (Sinclair and Seligman, 1996; also refer to Chapter 5).
2. Hypotheses will require input data either as parameters or driving variables. Therefore, hypotheses should be favored where there is observational data to quantify and implement the hypotheses.
3. The most useful models will be those that include parameters and state variables that can be readily determined by simple measurements or observations. Non-measurable, vague parameters that are estimated by "calibration" of the completed model need to be avoided, and used only with great caution.
4. Existing hypotheses/models are a useful resource, but they need to be evaluated in view of the objectives of the current model construction. While efficiency demands that successful approaches should not be ignored, a new objective may require new hypotheses or an old model may need to be extensively modified (Sinclair and Seligman, 1996).

In the example of modeling leaf area development for non-stressed conditions, the following hypotheses are identified:

1. Daily development of leaf nodes on the plant stem progress as a linear function of temperature.

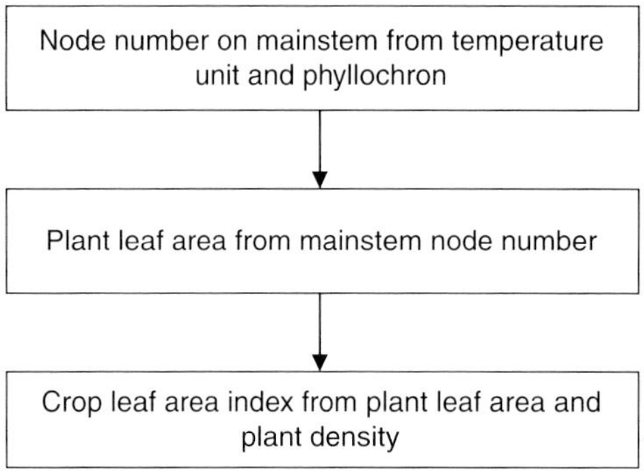

Fig. 2.2. Flow diagram of hypotheses and calculation sequence for sugarcane leaf area development model.

2. Leaf area of individual plants is dependent on the number of leaf nodes, i.e. number of leaves.
3. Crop leaf area is equal to plant leaf area multiplied by plant density.

These hypotheses can then be put into a descriptive flow diagram to illustrate how the hypotheses are linked. The linkage of hypotheses is a critical step. Uncertainty in how to link hypotheses is important in identifying inadequacies in developing the model. Figure 2.2 is a flow diagram illustrating the connections among the above three hypotheses for the leaf area development model.

Quantitative Description of Hypotheses

At this stage, qualitative hypotheses need to be expressed as mathematical functions. Equations need to be developed to express how each hypothesis can be expressed in the model system. The interaction of hypotheses also needs to be taken into account, which often introduces a whole new layer of assumptions. Commonly, it is assumed that there is no interaction among hypotheses other than what has been explicitly defined. In the example of leaf area development, the implicit assumption is that plant density does not influence the relationship between leaf area and leaf node number.

The quantification of the model can be the most challenging because it requires a thorough knowledge of the system being modeled, and an understanding of the most relevant relationships. That is, additional functions can always be added to the model, but do these functions enhance the performance of the model for the stated objective? Often, the most critical phase of the model construction will be selection of the quantitative functions that sufficiently and efficiently describe the model component being modeled. Assembling equations without understanding and evaluating their relevance to the objectives overlooks a critical aspect of modeling.

Construction of the quantitative functions, for the example of leaf area development from the hypotheses, is now presented. To track daily node development of the sugarcane plant, the phyllochron concept is used. A phyllochron defines the interval between the appearance of successive nodes on the plant stem. The phyllochron interval (PHYL) can be expressed as a function of cumulative temperature (°C), which is often referred to as "degree day" or "thermal time". However, the variable of cumulative temperature does not include time and these latter two terms are mathematically confusing due to the implicit inclusion of time. In this book, the term "temperature unit" will be used. For many crops, the value of PHYL is in the range of 40 to 120°C.

Therefore, the first step in calculating leaf node development is to calculate daily temperature unit (DTU). For temperate crop species, the simplest expression of DTU is the average daily temperature, which can be estimated as the average of the minimum and maximum temperature (TMIN and TMAX). Therefore, if the minimum temperature on a day was 20°C and maximum temperature was 30°C, the temperature unit for that day is 25°C.

Many crops with an origin not in temperate regions may develop nodes only when the temperature is above 10°C, i.e. a defined base temperature (TBD). Therefore, this base temperature is incorporated into the calculation of daily temperature unit by subtracting the base temperature. In sugarcane, the value of TBD is approximately 10°C (Sinclair *et al.*, 2004) so the estimate of DTU in the previous paragraph becomes 15°C (25–10 = 15°C). Therefore, the initial equation in the model is the following:

$$\text{DTU} = (\text{TMIN} + \text{TMAX}) / 2 - \text{TBD} \quad (2.1)$$

The progress in daily development of leaf nodes (INODE) can then be calculated from hypothesis (1) based on the calculated daily increase in phyllochron interval. That is, the progress in development of nodes is based on the fraction of the PHYL that is experienced each day by the crop.

$$\text{INODE} = \text{DTU} / \text{PHYL} \quad (2.2)$$

If the sugarcane variety of interest has a PHYL value of 100°C and DTU equals 15°C, the value of INODE for that day calculated from Eqn 2.2 would be 0.15. In this example, if the daily temperature remained constant, 6.7 days would be required to accumulate sufficient temperature units to reach the PHYL value of 100°C. Therefore, in this example of a constant temperature, every 6.7 days a new leaf node would be fully developed.

For accounting purposes, it is necessary to calculate the total number of leaf nodes that are present on the plant (MSNN_i) each day. This is easily achieved by adding INODE to the number of nodes that existed on the plant on the previous day (MSNN_{i-1}).

$$\text{MSNN}_i = \text{MSNN}_{i-1} + \text{INODE} \quad (2.3)$$

Hypothesis 2 can be invoked to calculate plant leaf area (PLA, m^2 per plant). Allometric relationships have been observed in many crops between leaf node number and plant leaf area during the major phase of leaf area development. Such allometric relationships have been reported in many crops (e.g. Sinclair, 1984; Wahbi and Sinclair, 2005; Soltani *et al.*, 2006c; refer to Chapter 9 for details). For sugarcane, the following relationship can be derived from the data presented for the sugarcane cultivar CP88-1762 by Sinclair *et al.* (2004) (Fig. 2.3):

$$\text{PLA} = a \times \exp(b \times \text{MSNN}) \quad (2.4)$$

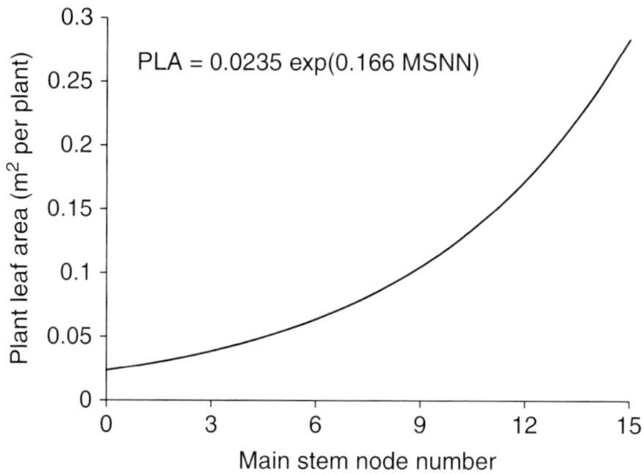

Fig. 2.3. Plant leaf area versus main stem node number in sugarcane described by Eqn 2.4 (T.R. Sinclair, unpublished data). In this example, *a* and *b* are 0.0235 and 0.166, respectively.

The final hypothesis (3) is that crop leaf area index, the ratio of leaf area to ground surface area (LAI, m^2 m^{-2}), can be obtained by multiplying PLA by plant density (PDEN, plants m^{-2}). For sugarcane, PDEN may be roughly 8 plants m^{-2}. Hence,

$$LAI = PLA \times PDEN \tag{2.5}$$

Finally, the qualitative flow diagram developed earlier needs to be redrawn to include the mathematical functions of the model. The development of the mathematical functions likely results in several steps in the calculations that are not fully illustrated in the original flow diagram. Therefore, the flow diagram showing the mathematical functions will likely be expanded over the original conceptual framework showing the basic hypotheses. Figure 2.4 shows the flow diagram for the leaf development model with the mathematical functions.

Programming

Once the hypotheses have been defined and quantified, the model is finally constructed into computer code. That is, the knowledge and insight about the system should have been captured and it should be a straightforward, even a trivial task, to translate the hypotheses into computer code. However, considerable care is required to accurately express the model in computer code. It is necessary to verify that computer algorithms and the codes are correct for the mathematical relationships defined. Preparing computer programs usually requires debugging to eliminate the errors that arise during programming.

The program code can usually be organized in individual sections as represented in the flow diagram. In more extensive models it is often useful to structure the code so that each section is placed into its own submodel, i.e. subroutine. The use of the submodels generally allows other submodels to be added relatively easily to the model if they become necessary.

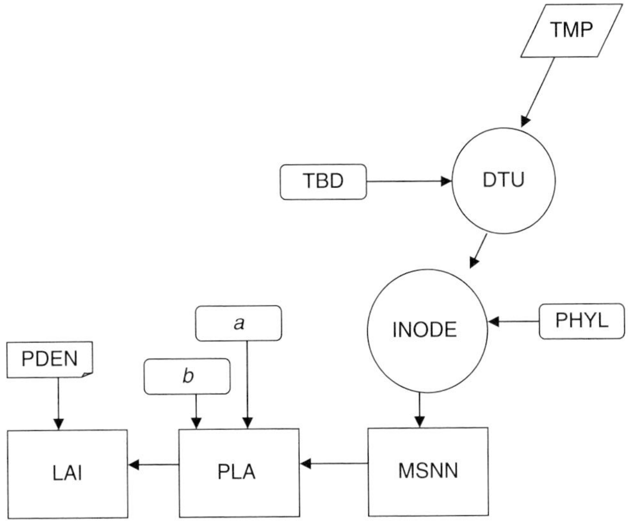

Fig. 2.4. Relational diagram of sugarcane leaf is a development model. TMP is the mean daily temperature (°C), DTU the daily thermal unit (°C), TBD the base temperature for development (°C), INODE the daily increase in main stem node number (day^{-1}), PHYL the phyllochron (°C per leaf/node), MSNN the main stem node number, PLA the plant leaf area (m^2 per plant), LAI the crop leaf area index (m^2 m^{-2}), *a* and *b* the coefficients describing the relationship between PLA and MSNN, and PDEN the plant density (m^{-2}). For more information refer to the text. Shapes are defined in Table 1.2.

Placing parameter values for the various functions in their own separate, initialization section can facilitate the use of the model in simulations using different parameters. That is, code the functions in the model using parameter names, and then define all the parameters at the beginning of the program, for instance. In this way, simulations of other conditions such as a different cultivar are facilitated by adjusting the relevant parameters in the initialization.

There is a choice of a number of computer languages in which to code the model. Computer languages are divided into four general categories (Fig. 2.5; Haefner, 2005):

1. Machine language is a binary code of 0s and 1s that are fed into the central processing unit (CPU) of the computer to compute the programmer's needs. This language is a lower-level language that is hard to learn but its capability is very high. Modern computer languages now make use of the binary code unnecessary.
2. Assembly language is a set of mnemonic instructions in English that code for machine language binary forms. This language manipulates the central processing unit of the computer directly and there is no need to remember sequences of bits because simple English language words are used.
3. General-purpose languages are most commonly used in crop modeling and scientific applications. In this language, direct access of the programmer to

the central processor of the computer is limited and simple English language instructions are used in programming. There are several general purpose languages that offer a great range of facilities for implementing algebraic operation and data manipulation. FORTRAN, BASIC and C are the most common general purpose languages.

4. Simulation languages allow users to work directly from a flow diagram to implement the model. These languages are usually simpler to use, but their capabilities are limited and may restrict the representation of the model hypotheses as desired. Stella and Fortran Simulation Translator (FST; van Ittersum *et al.*, 2003) are examples of these languages. FST is a product of Wageningen crop modeling group.

General-purpose languages are widely used in crop modeling, while the usage of simulation languages has been limited. Figure 2.6 compares a number of general-purpose languages and one simulation language with comments about their power and ease of use.

In this book, Visual Basic for Application (VBA) in MS Excel is used because many students are already familiar with Excel and it is possible to use graphical facilities of Excel without the need for more programming and hence avoiding big, complicated programs. Excel worksheets are used as the interface for input and output operations. Learning Basic is easy and can help in learning other languages.

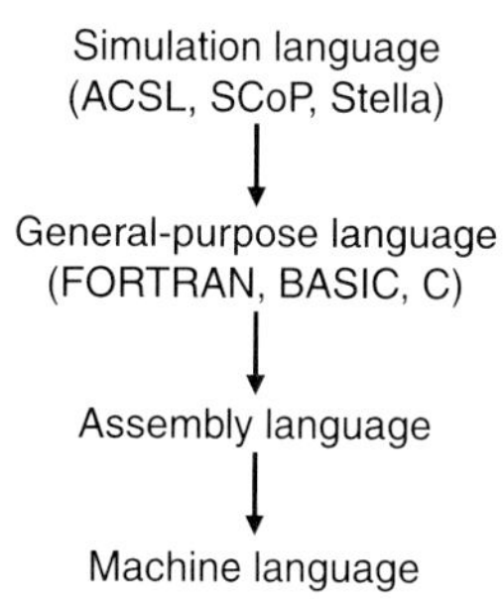

Fig. 2.5. A hierarchy of computer languages.

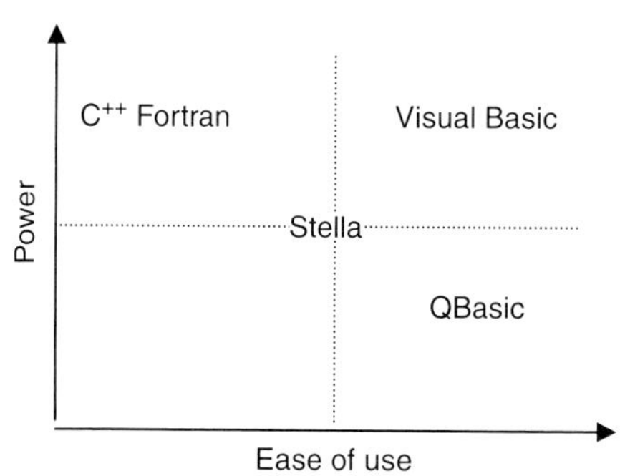

Fig. 2.6. Relationship between the power and ease of use for some computer languages (personal communications).

Estimating Parameters

There are two acceptable general methods for parameter estimation:

1. Literature. If the value of a parameter has been defined in research previously published, then these data are a strong starting point. This source of parameters is desirable because the research has been subjected to the rigors of peer review and the results are available to everyone.
2. Measurement. Experiments or observations need to be done by the person constructing the model or his/her collaborators. Special research protocols need to be established and executed to obtain information about the desired parameters. Careful statistical analysis of the results is needed to obtain parameters. This method is time-consuming, laborious, and expensive. Soltani *et al.* (2004a, 2006a–e) presented examples of parameter estimation procedures related to different aspects of crop growth and development.

Another method that is too frequently used for parameter estimation is "calibration" of parameters so that the final output of the overall model matches expected results. That is, the model is tested using different values for a specific parameter, then values are chosen that provide the closest match to the observations of the major outputs, frequently final yield. Adjustments in parameters to achieve closer matches with observations by the complete model are difficult because it is not clear what parameters need adjusting. A major problem is that parameter adjustments can compensate each other and the parameters diverge from how the plant operates.

The major limitation of the "calibration" approach is that it reduces the model to an exercise in empirically fitting the model to the observations under a particular set of circumstances. Parameters are adjusted to achieve end results rather than relying on understanding about how the plants are performing and defining specific functions for the individual processes.

Sometimes "calibration" is done be reserving a separate data set for the calibration procedure. This approach does not overcome the basic empirical problem in this approach. Success of "calibrated" parameters in matching a second set of data only indicates success in splitting the two data sets so that they represent the same population of data. That is, success with a calibrated model only assures that the calibration data set were adequate to empirically match the second set of data.

Appendix I includes a practical guide to identify which parameter estimate(s) might be the reason for disagreement between the observed data and predicted data of crop models.

Model Evaluation

Every model should be evaluated with respect to transparency and robustness. Transparency refers to how easy it is to understand a model and robustness refers to how closely the predictions of the model match with observed outputs from the system. Model evaluation is discussed in detail in Chapter 3 and here a brief explanation is provided.

There are no well-defined measures for transparency, but there are some guidelines that are explained in Chapter 3. However, after the above-mentioned steps in constructing a model, the model can be run and its output can be compared with measurements from the system. A test of model output against observations is especially necessary when a model is to be used in an application mode. The users of the model need to be given some notion of situations in which the model has proven useful, with a disclaimer for reliability in any other situations (Sinclair and Seligman, 1996). Model evaluation should be done based on predefined criteria established in the objective-definition stage. If the predictions are reasonably matched with the measurements from the system, based on predefined criteria, the modeling process is complete and the model is ready to use (Fig. 2.1).

If the results of the model evaluation stage are not satisfactory, it is necessary to reconsider the hypotheses, equations and the quantitative methods used to construct the model. If alternative methods and equations are identified, then model evaluation should be repeated (Fig. 2.1). If the predictions and the output are not still satisfactory, it is likely that there is some basic problem with the hypotheses. More experimental investigation is required and the modeling process should be set aside until observations allow improved hypotheses.

Exercise

Code the sugarcane leaf growth model and compare leaf area development for different climates and for different parameters.

3 Evaluation of Model to Meet Objectives

There are two critical criteria in evaluating the suitability of a model: transparency and robustness. Transparency means that model parameters, flow diagrams, and code can be readily understood by those that were not involved in its development. As much as possible, the functions are stand-alone descriptions of processes in the plant and crop. Transparency is facilitated by a minimum number of empirical coefficients, and these coefficients can be independently observed and measured. Robustness means that the model produces simulation results that compare favorably with observation. The judgment of "favorable" will depend directly on the original objectives for the model.

Transparency

The ability of others to digest and understand a model depends to a large extent on the complexity of the model. The model of how particles might interact offered by Newtonian physics, for example, is much more readily understood than models that include Einstein's laws of relativity. It is known that Newtonian physics is fundamentally flawed, but the erroneous assumptions only become important under certain conditions. For everyday experiences, Newtonian physics is completely adequate. Remember, Albert Einstein himself suggested "Everything should be made as simple as possible, but not simpler". This is certainly true in developing a transparent model.

What are the factors that determine the simplicity of a model, and hence its transparency? The complexity of a model emanates from the original statement of the objective of the model. If the objective is explicitly clear and well focused, then the key features of the model will be well-defined and superfluous elements of a model should become apparent. Increasing the numbers of processes modeled will necessarily increase the complexity of the overall model. Indices of the complexity of a model are the number of equations and parameters used to describe each process; the greater the number of equations and parameters, the greater the complexity of the model.

A challenge in modeling is to assemble the knowledge and understanding of a system to make judgments about the acceptability of various assumptions and simplifications. This is not a novel idea. Investigators must make similar judgments in designing and executing laboratory and field experiments. For example, experiments are frequently interpreted assuming no pest effects or non-limiting nutrients. In experiments, procedures are put in place to keep the crop free of pests and to eliminate disease limitations, but detailed analysis of the success in achieving these assumptions is rarely tested, especially in regards to possible confounding factors in the soil. In many cases, it is allowable to transfer such experimental assumptions to construction of a model if allowed by the objectives. On the other hand, we should not prepare models that are very simple and incomplete with respect to vital aspects of the system. In such cases, application of the model may result in incomplete or even incorrect understanding of the system (Monteith, 1996).

Real systems including crops are complex and to model them a number of assumptions are required; this is inevitable. Without these assumptions, modeling would not be possible. It is essential to remember that models are by definition a simple description of reality and are incomplete and imperfect descriptions (Teh, 2006). Adding complexity within a model does not necessarily move the model closer to reality. In fact, it is quite likely that including hypotheses without extensive experimental justification can easily increase the imperfection of the model (Sinclair and Seligman, 1996).

Sometimes complexity is unnecessary. Monteith (1996) has presented an example of such complexity: Boyle's Law is one of the most familiar models in the physical sciences. Boyle's Law states that the volume of a gas at constant temperature is inversely proportional to the pressure it exerts. It was demonstrated experimentally by Boyle. Later, Van der Waals showed that Boyle's Law is only an approximation of gas behavior, because it ignores intermolecular forces. To describe and explore the thermodynamic behavior of the atmosphere, meteorologists are content with Boyle's Law. Introducing the Van der Waals correction would make their work much more cumbersome without significantly improving predictions of atmospheric behavior.

Monteith (1996) argued that complexity is often a result of how researchers view the world whether they are physicists, engineers, or biologists. Those trained in physics see that existence is constrained by simple laws and these need to be included in models. Engineers start from the same standpoint as physicists, but often include extensive descriptions of various components of the system so that the model deals with all contingencies. This may be an outgrowth of the engineers' need to avoid catastrophic failures so models need to account for a large array of variables and parameters that may impact the system. Biologists commonly study the world from a reductionist view. The interpretation of any activity at one level can be explained by studying the next level of greater complexity. Applying the biologist's philosophy to modeling results in a proliferation of processes through various levels of complexity that need to be modeled and parameters that need to be quantified. Monteith (1996) suggested that these tendencies need to be resisted and that models should be simple enough to be comprehensible by others (i.e. transparent), but complex enough to be comprehensive in scope (Monteith, 1996).

A checklist of guidelines to keep in mind when evaluating the transparency of the model is summarized below.

1. Does the model strive to include the minimum possible number of equations and parameters? That is, are equations and parameters necessary to meet the objective for the specific model?
2. Are the equations based on established mechanistic relationships and the parameters directly observable?
3. Is the use of algorithms at a minimum and the structure of the model adaptable for use under new circumstances (Monteith, 1996)?
4. Does the model include only processes at the same organizational level or only one organizational level below the process defined in the objective? For example, if modeling at crop canopy level is desired, using equations and parameters at the community level or one level below, at plant level, would be sufficient and it is highly likely that there is no need to consider lower organizational levels, i.e. organ, tissue, cell, and enzyme. Increasing crop model complexity by adding lower level and peripheral processes or by involving cellular and molecular submodels is not likely to improve model performance or relevance. Rather the contrary: Excessive complexity will obscure and even distort benefit that may be gained about crop performance at the desired level of interest (Sinclair and Seligman, 1996).
5. Are summary expressions or conservative relationships incorporated into the model where it is appropriate and possible? There are several summary relationships that are sufficiently robust to express efficiently underlying empirical patterns or complex hypotheses. Examples of such summary relationships include an exponential function for describing radiation interception (Monsi and Saeki, 1953), radiation use efficiency to describe new carbon assimilation (Sinclair and Horie, 1989), transpiration–photosynthesis relationships to describe crop transpiration (de Wit, 1958; Tanner and Sinclair, 1983; Monteith, 1990), and maintenance and growth respiration (Penning de Vries, 1975).

Robustness

Robustness is an evaluation of the ability of a model to simulate reality. Not surprisingly from the previous discussion, the deviation between model predictions and real-world observations may increase with complexity of the model (Reynolds and Acock, 1985, cited in Passioura, 1996). The relationship between accuracy in prediction and complexity is illustrated in Fig. 3.1. Total error in prediction is the sum of two errors: structure error and error in estimating parameters. As a model becomes complicated, structure error declines because the model becomes closer to the system, i.e. it is a better surrogate of the system. However, structural errors can be reduced only to a certain limit (Fig. 3.1). This is especially true of biological systems where their structures are not

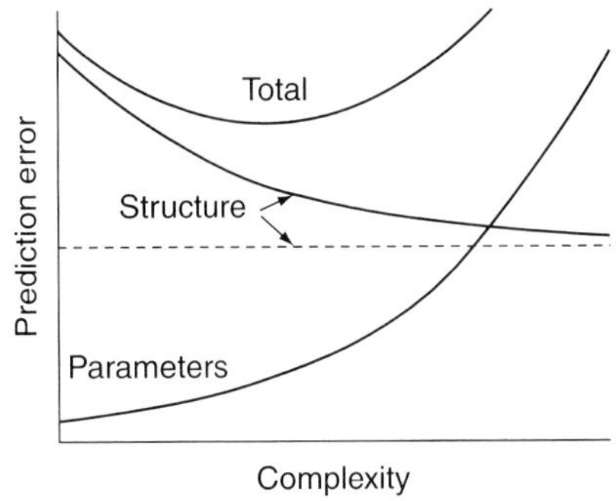

Fig. 3.1. Relationship between complexity and prediction error in biological (crop) models. Horizontal line represents irreducible structural error due to unknown structure of the system (Reynolds and Acock, 1985, as cited in Passioura, 1996).

completely known. Importantly, with increasing complexity, parameter errors increase. This results due to an increase in the number of equations and parameters, and to the inherent error related to estimating each parameter. Therefore, the greater the number of parameters, the greater will be the parameter error. Figure 3.1 illustrates that in principle there is an optimal level of complexity where the model predictions are accompanied with minimal error.

Statistical evaluation of a model can be done by comparing the results produced by a model with measurements from the real system. This type of evaluation is necessary if the model is to be used in an application mode rather than an heuristic mode, for example. Users of the model need to be given some notion of situations in which the model has proven useful, with a disclaimer for reliability in any other situation (Sinclair and Seligman, 1996). However, statistical robustness is not always essential in certifying the value of a model. If the purpose of the model is to integrate the existing knowledge and information, high predictive capability is less important than the need to identify weaknesses in conceptualization of hypotheses. A research model is more likely to be concerned with behavioral patterns than with precise quantitative predictions. Statistical tests of model robustness may not even be necessary in such cases (Sinclair and Seligman, 1996).

In many cases, however, the objectives of construction of a model include its application in simulating real-world behavior. In this case, robustness needs be tested based on predefined criteria in the objective-definition stage. If model predictions reasonably match measurements from the system and prove to be robust based on predefined criteria, then a model can be tentatively accepted for use.

Several cautions must still be remembered when applying a model.

1. Statistical evaluation of robustness can show only how well (or badly) a model performs in a particular circumstance. The evaluation cannot guarantee the performance of the model under any other environmental condition especially when the model has been "calibrated" to fit a specific circumstance or set of circumstances.

2. The statistical test of robustness needs to be placed within a scientific context of evaluation of a hypothesis, or a collection of hypotheses, i.e. a model. From the scientific perspective, a hypothesis can never be validated (Sinclair and Seligman, 1996, 2000). Simply finding a match between model output and observations does not prove that the collection of hypotheses in the model and the parameters used to describe the hypotheses are correct. In evaluating a collection of hypotheses, there is the real possibility that calculations in the model may counter-balance each other to give the expected outcome even though individual hypotheses may be applied incorrectly. Also, in the statistical test it needs to be remembered that there are experimental and observational errors in the data to which the model is being compared.

If the statistical evaluation of the performance of the model is not satisfactory, it is necessary to reconsider the equations and the quantitative methods and, if required, alternative methods and equations should be used (see Fig. 2.1). For example, a linear response of plant development rate to temperature may be substituted with a non-linear curve. After applying the new function, model evaluation should be repeated. As discussed in Chapter 2, if the predictions and the output are still not satisfactory, the assumptions may need reconsideration. For example, in the case of failure of the leaf development model based solely on temperature in Chapter 2, it may be necessary to include the effect of nitrogen by using appropriate equations. If the model predictions are still not acceptable, model objectives and expectations may be reconsidered, or the modeling process may need to be put aside until further understanding of the system results from experimental observation. Indeed, guidance to inadequacies in the understanding about a system can be one of the most valuable outcomes of a modeling effort!

Direct Evaluation of Robustness

Graphs for model evaluation

A very common and widespread way of evaluating the robustness of a model is drawing a plot of simulated output by the model as *y* versus measured output of the system as *x* (Fig. 3.2a). Often, a 1:1 line is also included in this graph (Fig. 3.2b). This line has a 45° angle with the *x*-axis, and if the model is perfect all model predictions will be equal to measurements of the system and lie on this 1:1 line. However, no model is reality and every observation of the system is accompanied by an error, so usually data are scattered around the 1:1 line even for robust models.

The plot of simulated versus observed results has an advantage that it readily highlights the comparison of model predictions and system measurements. This graph is especially common for output variables with a single value for each situation, e.g. days to maturity, crop yield, crop mass at maturity, and so forth.

Sometimes, divergence lines, e.g. ±15% lines, are added to the graph (Fig. 3.2c). A basis, of course, is needed to choose the amount of divergence for these lines. Probably, observed coefficient of variation (CV) for the variable under consideration is a good basis to draw these lines. If the model is robust, many data should be located between these divergence lines, say 80 or 90% of the points.

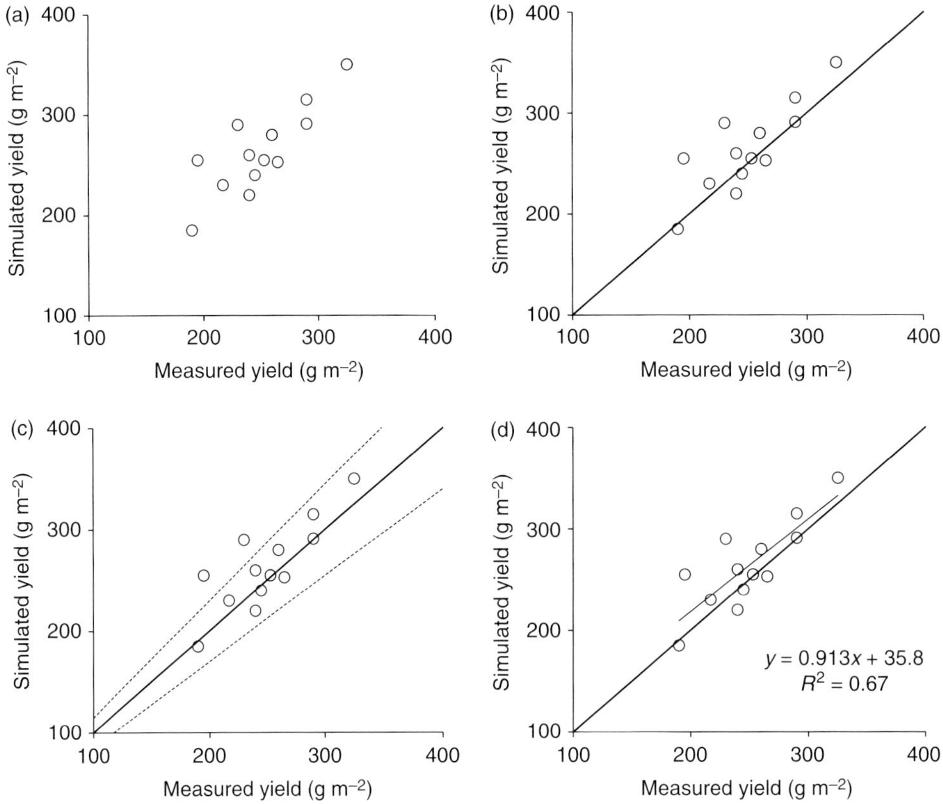

Fig. 3.2. Evaluation of model robustness using simulated versus measured graphs: (a) a simple graph of simulated versus measured variable, (b) simulated versus measured variable plus 1:1 line, (c) simulated versus measured variable plus 1:1 line plus ±15% discrepancy lines, and (d) simulated versus measured variable plus 1:1 line plus regression line.

A regression line between simulated and measured output is commonly included in the graph of simulated versus measured variable (Fig. 3.2d). This regression model will help to identify any bias in model prediction (see below) and determine correlation between model predictions and system measurements. However, inclusion of the regression line may be misleading because there might be a strong regression between simulated and measured variable while there is a bias in model prediction (Wallach, 2006). Figure 3.3 represents an example of this situation. From high R^2 (and r, also) one might conclude that the model is robust while it is not true and, in fact, the model resulted in an overestimation at lower values of the variable (less than 25 g m^{-2}) and underestimation at higher values of the variable.

A plot of simulated versus measured values can be drawn for output variables that have more than one value during the growing season for each situation or simulation exercise, e.g. crop leaf area, crop mass, nitrogen accumulation, and so on during the growing season. Figure 3.4 presents results of crop mass and crop nitrogen accumulation at various points in the growing season for a chickpea model (Soltani and Sinclair, 2011).

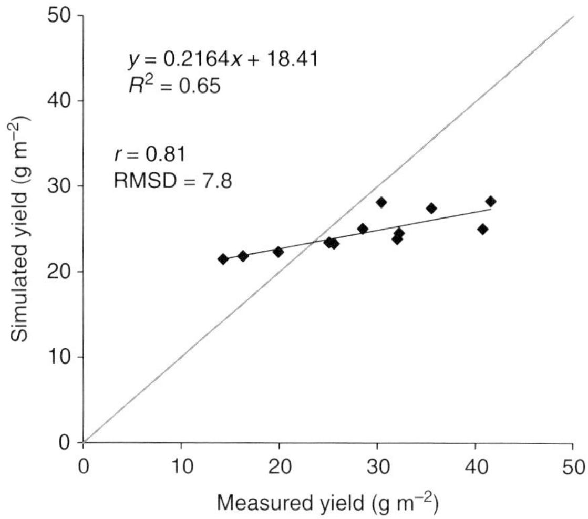

Fig. 3.3. A sample graph of simulated versus measured variable in which the linear regression between simulated and measured variable resulted in high R^2 and r, indicating that the model obviously is not a good simulator of the system.

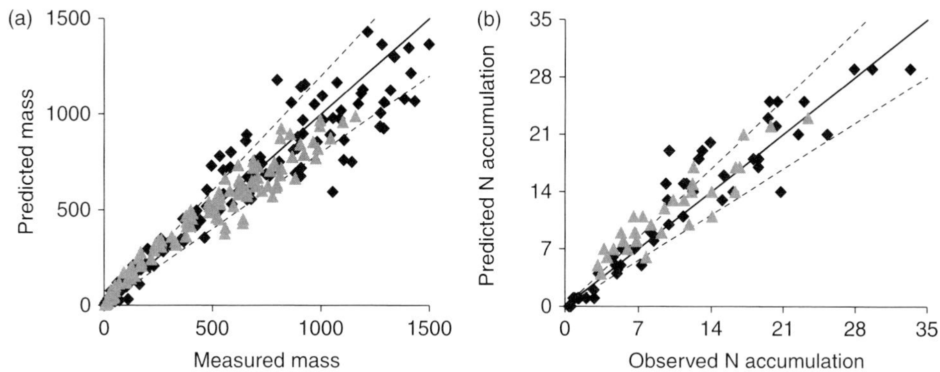

Fig. 3.4. Sample graphs of (a) predicted crop mass and (b) nitrogen (N) accumulation, both in g m^{-2}, during growing season in chickpea crops grown under a wide range of sowing date and density. Different symbols indicate two different experiments. The 18% ranges of discrepancy between simulated and measured values are indicated by dashed lines. Solid line is 1:1 line (Soltani and Sinclair, 2011).

Few modelers have used a graph of residuals (simulated – measured results plotted on the y-axis) versus measured data (plotted on the x-axis). Wallach (2006) stated that rare use of this type of graph is very unfortunate. This kind of plot is very popular in regression analysis and recommended as a part of any regression analysis (Montgomery *et al.*, 2001). One advantage of this type of graph is that deviation of the model from observations appears directly (Wallach, 2006). The graphs are thus easier to evaluate and to compare. Residual graphs are very important for bringing attention to systematic patterns in the errors. For example, Fig. 3.5 is a residual plot based on the data presented in Fig. 3.3. This graph clearly indicates model overestimation for lower values of measured crop yield and model underestimation for higher values of measured crop yield. The residual plot also reveals any bias in model predictions. Residual mean of zero indicates no overall bias, but negative and positive means indicate bias in model predictions (see below).

There is another common type of graphical evaluation of model robustness, i.e. plotting model predictions and system measurements for a given state variable over time. While there are model predictions for every day, measurements from the system might be occasional. Figure 3.6 indicates samples of this type of graph for leaf area index and cumulative organ mass for chickpea through a growing season (Soltani and Sinclair, 2011).

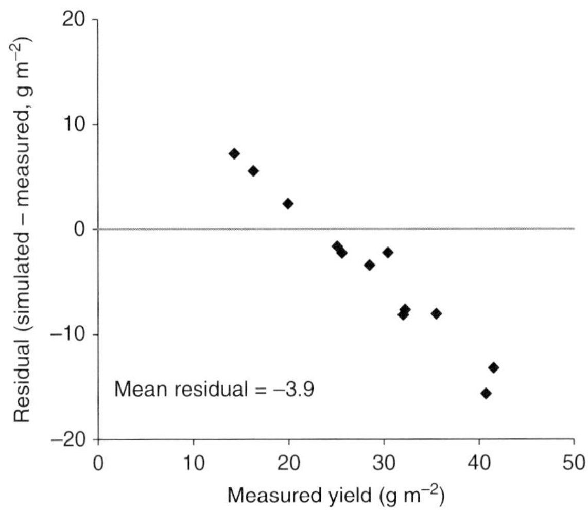

Fig. 3.5. A sample graph of residual (simulated – measured) versus measured variable derived from Fig. 3.3. This example indicates an obvious bias in model prediction. Mean bias (residual mean) is –3.9 g m^{-2}.

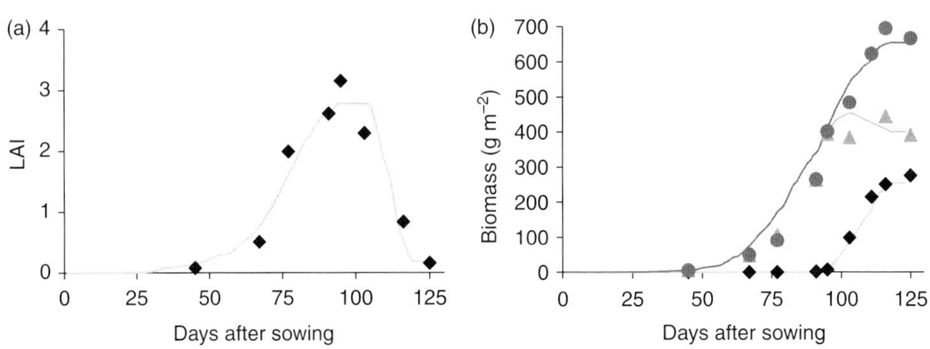

Fig. 3.6. Simulated and measured values of (a) leaf area index (LAI) and (b) vegetative, seed, and total dry matter by a chickpea crop (Soltani and Sinclair, 2011).

Statistics for model evaluation

Various kinds of statistics can be used to test the robustness of a model. These statistics can be classified in two categories:

1. Differences between the predictions and the measurements.
2. Correlation between the model outputs and the measurements from the system.

Deviation-based statistics have often been used with correlation-based statistics. Although these different statistics may represent different aspects of the model-measurement discrepancy, the deviation-based statistics (e.g. root mean square of deviation) and the correlation-based statistics (e.g. correlation coefficient) are not really consistent with each other in their assumptions (Kobayashi and Salam, 2000).

There has been a controversy about using statistics in model evaluation. For example, Harrison (1990) challenged using regression-based tests of model evaluation. Later, Mayer *et al.* (1994), using a Monte-Carlo experiment, showed that this test performs well, except in situations where the errors are auto-correlated. However, Analla (1998) again challenged it. Kobayashi and Salam (2000) argued that the correlation coefficient and linear regression are not entirely satisfactory for model evaluation and suggested that mean square of deviation and its components are more informative. More recently, Gauch *et al.* (2003) proposed a partitioning of mean square deviation which complements correlation and linear regression in evaluating the predictive accuracy of models. For more information about statistics in model evaluation refer to Wallach (2006). Fila *et al.* (2003) developed software that can be used in the calculation of different statistics.

A few statistics of both categories are presented below. These statistics are very common in modeling work and are usually used together.

Perhaps the most important statistic from the first category is Root Mean Square of Deviation (RMSD), which is calculated from the following equation:

$$\text{RMSD} = \left(\frac{\sum (X_i - Y_i)^2}{n-1} \right)^{0.5} \tag{3.1}$$

where X_i is the measured value, Y_i the simulated value and n the number of pairs of measured and simulated values that are used in calculating RMSD. RMSD is usually reported and discussed as a percentage of average measured performance of the system.

Bias is obtained as average deviation between model predictions and system measurements:

$$\text{Bias} = \Sigma\,(Y_i - X_i)/n \tag{3.2}$$

In the second approach the correlation coefficient between measured and simulated values are calculated. Correlation coefficient (r) can be obtained as:

$$r = \frac{\sum_{i=1}^{n}(X_i - \overline{X})(Y_i - \overline{Y})}{\sqrt{\sum_{i=1}^{n}(X_i - \overline{X})^2}\sqrt{\sum_{i=1}^{n}(Y_i - \overline{Y})^2}} \tag{3.3}$$

where X_i is the measured value, Y_i the simulated value, X and Y the mean value of X_i and Y_i and n the number of pairs of measured and simulated values that are used in calculating r.

Simple, linear regression is another method that is commonly used to evaluate model robustness. A simple linear regression model ($Y = a + bX$) is fitted to data of simulated (as y-axis) versus measured (as x-axis) variable values. If the intercept (a) is not significantly different from 0 and slope (b) is not significantly different from 1, the model is assumed to be robust.

Exercise

The table below includes observed grain yield and simulated yield by a crop model, both in g m^{-2}. Using the data, try to evaluate model robustness using graphical and statistical measures discussed in this chapter.

Observed grain yield and simulated yield by a crop model, both in g m^{-2}.

Simulated	Measured
175	180
175	160
180	190
185	180
205	180
210	200
218	232
230	230
235	250
250	273
250	240
258	267

4 Applications of Crop Models

Using crop models, different questions related to the system (from which the model is built) can be answered. For example, how can crop yield be increased in a given environment via manipulation of crop traits? How important is the amount of soil water at sowing in determining final crop yield? What will be the impact of climate change on crop yield and water use? How can nitrate leaching from a field be decreased?

It is obvious that crop models alone cannot answer all the above questions. However, when crop models are integrated with experiments and measurements from the system and their application is accompanied by physiological/ecological understanding of the system, they become powerful tools in answering questions about the systems. Nevertheless, crop models, like other tools, have various strengths and limitations. Therefore, achievement of desired results from application of a crop model needs a proper understanding of the capabilities and limitations of crop models. In the application of crop models the following points need to be remembered.

1. Model users need to understand the structure of the crop model they want to use before model application. They need to be aware of limitations in the model and their inputs/parameters requirements. As models are incomplete surrogates of the real-world system, model limitations are normal and unavoidable results of simplification.
2. During the juvenile stage of crop modeling (Sinclair and Seligman, 1996; also see Chapter 5), it was thought that crop models could substitute for (eliminate) field experiments. As indicated in Chapters 2 and 3, this is not a correct perception. At best, crop models supplement field experiments helping to identify the processes and parameters that are especially critical in understanding the performance of the system. In this way, models can help to prioritize experiments; they can be used to extrapolate and check the results of the experiments over years and locations, which will increase the effectiveness of the experiments.

3. Models are developed by individuals who are trained in specialized disciplines and they may have less competence in other disciplines that might be components of the model. It is useful to take a skeptical viewpoint when examining an existing model. What are the hypotheses and assumptions? Do specific submodels match with the experimental experience about the new system of interest?
4. Application of crop models is limited by availability and quality of required inputs/parameters. The quality of model outputs depends on the quality of input data used. Some reasons for poor quality of input data (Boote *et al.*, 1996) are: (i) cost of obtaining data, e.g. high prices of research instruments for measurement; (ii) spatial variability, which might be important for field applications where soil properties vary considerably within a single field; (iii) the technical knowledge required for some inputs; (iv) temporal variability due to pest outbreaks; and (v) data quality as a result of poorly calibrated sensors. In field crops, a large quantity of data exists for phenology and biomass accumulation. However, it is possible that phenological stages are recorded using different criteria (keys). For other parameters, the observational data might be very limited, e.g. leaf area development, root growth, and crop nitrogen accumulation. In addition, there are commonly difficulties in weather data. Weather records might be incomplete or distance between the weather station and the field might be large. Weather stations usually do not measure solar radiation.
5. When a crop model is used in a new situation, e.g. in a new environment or for a new cultivar, model parameters need to be estimated for the new conditions and model performance needs to be rechecked.

Crop models have different applications. The applications have been classified by different researchers (e.g. Whisler *et al.*, 1986; Boote *et al.*, 1996; Sinclair and Seligman, 1996). The classification used here is a combination of that of Boote *et al.* (1996) and Sinclair and Seligman (1996) with modifications. Generally, model applications are in research, in management, and in teaching. Table 4.1 summarizes the classification.

Table 4.1. Summary of different types of models' applications.

Using crop models in research
- Integration of research knowledge
- Integration of knowledge across disciplines
- Improvement in experiment documentation
- Crop genetic improvement
- Crop management improvement
- Yield analysis
- Response to climate variability and change
- Environmental impact of crop production

Using crop models as a tool in crop management
- Best management practices
- Pre-sowing and in-season decision aid for farmers
- Site-specific or precision farming
- Pre-harvest yield forecasting

Using crop models in education
- Student education
- Farmer education

Research Applications

Many modeling groups and modelers have used their crop models in a research mode. Application of crop models in research can be classified into subclasses.

Integration of research knowledge

Crop models are valuable tools for integration of research findings resulting from research on different aspects of crop processes. In fact, crop models may be the only way to integrate research findings that have been obtained from studies of different processes in isolation (Boote *et al.*, 1996). In constructing and evaluating the model, the key factors influencing the performance of the system are likely to be identified.

Crop models can also reveal those aspects of crop growth where little information exists or the understanding is incomplete. Such revelation can occur when a model is built or when the response of a crop model to an environmental and/or management factor is unexpected (Boote *et al.*, 1996). For instance, development of a chickpea model by Soltani *et al.* (1999) indicated scarcity in quantitative information about some processes. Later, Soltani *et al.* (2005, 2006a–e) conducted detailed experiments to obtain the quantitative information.

Integration of knowledge across disciplines

Developing and application of crop models to simulate crop production under different environmental conditions needs integration of knowledge from different disciplines such as agronomy, soil science, pest control, economy, etc. Models are thus valuable interdisciplinary research tools that integrate disciplinary knowledge and relationships to produce a descriptive tool for application beyond the individual science discipline (Boote *et al.*, 1996).

Improvement in experiment documentation

Development, parameterizing, evaluation, and application of crop models require gathering and organizing experimental data. Unfortunately, not all experiments are likely to be done by individuals familiar with the hypotheses of the model. Therefore, guidelines for experiments that document the desired information must be developed for ready access by others. In an effort to increase utilization of cropping system data, the International Consortium for Agricultural Systems Applications (ICASA) has developed an Internet-based system that provides a forum for documenting, archiving, and exchanging cropping system experiment and/or weather data sets (Bostick *et al.*, 2004).

Crop genetic improvement

Crop models can assist in identifying traits for genetic improvement through different applications. Crop models can be used in environmental characterization and evaluation of putative value of different traits for plant breeding. Environmental characterization is considered an important step in understanding genotype–environment interactions, which limits genetic gain for complex traits such as tolerance to drought (Chapman *et al.*, 2002). Crop models can be used to characterize broadly (large geographic area, long-term period) and locally (field experiment) drought-related environmental stresses, which enables breeders to analyze their experimental trials with regard to the broad population of environments of interest (Muchow *et al.*, 1996; Chapman *et al.*, 2002; Chenu *et al.*, 2011).

An example of using a model for geographical characterization was presented by Chenu *et al.* (2011). They used the APSIM models to do simulation for northeastern Australia for representative locations (25), soils, and management systems using more than 100 years of historical climate data. They identified three major water-deficit patterns as indicated in Fig. 4.1. The first environment type was comprised of situations where plants were effectively not limited by water or experienced only short-term water deficit; the second environment type was characterized by mid-season water deficit, starting during the vegetative period and relieved by rainfall events during grain filling; and the third environment type also had a mid-season water deficit beginning at around the same time as in the second environment type, but continuing through to plant maturity.

Genetic improvement programs dealing with physiological traits have three steps (Jordan *et al.*, 1983): (i) trait or traits that promote genetic yield potential must be identified; (ii) genetic variation and its nature for the traits must be assessed and superior genetic resources must be identified; and (iii) the gene or genes governing the traits must be incorporated into the current good cultivars. Therefore, before physiological traits are proposed for inclusion in breeding programs, their benefit for grain yield should be assessed in terms of the components of yield and the determinants of survival (Ludlow and Muchow, 1990).

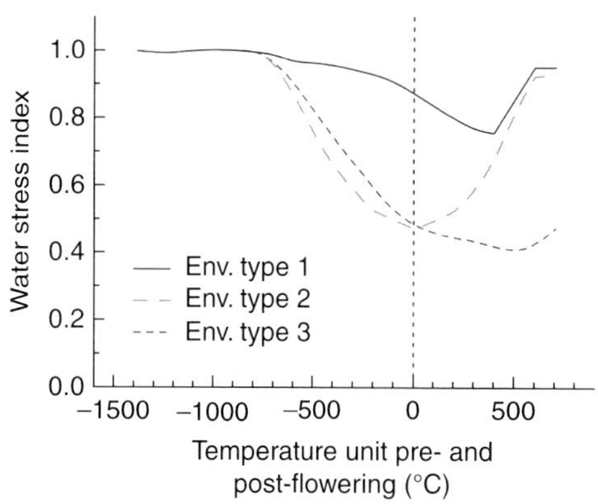

Fig. 4.1. Water-stress index throughout the crop cycle, represented as a function of temperature unit relative to flowering. A water-stress index of 1 indicates no water stress, while a ratio of 0 corresponds to a full stress, with no water available to the crop (Chenu *et al.*, 2011).

Unless they make a contribution to one or more yield component or determinant, there seems little use in breeding for the trait. A common approach for assessing the value of traits is the comparison of yield of isogenic or near-isogenic lines or populations. Another alternative approach is to use simulation modeling (Sinclair and Muchow, 2001; Sinclair, 2011). Modifying crop parameters in a crop model is analogous to the creation of genetic isolines. Simulation models have been used for determining critical traits for higher yield or prolonged survival in a number of crops and cropping situations (Muchow and Carberry, 1993; Aggarwal *et al.*, 1997; Aguera *et al.*, 1997; Sinclair, 2000; Soltani *et al.*, 2000; Sinclair and Muchow, 2001; Soltani and Galeshi, 2002; Sinclair *et al.*, 2005a, 2010; Manschadi *et al.*, 2006).

Recently, Sinclair *et al.* (2010) used a soybean model to assess the yield benefits from altered drought traits across the USA. The traits they evaluated were: rooting depth extension, rate of leaf area development, decreased stomata conductance at high soil water content, limited maximum transpiration rate, and drought-tolerant symbiotic nitrogen fixation. Simulations were done for 50 years of weather data for 2655 grid locations of 30 km by 30 km size in the USA. They found that a fast rate of rooting development was a neutral or negative trait in most locations. Slow leaf area development proved beneficial in less than half the years and in wetter years it resulted in yield losses. Water conservation both by early decrease in stomata conductance with soil drying and by imposing a maximum transpiration rate resulted in yield increases in many locations in 70% or more of the years. Both traits resulted in only small yield decreases in the wet years. Drought-tolerant nitrogen fixation had the greatest benefit of all traits with a yield gain in more than 85% of the years at nearly all locations, and in those cases with no yield increase there was only a very small yield loss. Figure 4.2 presents yield gain predicted for drought-tolerant nitrogen fixation.

Crop models have proved to be useful in decreasing the number of field experiments and duration of selection period in field crops. Currently, a long series of pre-testing trials of promising cultivars at many locations is a prerequisite for selection and release of a crop cultivar, involving large investments in time and effort. Palanisamy *et al.* (1993) used a rice model to predict the performance of pre-release genotypes at different locations in India to minimize the requirement for field study. However, additional successful examples illustrating this use of models are scarce.

Application of crop models in integration with molecular genetics and genomics has also been used to help plant breeding (Hunt *et al.*, 2003; Yin *et al.*, 2003b). Wight and Hoogenboom (2003) discussed the experience of gene-based modeling and its future.

Crop management research

Management decisions regarding cultural practices and inputs have a major impact on yield. Simulation models that allow the specification of management options offer a relatively inexpensive means of evaluating a large number

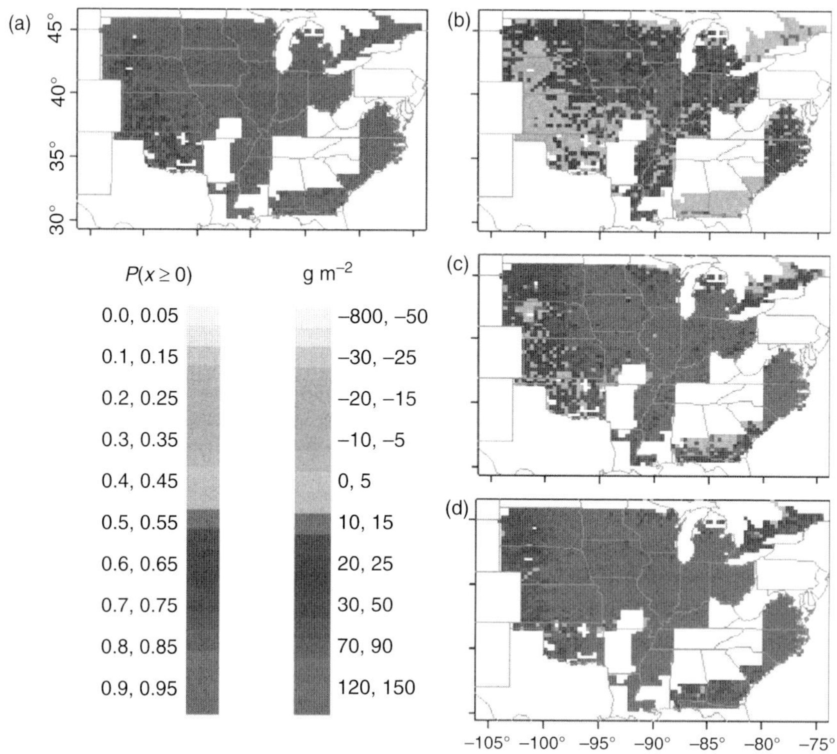

Fig. 4.2. Regional patterns of yield variation due to change in sensitivity of nitrogen fixation to drying soil. (a) Probability of yield for simulation with altered trait being greater than the standard simulation. Absolute yield difference between the simulation with the altered trait and the standard simulation presented for each grid location for the 75%, median, and 25% quartiles among all years simulated (b to d) (Sinclair *et al.*, 2010).

of options that would rapidly become too expensive if the traditional experimentation approach were to be adopted (Cheeroo-Nayamuth, 1999).

Many publications are available describing the use of simulation models with respect to crop management such as sowing date and density, water and fertilizer management, pest and disease management, precision agriculture and so on. In this kind of application, crop models alone or in combination with field experiments are used to study response of crop yield and other crop variables to management options. As a result, better management options are more expeditiously identified.

Some examples of application of crop models in optimizing cultural management are Egli and Bruenning (1992), Muchow *et al.* (1994), Anapalli *et al.* (2005), Heng *et al.* (2007), and Monzon *et al.* (2007). Similarly, crop models have been used to optimize water and fertilizer management (O'Leary and Connor, 1998; Soltani *et al.*, 2001a; Rinaldi, 2004; Lobell and Ortiz-Monasterio, 2006; Benli *et al.*, 2007; Salado-Navarro and Sinclair, 2009).

Soltani *et al.* (2001a) used a crop model and long-term weather data to analyze the response of chickpea to limited irrigation. They showed that in northwest Iran, chickpea experiences terminal drought stress that is started at a time between flowering and beginning seed growth and this stress severely reduces grain yield by 67% compared to irrigated conditions (from 2766 to 909 kg ha^{-1}) (Table 4.2). Simulated grain yield showed a large response to limited irrigation. Table 4.2 shows the best results were obtained when the first irrigation was applied at the beginning of seed growth P1(25) resulting in a yield increase of 81% (739 kg ha^{-1}) compared to rainfed conditions. For this irrigation, 124 mm water was required, assuming an irrigation efficiency of 100%.

Rinaldi (2004) using DSSAT-CERES-Wheat found in a Mediterranean environment that crop-available water at sowing is important for winter durum wheat productivity (Fig. 4.3). They also found that nitrogen fertilizer is important for grain yield and the optimal nitrogen fertilizer amount for durum wheat was about 100 kg N ha^{-1} from both productive and economic points of view.

Geographical yield analysis

Crop models have successfully been used in evaluation and determination of potential production on different scales (Meinke and Hammer, 1995; Wilson *et al.*, 1995; Caldiz *et al.*, 2001; Wu *et al.*, 2006; Binder *et al.*, 2008). On a regional scale, a crop model is usually used in combination with a Geographical Information System (GIS). Some models have been used to quantify yield gap (the difference between attainable yield and actual yield) and indicate possible explanations for the gap (Affholder *et al.*, 2003; Calvino *et al.*, 2003; Kalra *et al.*, 2007). Figure 4.4 presents potential production of wheat in the North China Plains under irrigated and rainfed conditions as quantified by Wu *et al.* (2006) using WOFOST-wheat model.

Table 4.2. Effect of limited irrigation with one irrigation at flowering (F1) or beginning of seed growth (pod filling, P1) on grain yield (YLD, kg ha^{-1}), coefficient of variation (CV) of grain yield, crop evapotranspiration (ET, mm), water use efficiency (WUE, kg ha^{-1} mm^{-1}), applied irrigation water (APLDW, mm) and applied-water use efficiency (EAW, kg ha^{-1} mm^{-1}). Rainfed and full-irrigated variables are included for comparison. Numbers in parentheses show plant density (plants m^{-2}) (Soltani *et al.*, 2001a). The numbers with different letters indicate significant differences at 5% level of probability.

Treatment	YLD	CV	ET	WUE	APLDW	EAW
Rainfed	909d	29	228g	3.99c	0	0
F1 (25)	1442c	16	296f	4.90b	68d	7.84a
F1 (38)	1503c	19	305ef	4.93b	74c	8.01a
F1 (50)	1524c	19	310de	4.93b	78c	7.94a
P1 (25)	1648b	18	323cd	5.12b	124b	5.99bc
P1 (38)	1701b	19	334bc	5.10b	127b	6.25bc
P1 (50)	1721b	20	338b	5.10b	128b	6.34b
Irrigated	2766a	4	441a	6.31a	325a	5.71c

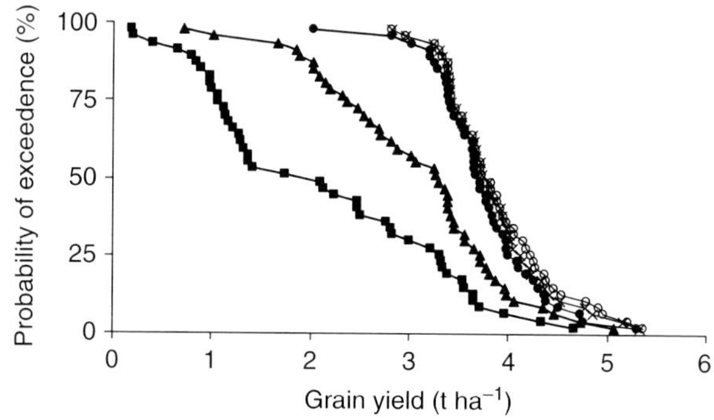

Fig. 4.3. Cumulative probability, using a long-term simulation (47 years), to exceed a grain yield value of durum wheat as function of different crop available water (CAW) at sowing of durum wheat: (■) CAW 0%; (▲) CAW 20%; (●) CAW 40%; (×) CAW 60%; (o) CAW 80% (Rinaldi, 2004).

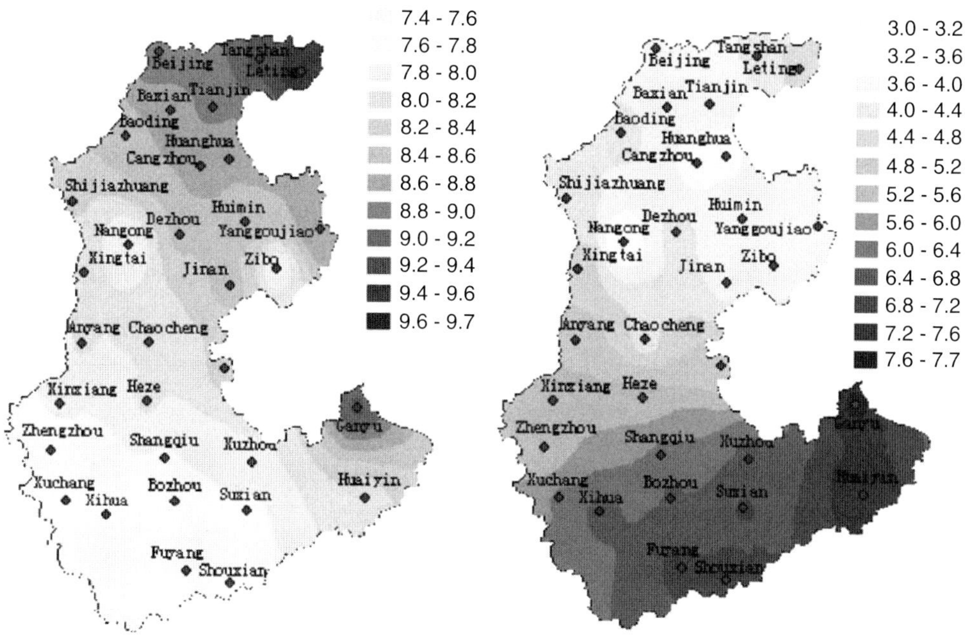

Fig. 4.4. On the left-hand side the simulated irrigated (potential) yields (Mg ha^{-1}) of winter wheat, and on the right-hand side the winter wheat yields under rainfed (water-limited) conditions in the North China Plains (Wu *et al.*, 2006).

Using simulation modeling, quantification of yield reductions caused by non-climatic causes (e.g. delayed sowing, soil fertility, pests, and diseases) becomes possible. Simulation models have also been reported as useful in separating yield gain into components due to changing weather trends, genetic improvements, and improved technology (Bell and Fischer, 1994, cited in Boote *et al.*, 1996). Brisson *et al.* (2010) used a crop model (STICS) along with other tools and data sources to analyze yield trends in France and evaluated the reasons why wheat yields are stagnating in Europe.

Agro-ecological zoning is a data inventory of environmental resources, identification of homologous environment, determination of agricultural production of a region, planning for regional development, and identification of research priorities (Aggarwal, 1993). Conventional methods are based on overlaying maps and statistical techniques. An additional approach is to use crop models to integrate the effect of climatic, soil, and crop management factors in determining crop yield (Aggarwal, 1993; Caldiz *et al.*, 2001).

Response to climate variability and change

Irrespective of technological development, climate is still one of the most important determinants of crop production. In some studies, crop models have been used to evaluate crop responses to within-season and among-season variability in weather data. Some examples are Semenov and Porter (1995), Mearns *et al.* (1996), and Soltani *et al.* (2004b).

Global climate change, i.e. elevated CO_2 concentration, increased temperature, and altered rainfall patterns, may have serious impacts on crop production in the future. Assessing the effects of these changes on crop yield is important. Despite some objections, it seems the simulation approach still remains the best tool for quantifying these effects (Cheeroo-Nayamuth, 1999). A number of crop growth models have been used to evaluate consequences of global climate change on agricultural production, regional economies, and other topics (Sinclair and Rawlins, 1993; Boote *et al.*, 1996; Haskett *et al.*, 2000; Reyenga *et al.*, 2001; Soltani *et al.*, 2001b; Koocheki *et al.*, 2006; Ludwig and Asseng, 2006; Sinclair, 2010).

In these studies, long-term sequences of historical weather are needed. Temperature and rainfall (and probably solar radiation) data are then modified proportionately to correspond to the monthly temperature and rainfall changes predicted by Global Circulation Models (GCM) for a higher CO_2 concentration, e.g. doubling CO_2. The next step is to use a crop model to evaluate crop yield and other variables in response to the changed weather and CO_2 concentration. The crop model used must include the impact of higher CO_2 on crop key processes, including radiation use efficiency (Chapter 10) and transpiration efficiency coefficient (Chapter 14).

With climate change, the weather is proposed to be more variable. Weather generators can be applied to create sophisticated weather data for future climate that include both change in averages and variability. LARS-WG is one of the generators developed by Semenov and co-workers (e.g. Semenov, 2008, 2009; Semenov and Stratonovitch, 2010). LARS-WG version 5.0 includes climate scenarios based on the 14 GCMs that have been used in the latest IPCC (2007). Unfortunately, there is considerable disagreement over the degree of temperature increase and rainfall variation projected by GCMs. Because of this, some researchers prefer to simply change historical weather data to create future climate data (e.g. Ludwig and Asseng, 2006).

Chavas *et al.* (2009) examined potential climate change impacts on the productivity of five major crops in eastern China using the EPIC model. They used weather data from 1961 to 1990 as a baseline and changed it for future conditions (2071 to 2100). Simulations were performed with and without the enhanced CO_2-fertilization effect. Figure 4.5 summarizes their results; aggregate potential productivity (i.e. if the crop is grown everywhere) increases 6.5% for rice, 8.3% for canola, 18.6% for maize, 22.9% for potato, and 24.9% for winter wheat, although with significant spatial variability for each crop. However, without the enhanced CO_2-fertilization effect, potential productivity declines in all cases ranging from 2.5 to 12%.

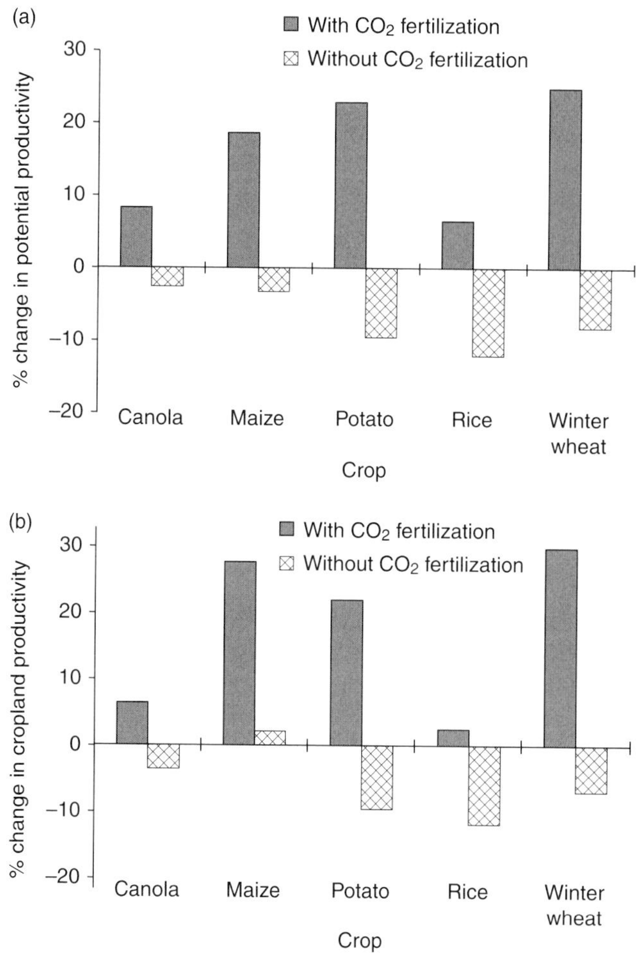

Fig. 4.5. Change in productivity between baseline (1961–1990) and the future (2071–2100) with and without inclusion of CO_2 fertilization. (a) Change (%) in aggregate potential production (i.e. if the crop is grown everywhere in the domain). (b) Change (%) in aggregate production weighted by where the crop was actually grown in 1990 (Chavas *et al.*, 2009).

Environmental consequences of crop production

Intensification of agriculture, including crop production, and the prospects of future intensification, may have major detrimental impacts on the environment (Tilman, 1999). Crop production, similar to other human activities, results in considerable environmental burdens and hence environmental impacts. Soil erosion, nitrate and pesticide leaching, gaseous losses of nitrogen to the atmosphere, phosphorus loss, soil organic carbon loss, energy use, and greenhouse gases emissions are examples of environmental burdens related to crop production (Nemecek and Kägi, 2007).

Crop models can be used as tools to estimate the amount of the environmental burdens. Environmental impact assessment methods, e.g. life cycle assessment, need these estimates. However, crop models have not been used to any great extent in life cycle assessments, and there is a great potential that can be exploited in the future. Crop models have been used to evaluate carbon sequestration (Doraiswamy *et al.*, 2007), nitrate and pesticide leaching and their resultant environmental pollution (Dueri *et al.*, 2007), and soil loss and erosion (Wang *et al.*, 2008).

Peralta and Stockle (2001) using CropSyst, performed a long-term analysis of the potential for nitrate leaching that might result from a typical irrigated potato system in the Quincy–Pasco area of the state of Washington. They showed that when scenarios with fertilization rates above current recommended rates were simulated, potato had the largest nitrate leaching amounts during the growing season (Table 4.3). When fertilization approached recommended rates, the simulated nitrate leaching during the potato growing seasons was low and not different from that of maize growing seasons. Nitrate leaching was found to be more significant during no-crop periods (fall and winter), particularly following potato.

Lugato *et al.* (2010) used the DNDC model (Li *et al.*, 1992, 2006) to estimate greenhouse gas (GHG) emissions from Italian agricultural areas. They also simulated the effect of alternative management practices on GHG emissions to assess the potential mitigation effects or impacts of the strategies adopted.

Table 4.3. Thirty-year simulated mean of nitrate leaching (kg N ha^{-1}) for six periods of the rotation and all the irrigation and nitrogen fertilization treatments (irrigation and nitrogen fertilization combinations). Irrigation levels are: excess irrigation (E), normal irrigation (N), and deficit irrigation (D). Nitrogen fertilization rates are: low fertilization (F1), medium fertilization (F2) and high fertilization (F3) (Peralta and Stockle, 2001). "Crop" is average of the six rotations.

Period	EF1	EF2	EF3	NF1	NF2	NF3	DF1	DF2	DF3
Wheat–Maize	7.1	23.4	26.8	2.8	18.9	27.8	8.5	20.8	32.8
Maize	6.0	15.6	20.7	3.0	20.4	28.2	5.6	13.8	18.7
Maize–Potato	5.2	27.2	87.6	2.3	23.0	29.5	17.8	43.4	49.7
Potato	5.2	46.7	68.6	3.1	21.5	34.0	5.9	14.3	27.7
Potato–Wheat	4.7	25.3	35.6	9.6	82.4	115.9	59.7	143.5	173.2
Wheat	0.7	0.9	0.8	0.0	0.3	0.4	0.0	0.0	0.0
Crop	3.9	21.0	30.0	2.1	14.1	20.9	3.8	9.4	15.4
No-crop	5.7	35.3	50.0	4.9	41.4	57.7	28.7	69.2	85.2

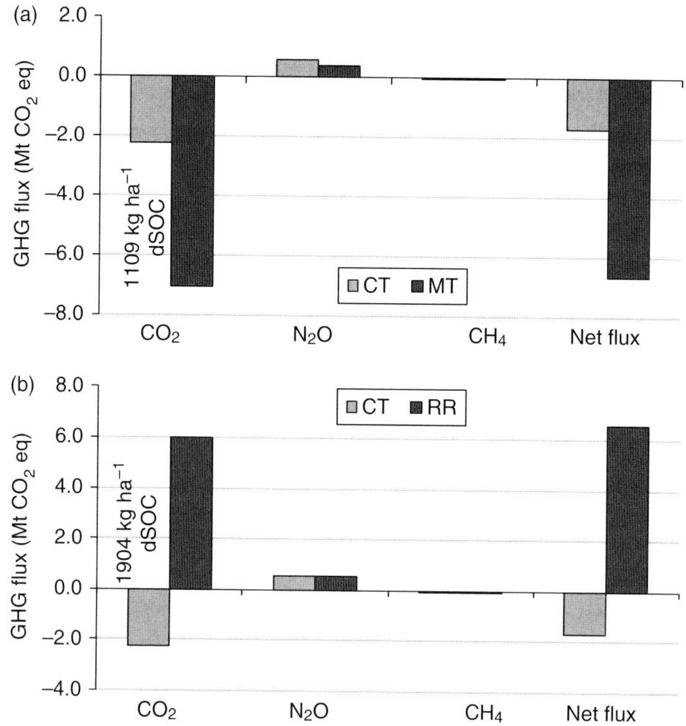

Fig. 4.6. Annual net greenhouse gas (GHG) fluxes and overall GHG balance of the grain maize with alternative management practices: (a) conventional (CT) versus reduced tillage (MT) and (b) incorporation (CT) versus removal of residues (RR) (Lugato *et al.*, 2010).

Figure 4.6 illustrates this in grain maize crop including the adoption of a minimum tillage (MT), and the residue removal (RR) from the field, both compared with the current management based on plow and residue incorporation (CT). The conversion of all the grain maize area to minimum tillage allowed the accumulation of about 1100 kg C ha^{-1} of soil organic carbon on average, saving more than 4 t ha^{-1} of CO_2 equivalent including the lower N_2O emissions. On the contrary, the residue removal could strongly impact the soil organic carbon balance leading to a very high soil C depletion (1904 kg C ha^{-1} on average) and consequently increasing the GHG emissions (8.2 t ha^{-1} of CO_2 equivalent).

Crop Management Applications

Crop models can be used directly to assist crop producers. In this mode, a crop model is used as a decision support system to help farmers in making decisions about how to manage their farms. That is, simulations from crop models are used to take pre-sowing, sowing, or in-season decisions and adjustment of crop management practices. In some cases, government and private agencies can also benefit from the simulations.

Best management practices

Crop management decisions might have a major impact on crop yield. However, many agriculturally important locations do not have local research stations to advise farmers even though best management practices often depend on local conditions. Thus, a crop model along with long-term weather data and some soil information can help to find the best management practices for a specific location. For this application, crop models must allow the specification of management options. In this mode, a crop model offers a relatively inexpensive means for the producer to evaluate a large number of options. Crop models can be used to find for example the best sowing date and plant density, harvest time, optimal nutrients, and irrigation management (amount and time of application), and selection of crop and/or cultivar that results in optimal crop yield and income in long-term analysis (Boote *et al.*, 1996; Reddy *et al.*, 2002; Soltani and Hoogenboom, 2007). Economic consequences of the decisions for the farmers can be evaluated as well.

Models that include chemicals leaching or erosion components have been applied to determine the best crop management practices over the long-term to reduce leaching and erosion (Boote *et al.*, 1996).

Pre-sowing and in-season decision aid for farmers

In this type of model application, a crop model along with weather data is used to find the best practice for the current season. Usually some forecast of weather data for future weeks or months is required. Then, farmers are advised to adjust their pre-sowing and/or in-season decisions including selection of crop and cultivar, and water and fertilizer management. Software programs, e.g. Weather Analogue (Bannayan and Hoogenboom, 2008), are available to provide prediction of future weather data. Reddy *et al.* (2002) described how GOSSYM/COMAX has been used to help farmers with pre-season and in-season decisions. GOSSYM is a cotton crop model and COMAX is a crop management expert system. FARMSCAPE (Farmers', Advisers', Researchers', Monitoring, Simulation, Communication And Performance Evaluation) is another example from Australia (Carberry *et al.*, 2002). FARMSCAPE is a program of participatory research with the farming community of northeast Australia. Its goal is to benefit farmers from tools such as soil characterization and sampling, climate forecasts and, in particular, simulation modeling.

WebGro is an example of such an application in which a web-based soybean decision support system is based on the CROPGRO-soybean model (Paz *et al.*, 2004). The aim of WebGro is to help soybean producers in midwestern USA understand how different stresses interact to limit soybean yield in their fields. Stresses include water, soybean cyst nematode (*Heterodera glycines*), herbicide injury, rhizoctinia root rot disease (*Rhizoctonia solani*), and hail damage. The user can set up a field scenario by selecting cultivar, sowing date, plant population, soil type, and the nearest weather station, using a web form. Different stress levels can then be entered, and the model can be run interactively by simulating the effect of one or more stress at a time. WebGro can be used either as a planning tool to determine how to best decrease the adverse impacts of different plant stresses during the upcoming year or as a diagnosis tool to estimate the cost of interactions during the past year. Table 4.4 includes a sample summary table from WebGro simulations (Paz *et al.*, 2004).

Site-specific or precision farming

In some conditions, maximum income from the field is obtained if the field is divided into smaller parts and each part is managed for its own optimum performance. For example, each individual sub-part receives the appropriate amount of suitable fertilizers and pesticides depending on soil conditions and pest occurrence. This practice is called precision farming and its aim is to enhance the economic production of field crops and to protect the environment through more precise timing and usage of inputs like seed, chemicals, fertilizers, and irrigation water (McKinion *et al.*, 2001). Interest in precision farming is driven by the availability of field combine monitors, global positioning sensors (GPS), and GIS that can produce maps of field yield data (Paz *et al.*, 2001a, b).

Precision farming requires a description of the basis of yield variability, which can be exploited for each management unit within fields to reduce costs and increase profits. Addressing these issues will be an interesting way to involve producers in crop modeling (Boote *et al.*, 1996). Crop models have been coupled to GIS, to produce maps of predicted yield, both on a regional scale and for single fields with small subunits. This application will require costly input information to characterize field units and many simulation trials to optimize management for the entire field. For examples of application refer to Paz *et al.* (2001a, b) and McKinion *et al.* (2001), which are based on DSSAT and GOSSYM/COMAX models, respectively.

McKinion *et al.* (2001) studied precision farming options using crop simulation models and GIS technology for cotton production system in the

Table 4.4. Summary table of soybean yield response as affected by different levels of plant stresses in Nashua, Iowa, USA. Planting date was 1 May 2003 and plant population is 18 plants m^{-2}. For the relevant simulations, date of hail damage was 21 June 2003 and date of herbicide injury was 21 July 2003 (Paz *et al.*, 2004).

SCN egg count	Herbicide injury level	Hail defoliation damage (%)	Rhizoctonia root rot severity	Water stress	Grain yield (kg ha^{-1})
0	0	0	0	N	4286
0	0	0	0	Y	3668
5000	0	0	0	Y	2665
5000	0	50	0	Y	2625
5000	50	50	0	Y	2274
5000	50	50	50	Y	1938

SCN = soybean cyst nematode.

Table 4.5. Comparison of total actual inputs made by grower and inputs recommended by simulated precision agriculture (McKinion *et al.*, 2001).

	Nitrogen input (kg N ha^{-1})	Irrigation applied (mm water)	Yield (kg ha^{-1})
Actual	159.4	83	1088
Simulated	159.4	83	1123
Precision	124.3	109	1410

mid-south USA. They applied GOSSYM/COMAX in their analysis. Simulations were done using an expert system to optimize water and nitrogen input on 1 ha subunits in the field. The model predicted that an increase of 322 kg ha^{-1} could be obtained by using only an average increase of 26 mm of irrigation ha^{-1} and an average decrease of 35 kg N ha^{-1} (Table 4.5).

Pre-harvest yield forecasting

Pre-harvest yield forecasting, especially over large areas, is important to the producers (harvesting and transport), the processing agents (scheduling of transportation and processing), as well as the marketing entities. Reliable yield forecasting within the growing season would enable improved planning and more efficient management of grain production, handling, and marketing. There are some successful examples where crop models have been used to predict crop yield well before harvesting (Hodges *et al.*, 1987; Bannayan *et al.*, 2003). In this application, a crop model and existing weather data are used to predict crop growth until the current point in time, e.g. anthesis. Then, stochastically generated or historical weather data are used to predict crop growth during the remaining growing season and final yield. For example, Bannayan *et al.* (2003) were able to successfully predict final wheat yield at anthesis without any significant difference with measured yield. They used CERES-Wheat model, actual weather data until anthesis and generated weather data after anthesis.

Educational Applications

It has been said that "one of the best ways to promote deep conceptual understanding of the real world is through the investigation of simulation models" (cited in Graves *et al.*, 2002). Crop models have successfully been used in laboratory and classroom-based education and in farmers' education (Sinclair and Seligman, 1996; Graves *et al.*, 2002). Graves *et al.* (2002) nicely reviewed the application of crop modeling in education. They pointed out that the first educational use of crop models dates backs to 1981 when McLaren and Craigon (1981) developed and used TRITIGRO. Graves *et al.* (2002) synthesized the positive and negative experiences in education to provide guidelines for using crop models in computer laboratories and the classrooms. They identified the major benefits to students as the following:

1. Crop models reduce the time required for experimentation and observation because they provide rapid simulation of the whole growing season, or even the simulation of several seasons. Class time is limited for students and is a major constraint in conducting traditional field or laboratory experiments. Hence, such experiments are infrequently incorporated into teaching programs. Therefore, crop models may offer an effective surrogate in teaching for real-life experiments.
2. Crop models provide increased control over plant and environmental variability (e.g. weather, soil, pests, and diseases) that frequently confounds field experiments. This variation can negatively affect the anticipated learning outcomes. Using a crop model, however, allows these environmental factors to be controlled and the impact of the treatment can be isolated.
3. Certain obscure biophysical processes cannot be effectively observed in the laboratory or can only be observed using expensive equipment. A simulation model can demonstrate these processes and allow greater insight into cause and effect than would be possible in simple observational experiments.
4. Complex concepts and relationships in crop science are more easily conveyed to students through crop models than through traditional means.
5. Crop models help in integration of fragmented knowledge. Typical reductionist scientific approaches fragment knowledge while for students, an understanding of a whole picture and how various disciplines interact and interrelate is very important. Crop models are a means to integrate disciplines, knowledge, and relationships beyond the individual disciplines.
6. Crop models are able to promote heuristic learning. Lessons in good crop management that might have taken months or years to learn in the real world are demonstrated quickly with models.

Using crop models, like other tools, is accompanied by limitations. Major limitations (Graves *et al.*, 2002) include the following:

1. Excessive use and reliance on crop models can lead to the loss of field and laboratory skills. Students must learn the practical skills required for the measurement and recording of data produced by scientific experimentation.
2. Crop models may separate students from the real-world phenomena on which they are working. This might lead to an inappropriate view that crop models are "reality" (Chapters 1 and 2).
3. Crop models may be used by students for simulations outside the geographical and environmental range of the model, giving misleading results and understanding.
4. Using crop models may be frustrating and boring due to difficulties with software.

Graves *et al.* (2002) concluded that crop models are valuable tools in education. However, they must be properly integrated into the teaching program and appropriately used by instructors. They stated that the following factors determine the success of using crop models in education:

1. Crop models should be used as an adjunct to, rather than as a substitute for, other teaching methods.
2. Instructors should maintain sufficient dialogue with students.

3. Exercises developed for use with crop models should encourage cognitive advances by the student.
4. Visual appeal and clarity of crop models should be ensured with standardized interfaces and graphical, dynamic representation of results.
5. Models must be transparent because transparent models are usually the most useful in education as they facilitate the process of learning. This does not mean that black-box models cannot be used in education. Students might learn from the models by evaluation of their outputs produced as a result of given inputs.
6. Input and output values should be in units appropriate to the topic of study or the country of use.
7. Crop models need to come with default values for parameters.
8. Online help, explaining the science behind the crop models, is important.
9. Students should find using the software clear, simple, intuitive, and flexible. There have been examples where models developed for other purposes have been used in education and students have struggled with the interface. Students then start to mistakenly perceive that the objective of the exercise is to get the model to function.

Exercise

Try to find and review more examples of applications of crop models (one paper for each application).

5 Status of Crop Modeling

Before launching into describing options for modeling the various processes of crop development, growth, and yield, this chapter reviews the current status of crop modeling. In this chapter, we take an historical perspective on the progress made in the construction of crop models. Sinclair and Seligman (1996) described the progression in crop modeling using the analogy of the stages in organisms' life cycles. The stages they described were infancy, juvenility, adolescence, and maturity.

Infancy

After World War II, systems analysis and computer science developed to the stage where these technologies provided convenient and relatively user friendly techniques to simulate the interaction of components in complex systems. This development was influenced by the Cold War and space exploration. Crop modeling was born in this era. The first attempts at crop modeling were models developed to estimate light interception and photosynthesis in crop canopies (Loomis and Williams, 1963; de Wit, 1965; Duncan *et al.*, 1967). These models made it possible to assess the sensitivity of crop photosynthetic rates to sun angles, leaf angle distribution, and the latitudinal position of the crop and to calculate potential production. An important outcome of these early models was that more information was needed to better understand assimilate partitioning, ontogeny, product quality, and genetic control of crop characteristics. This also marked the end of the infancy stage of crop modeling.

Juvenility

Childhood is accompanied by widening horizons. Development of the new technologies and advances in computer science, innovation and production of new equipment for field experimentation and data logging during the late 1960s and early 1970s opened new vistas in understanding crops. It seemed straightforward to proceed with modeling the many factors that influence crop yield: weather, soils, crop genetics, plant physiology, and pest damage.

Important advances in modeling stomatal conductance and leaf gas exchange (e.g. Cowan, 1977), and growth and maintenance respiration (Penning de Vries, 1975) were made during this period. As the timeframe of models was lengthened to include the entire growing season, crop phenology and partitioning of assimilate among various tissues became an important consideration.

Some modelers thought that it was possible to develop universal models by quantifying individual plant processes and their responses to the environment (Whisler *et al.*, 1986). Therefore, the complexities of crop models increased as various details were incorporated into the models. Ultimately, a number of models with tremendous complexity, such as GOSSYM (Whisler *et al.*, 1986) and SOYGRO (Wilkerson *et al.*, 1983) were developed.

The development of these complex models was accompanied by some of the stresses and strains frequently associated with juvenility. As models became more complex, the number of parameters required to describe the system increased greatly. Many coefficients were needed to describe cultivar characteristics and estimation of these coefficients included the inevitable experimental errors. Some parameters were included in models that could not be measured directly in experiments. These parameters were quantified by "calibrating" the whole model to achieve outputs that matched relevant observations. All these complications greatly limited transparency due to ambiguous interconnections between processes and inhibited implementation of the models. It was realized that the role and function of models in solving engineering problems did not transfer directly to biological systems (de Wit, 1970). In an engineering model, unlike a biological model, all components are defined and have clear specifications.

Biological systems are composed of a vast number of components and processes interacting over such a very wide range of organizational levels. It is impossible to identify all possible factors for all situations that may influence their performance (Mayr, 1982; Pease and Bull, 1992). Crop models are highly simplified surrogates of the cropping system, even when defined in great detail. The discrepancy between the model and the actual system is inevitable and one-to-one correspondence is, therefore, unattainable. Consequently, the dilemma presented by the attempt to model the complexity of the crop system while avoiding the danger of sinking into a "madness of detail" became increasingly acute (Sinclair and Seligman, 1996).

Adolescence

Adolescence is commonly associated with considerable confusion and turmoil. In this stage of transition from juvenility to adulthood, basic assumptions are questioned and perspectives are changed. The unbounded possibilities of earlier developmental stages shrink as the realities of limited resources and possibilities encroach on original expectations. So too with crop modeling the original tenets need to be reevaluated in the light of accumulated experience (Sinclair and Seligman, 1996). Three of the original, basic tenets of crop models have been discredited (Sinclair and Seligman, 2000): models are not necessarily improved by extensive reductionism, universal crop models cannot be constructed, and models cannot be validated. A new perspective on the construction and benefits of crop models was necessary.

Extensive reductionism

Many modeling efforts were based on the assumption that a powerful model, from a scientific point of view, is the model in which processes are quantified in detail based on their physical, chemical, and physiological principles. As a result, extensive reductionism was applied and complex models were developed. In many cases, however, increasing reductionism forced the use of expressions that were often no more than weakly supported hypotheses made by the model builder. For example, the processes that determine how materials are partitioned within the plant are not well understood. In order to describe these processes within the plant, model builders have defined hypothetical pools of compounds that responded to supply and demand. Such reductionism when inappropriately applied can be misleading. When a high level of plant organization is being modeled, its use may well give a more distorted representation of organ growth than the use of conservative allometric relationships.

There are a number of examples where detailed, reductionist models are less reliable than simpler models for simulating observations or predicting yields. A simple water balance model was found superior to COTTAM and GOSSYM in approximating crop water stress and field water balance (Asare *et al.*, 1992). An empirical equation was found superior to CERES in predicting annual potential wheat yields in Mexico (Bell and Fischer, 1994). Goudriaan (1996) compared the performance of 14 mechanistically based wheat models representing a full range of complexity from very simple models involving only a few lines of code to a model of such great complexity that a CRAY supercomputer was required for simulations. The models were given the same input data for two locations to simulate wheat growth and yield. Good yield predictions were not associated with model complexity, and increasing reductionism in models did not result in less variability in predictions among the complex models.

Universal models

It has been found that in crop modeling each new season or new location brings new challenges that were likely not foreseen in the original model, and the expectation of universality fails. For example, attempts to use existing crop models developed for higher latitudes failed when an attempt was made to simulate crops in the semi-arid tropics of Australia (Carberry and Abrecht, 1991).

Important deficiencies were found in each of three complex wheat models even after they had been calibrated for a new set of conditions in New Zealand (Porter *et al.*, 1993). Considerable effort and model modification is required to make models account for discrepancies that derive from changes in cultivars, cropping conditions, and peculiarities of the application environment. The practical consequence is that it is impossible to create universal crop models (Spitters, 1990).

Validation

Finally, there is the tenet that crop models must be validated. There is a fundamental difficulty with this concept as discussed previously in Chapter 3.

Maturity

With developing awareness not only of the limits in modeling system behavior but also of the nature of the essential limiting factors, a stage of maturation started in the early 1990s (Sinclair and Seligman, 1996). The limits of crop models as surrogates for reality were being recognized and accepted as inevitable consequences of simplification. The role of models as heuristic tools to aid interpretation of reality was recognized as an important goal of crop modeling (Wullschleger *et al.*, 1994). Crop models could be used effectively to discover (or uncover) faulty reasoning or interesting implications about crop development and growth. Therefore, models as discussed in Chapter 4 became very useful tools in teaching, research, and applied modes as powerful aids to reasoning about the performance of a crop or about the relative benefits of alternative plant genetics and management strategies.

In this maturity phase several major modeling approaches and frameworks have evolved. The group working in Wageningen, the Netherlands, has a long tradition in developing and applying crop models, based on the pioneering work of C.T. de Wit (van Ittersum *et al.*, 2003). Modeling activities of this group began with calculation of photosynthesis of plant canopies in the mid-1960s and is still continuing. Bouman *et al.* (1996) reviewed the pedigree of Wageningen models from 1965 to 1995 (Fig. 5.1). In the 1960s and 1970s the main aim of these modeling activities was to obtain understanding at the crop scale based on the underlying processes. de Wit and co-workers developed the model BACROS and evaluated components of the model with field experiments (Penning de Vries *et al.*, 1974; van Keulen, 1975; Goudriaan, 1977; de Wit, 1978). In the 1980s a wide range of scientists in Wageningen became involved in development and application of crop models. The generic crop model SUCROS for potential production situation was developed, which formed the basis of Wageningen crop models such as WOFOST, MACROS, ARID CROP, SAHEL, PAPRAN, INTERCOM and ORYZA during the 1980s and 1990s (van Ittersum *et al.*, 2003). Spitters and Schapendonk (1990) developed a simple model (LINTUL) in which calculation of photosynthesis and respiration was replaced by radiation use efficiency.

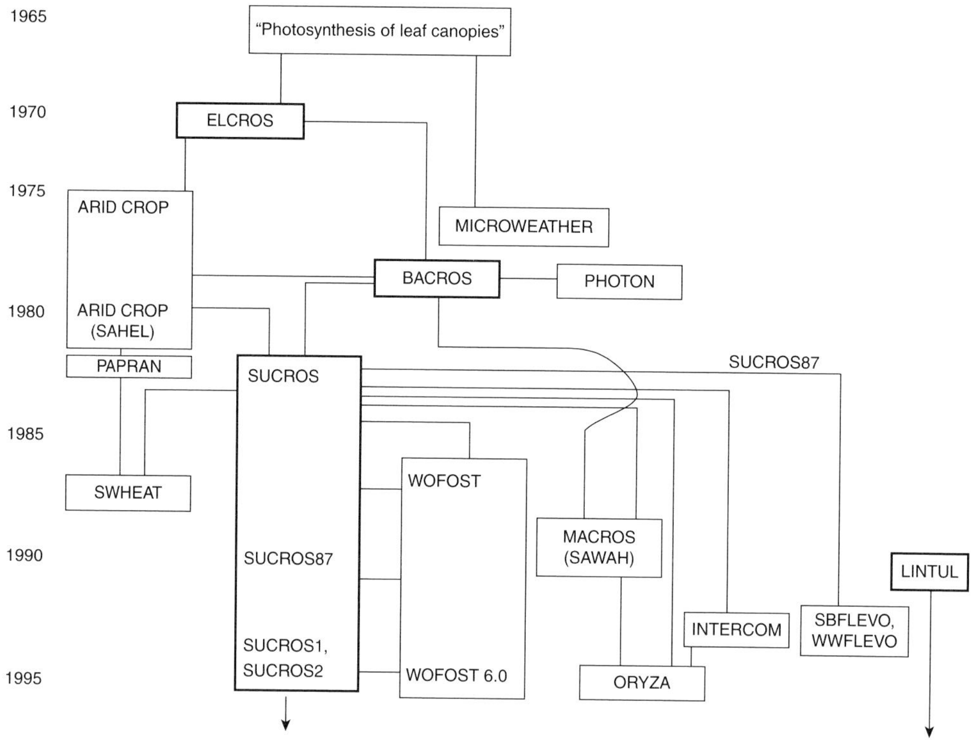

Fig. 5.1. Pedigree of crop models developed by Wageningen group, 1965–1995. Models in bold boxes have been "lead" models for the development of other crop models (Bouman *et al.*, 1996).

Later work by the Wageningen group resulted in development of ORYZA 2000 (Bouman *et al.*, 2001), GECROS (Yin and van Laar, 2005), and LINTUL3 (Shibu *et al.*, 2010). These modeling approaches have served as basis and inspiration for modeling groups around the world (van Ittersum *et al.*, 2003).

Another well-known modeling group is DSSAT (Decision Support System for Agrotechnology Transfer), which was originally developed by an international network of scientists, cooperating in the International Benchmark Sites Network for Agrotechnology Transfer project (IBSNAT) to facilitate the application of crop models to agronomic research using a systems approach. Its initial development was motivated by a need to integrate knowledge about soil, climate, crops, and management for making better decisions about transferring production technology from one location to others where soils and climate differed (Jones *et al.*, 2003).

DSSAT has been in use since 1989 by researchers worldwide. It incorporates models of more than 15 different crops with software that facilitates the evaluation and application of the crop models for different purposes. Recently, the DSSAT crop models have been redesigned and programmed to facilitate more efficient incorporation of new scientific advances, applications, documentation, and maintenance. Now, it is a collection of independent programs that operate together; crop simulation models are at its center (Fig. 5.2). Databases

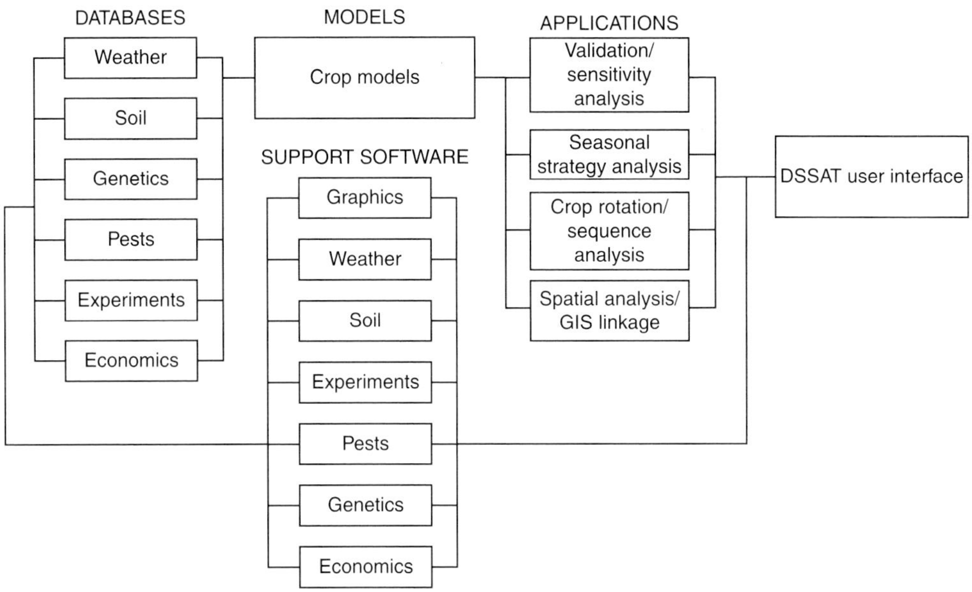

Fig. 5.2. Diagram of database, application, and support software components and their use with crop models for applications in DSSAT (Jones *et al.*, 2003).

describe weather, soil, experiment conditions and measurements, and genotype information for applying the models to different situations. Software aids potential users to access these databases and compare simulated results with observations to give them confidence in the models or to determine if modifications are needed to improve accuracy. In addition, programs contained in DSSAT allow users to simulate options for crop management over a number of years to assess the risks associated with each option. DSSAT has been used in many studies for various applications. For more information refer to Tsuji *et al.* (1998) and Jones *et al.* (2003).

Modelers in Australia developed APSIM (Agricultural Production systems SIMulator), which is a modular modeling framework. This model is based on earlier work of these modelers (e.g. Carberry and Abrecht, 1991; Chapman *et al.*, 1993; Muchow *et al.*, 1994; Hammer *et al.*, 1995). APSIM was developed to simulate biophysical processes in farming systems, in particular where there is interest in the economic and ecological outcomes of management practice in the face of climatic risk (Keating *et al.*, 2003; Hammer *et al.*, 2010). APSIM has many different plant, soil, and management modules, including a diverse range of crops, pastures, and trees, soil processes including water balance, nitrogen and phosphorus transformations, soil pH, erosion, and a full range of management controls (Fig. 5.3). APSIM has been used in a broad range of applications, including support for on-farm decision making, farming systems design for production or resource management objectives, assessment of the value of seasonal climate forecasting, analysis of supply chain issues in agribusiness activities, development of waste management guidelines, risk assessment for government policy making, and as a guide to research and education activity.

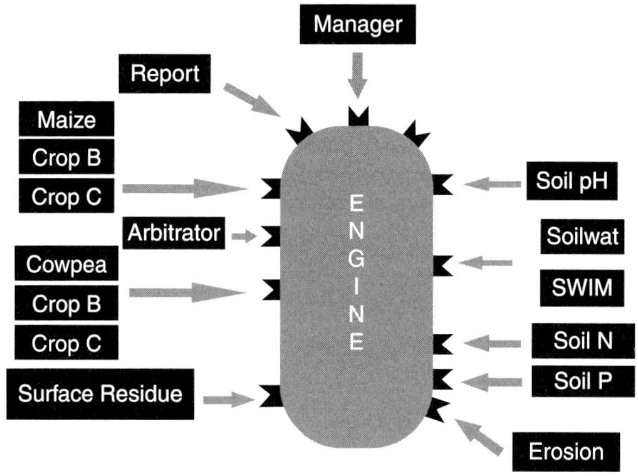

Fig. 5.3. Diagrammatic representation of the APSIM simulation framework with individual crop and soil modules, module interfaces, and the simulation engine (Keating *et al.*, 2003).

The CropSyst group is based in Washington State University, USA (Stockle *et al.*, 2003). The development of CropSyst started in the early 1990s (Stockle *et al.*, 1994). The motivation for its development was based on the observation that there was a niche in the demand for cropping systems models, particularly those featuring crop rotation capabilities, which was not properly served. Efficient cooperation among researchers from several world locations, a free distribution policy, active cooperation of model developers and users in specific projects, and careful attention to software design from the outset allowed for rapid and cost-effective progress.

The CropSyst model has been used as an analytical tool to study the effect of climate, soils, and management on cropping systems productivity and the environment. CropSyst simulates the soil water and nitrogen budgets, crop growth and development, crop yield, residue production and decomposition, soil erosion by water, and salinity. CropSyst has several utility programs including a weather generator, GIS application, and a watershed model. CropSyst has been applied to perform risk and economic analyses of scenarios involving different cropping systems, management options, and soil and climatic conditions. For more information refer to Stockle *et al.* (2003).

Researchers at the USDA Blackland Research Center, Texas, USA, have had an influential impact on crop modeling by developing models like ALMANAC (Kiniry *et al.*, 1992; Debaeke *et al.*, 1997; Xie *et al.*, 2003) and EPIC (Williams *et al.*, 1984, 1989; Ko *et al.*, 2009). CERES-Maize, which is part of the DSSAT package now, was first developed at this center (Jones and Kiniry, 1986). Both ALMANAC and EPIC are able to simulate crop growth and yield using simple algorithms and include many utilities to facilitate model application. In addition, ALMANAC is able to simulate crop–weed competition. Both models have been used around the world and have been implemented in other models such as the watershed model, SWAT (Arnold *et al.*, 1998).

Besides the crop modeling groups described above, there have been many individuals who have had a prominent impact on development of crop modeling. J.L. Monteith developed some important concepts in crop modeling, e.g. radiation use efficiency (Monteith, 1977) and soil water extraction by roots (Monteith, 1986). He and co-workers developed a few simple models (RESCAP and PARCH) and applied them in simulation analyses (Monteith *et al.*, 1989; Bradley and Crout, 1994).

The paper on calculation of crop potential growth by Loomis and Williams (1963) is considered one of the pioneering works in crop modeling. Loomis and co-workers also developed detailed models to simulate crop growth and yield in sugarbeet (Fick *et al.*, 1978) and potato (Ng and Loomis, 1984).

T.R. Sinclair and colleagues developed a range of simple crop models for different crops including soybean (Sinclair, 1986), maize (Sinclair and Muchow, 1995), sorghum (Sinclair *et al.*, 1997), wheat (Sinclair and Amir, 1992), barley (Wahbi and Sinclair, 2005), peanut (Hammer *et al.*, 1995), and chickpea (Soltani and Sinclair, 2011). These models were successfully applied in simulation analysis with the objectives of crop genetic improvement and increased crop yield. T.R. Sinclair has been involved in development of some basic concepts in crop modeling, e.g. radiation use efficiency and its determinants (Sinclair and Horie, 1989), linkage between dry matter production and transpiration (Tanner and Sinclair, 1983), nitrogen limitation in yield formation and resultant leaf senescence (Sinclair and de Wit, 1976), yield formation based on linear increase in harvest index (Speath and Sinclair, 1985), and quantifying crop responses to soil water deficit (Sinclair and Ludlow, 1986; Sinclair, 2005).

A joint modeling team from New Zealand and the UK developed a wheat model, Sirius (Jamieson *et al.*, 1998; Jamieson and Semenov, 2000), which includes interesting details and innovations with regard to phenology, leaf area development, and nitrogen uptake and distribution.

The Future of Crop Modeling

Hammer *et al.* (2003) discussed two important opportunities for crop modeling activities in the future. One of these opportunities is wider applications of crop models in their heuristic role (Sinclair and Seligman, 1996) to support scientific research, to facilitate farm management decisions and help in education and training. They stated that this aspect of crop modeling activities could also extend to broader environmental and ecological issues in crop production. Recent works of Rinaldi (2004), Wessolek and Asseng (2006), and Brisson *et al.* (2010) are examples of such applications (see Chapter 4).

A second opportunity for future uses of crop models is to describe and evaluate expectations for crop growth and yield resulting from genetic alterations of specific crop traits. As genetic studies of plant traits at the molecular level become more removed from easily visualized crop behavior, models can aid in extrapolating the genetic modification to actual impact on crops grown over a range of conditions. Physiological dissection and modeling of traits provides an avenue by which crop modeling could contribute to enhancing integration of molecular genetic technologies in crop improvement (Hammer *et al.*, 2003). Recent studies by Manschadi *et al.* (2006) and Sinclair *et al.* (2005a, 2010) are examples of such applications (see Chapter 4).

Indeed, the future is bright for progress in developing and using crop models. The increasing challenges and complexity in meeting the human demands for food and fiber will require greater insight about how the plant–environment system operates. What genetic modifications of various crop species have the potential for increasing yields? What management practices can be introduced to increase yield and minimize environmental impacts of growing crops? How do cropping practices interact with other systems on which humans depend? Crop models are on the cusp of making important future contributions to a number of critical issues in crop and environmental sciences. While developing quantitative relationships and constructing model code can be tedious, simplifying a complex system (such as a wheat field) into a computer program and then using it to analyze the system can be quite rewarding. It is an exciting time to develop the knowledge and skills to apply the crop modeling tool to future challenges in crop science.

The remaining chapters of this book offer the reader critical perspectives on the physiology of crop plants to model plant development, growth and yield under potential and water- and nitrogen-limited conditions. A balanced approach has been attempted in dealing with various processes so that each process is modeled at equivalent levels of complexity. All programs and models described in the book can be downloaded from the book's website (https://sites.google.com/site/cropmodeling). Thus, readers are free to explore the use of these models.

Exercise

The table below includes different stages of crop modeling as described by Sinclair and Seligman (1996). Discuss and try to fill out the table with the approximate beginning and ending dates for each stage and its duration.

Stage	Beginning	Ending	Duration
Infancy			
Juvenility			
Adolescence			
Maturity			

6 Phenology – Temperature

Phenology is the study of the progress in development stages in plant cycles, and how these are influenced by environmental factors. Modeling of phenological development is predicting the time of occurrence of different plant developmental stages such as emergence, flowering, or maturity. In Chapter 2, a simple model was presented to predict the development of successive leaves on sugarcane plants after the plant emerged from the soil up to the cessation of leaf production. To have a complete model, it is necessary to also predict the two critical stages of the beginning and ending of leaf development. Hence, a complete crop simulation model needs to include the capability of predicting the timing of various phenological events in the crop growth cycle.

As it turns out, modeling phenological development is likely to be critically important in modeling crops because unlike many other plant processes, there are large differences in phenological development among species and cultivars within a species. Without accurate predictions of phenology, the model will inaccurately simulate growth processes because the timing of events and the environmental conditions during these events will be in error (Hodges, 1991). Indeed, much of physiological improvement of crops to achieve greater yields has been a result of modification of phenological development.

Predicting phenological development in a crop model in itself can have important applications (Hodges, 1991; Ritchie and NeSmith, 1991; McMaster *et al.*, 2009).

1. The ability to predict the stage of crop development is important for such management decisions as timing pesticide application, scheduling the orderly harvest of crops, or synchronizing the flowering of cross-pollination crops for hybrid seed production.
2. Predicting crop growth duration is necessary to find cultivars with a desired growth period that enables farmers to optimize yields. In many regions, growth of

crops needs to be adjusted to minimize the impact of episodes of drought, heat, or cold. For optimum yield, it is essential that cultivars and sowing dates be selected so that stages critically affecting economic yield will occur during periods of optimum growing conditions and so that full use will be made of these periods.
3. Temperature increases due to global increases in atmospheric CO_2 concentration and other greenhouse gases will likely require changes in phenological responses via new crop cultivars or new crop types. To optimize phenological responses to changing temperatures, it will be necessary to have modeling tools to predict responses to evolving changes in local climates.

Several environmental factors affect phenological development of plants, including temperature, photoperiod, drought, and nutrition. Among them, temperature and photoperiod are commonly the most important factors. For example, unfavorable temperature and photoperiod may delay flowering time of chickpea from 40 to 250 days (Soltani *et al.*, 2006b). In Chapter 7 the combined influence of temperature and photoperiod are considered. The impacts of drought (Chapter 15) and nutrient limitations usually directly influence phenological development only for a few days. In this chapter, the basics of modeling phenological development are developed by considering only the impact of temperature.

Background

Farmers and agricultural scientists have long realized that calendar time is not a suitable measure to characterize plant age, and that temperature has a large influence (Gilmore and Rogers, 1958). For example, at 30 days a plant may have ten leaves in one year, but six leaves in another year. To resolve this problem, calendar age has been replaced by plant development age.

One of the first approaches to describing development age was plastochron index, which is the number of meristematic leaves formed in the growing tip by a specific time (Loomis and Connor, 1992). Plastochron is the interval between the initiation of successive leaves. Using this index requires destructive sampling of the plant to identify at the microscopic level when a new leaf is initiated. Plastochron index has been generally replaced with the more easily observed progress in development of visible leaves or number of nodes on the main stem of the plant, which is identified as the phyllochron index.

Phenological development of plants is now described for most crop species using standard developmental keys, e.g. maize (Ritchie *et al.*, 1992) and rapeseed (canola) (Sylvester-Bradley and Makepeace, 1984). A summary of the standard keys for wheat and soybean is given in Tables 6.1 and 6.2. Under field conditions, the identification of the developmental stages can be a bit challenging because of variation among individual plants. When identifying standard phenological stages in the field, a specific stage is assumed to occur if 50% of the individual plants show that stage.

The cumulative temperature unit concept has been extensively used to quantify phenological development (McMaster and Wilhelm, 1997; Bonhomme, 2000; Sinclair and Weiss, 2010). Various terms have been used by different

Table 6.1. Important phenological stages in wheat based on a combination of Feeks (Large, 1954) and Zadoks (Zadoks *et al.*, 1974) methods.

Stage	Definition (observed in 50% of plants)
Emergence (EM)	Appearance of coleoptile (any part) at soil surface
Beginning of tillering (TL)	First tiller visible
Stem elongation (SE)	First node visible
Termination of leaf production on main stem (TLM)	Ligule of flag leaf visible
Ear emergence (EE)	Awns or ear visible from flag leaf sheath
Anthesis (ANT)	Half of anthers opened in spikes
Physiological maturity (PM)	Spike and plant have lost their green color
Harvest maturity (HM)	Plant is dry and grains are hard and dry

Table 6.2. Important phenological stages in soybean based on a Fehr and Caviness (1977) method.

Stage	Definition (observed in 50% of plants)
Emergence (VE)	Cotyledons above the soil surface
Beginning flowering (R1)	One open flower at any node on the main stem
Beginning pod (R3)	Pod 5 mm long at one of the four uppermost nodes on the main stem with a fully developed leaf
Beginning seed (R5)	Seed 3 mm long in a pod at one of the four uppermost nodes on the main stem with a fully developed leaf
Beginning maturity (R7)	One normal pod on the main stem that has reached its mature pod color
Full maturity (R8)	95% of the pods have reached their mature pod color. (5–10 days of drying weather are required after R8 for the soybean moisture levels to be reduced to less than 15%)

researchers, e.g. degree-days, growing degree-days, heat units, heat sums, thermal units, and thermal time. The term "degree-days" is rejected here because the term should not include "days" as a part of its definition. As these terms all refer to a summation of temperature and only have the units of Celsius, the term "temperature unit" is used in this book.

The traditional method for calculation of temperature unit was described in Chapter 2 (Eqn 2.1). In this method, each day a base temperature is simply subtracted from daily mean temperature to obtain the effective temperature experienced by the crop on that day, i.e. daily temperature unit. By summing daily values of temperature unit, cumulative temperature unit is obtained to assess crop progress in moving through one developmental stage to another. As we will see, temperature unit calculations can become more complicated if there is a maximum development rate at high temperatures, or under very high daily temperatures where development rate might actually decrease.

The traditional method of calculating temperature unit is reliable under the following conditions (Ritchie and NeSmith, 1991):

1. The temperature response of development rate is linear over the range of temperatures experienced by the crop.
2. The daily temperature does not fall below the base temperature for a significant part of the day.
3. The daily temperature does not exceed an upper threshold temperature for a significant part of the day.
4. The developing meristem of the plant is exposed to the same mean temperature as the average daily air temperature. This assumption can be violated, for example, in the early development stages when the meristem is still located in the soil and is exposed to soil temperature rather than air temperature (Vinocur and Ritchie, 2001).

Basics

In this chapter, a method for quantifying and predicting phenological development in grain crops such as wheat and soybean as a function of temperature is provided. The method is based on the temperature unit concept, which can be calculated from standard weather reports that include temperature data. It is assumed the response of development rate to temperature does not change during the crop life cycle. A specific stage is predicted to occur when the cumulative temperature unit reaches or surpasses a value that is required for completion of that specific stage.

The temperature-based method used here will give appropriate predictions under the following two conditions:

1. Crop or phenological phase is not sensitive to changes in photoperiod. (In Chapter 7, a simple method is presented to quantify the combined effects of both temperature and photoperiod.)
2. The crop does not encounter drought, nutrient deficits, insects, diseases, and weeds during growing season, or if it does these factors do not affect development rate.

To begin development of the phenological-development model, the following questions need to be answered:

1. Which phenological stages need to be predicted, e.g. emergence, flowering, and maturity? Selection of these stages depends on the objective(s) of the crop model as well as the crop species and the type of methods that are used to predict other processes.
2. What environmental factors control each phenological stage? In this chapter, it is assumed that temperature is the dominating factor in phenological development.
3. What is the response function of development rate to temperature and what are the values of the parameters in this function?

Which phenological stages need to be predicted?

Selection of phenological stages that should be predicted depends on the objectives of the model and the need for prediction of phenological stages to simulate other processes. In the next chapters, several key phenological stages are required to simulate physiological processes of growth and yield in the plant because major phenological events switch on or off other growth processes. These phenological events are:

- *Emergence*: the date when 50% of plants emerged from the soil.
- *Termination leaf growth on main stem*: the stage at which *effective* leaf or node production on main stem terminates (discussed further in Chapter 9).
- *Beginning seed growth*: the stage when grains *effectively* begin to grow. Linear increase in harvest index starts at this stage (Chapter 11).
- *Termination seed growth (physiological maturity)*: the stage when *effective* growth of grains terminates. Or, linear increase in harvest index ceases (Chapter 11).
- *Harvest maturity*: the time when grain moisture has decreased following physiological maturity to a level such that grains are ready for machine harvesting.

Additional stages might be added to the list above that are important for practical crop management purposes, for instance, timing of application of fertilizers or pesticides or estimating crop damages due to adverse weather for an insurance company.

What are the temperature response functions?

Plant response to temperature is calculated on a daily basis. The daily calculations are simplified by basing the response on the mean temperature (TMP, °C) for the day. The value is commonly calculated as the mean of the daily minimum and maximum temperatures. Consequently, the daily temperature response can be determined from TMP using a curvilinear response curve as illustrated in Fig. 6.1.

The various parts of the curvilinear response represent different sensitivities in the plant. The response to sub-optimal temperatures (region A in Fig. 6.1) is classically described by the Boltzman energy distribution (Sinclair, 1994). Using this distribution, the reaction rate of a system is best described by an exponential function of temperature. Segments of the exponential response function can be adequately represented by linear approximations (Sinclair, 1994). Region B in Fig. 6.1 is related to enzyme saturation or substrate limitations response as a linear function of temperature. At high temperatures, there are likely injuries to critical enzymes associated with the decrease shown in region C (Loomis and Connor, 1992), or losses of membrane integrity.

To simplify simulations and facilitate model transparency, the curvilinear response (Fig. 6.1) is represented by linear segments for each region of the curve. It needs to be remembered, however, that the linear approximation represents

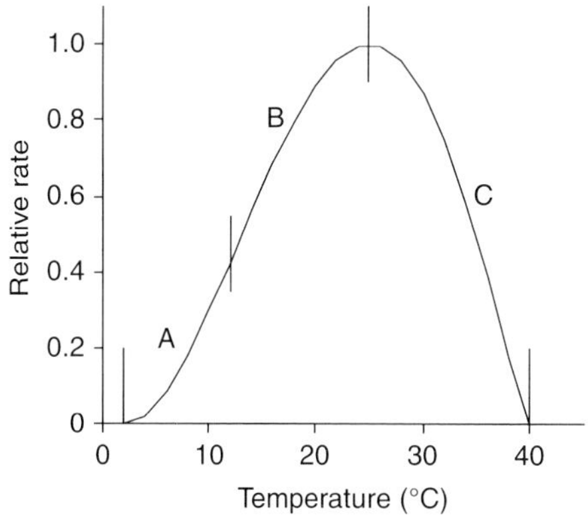

Fig. 6.1. Curvilinear response to temperature of plant processes. Various response regions (A, B, and C) are labeled.

only the temperature range for which the data were collected. Due to the inherent curvilinearity of the temperature response, a different linear approximation may result when the temperature range used in obtaining the data for the approximation is changed. Therefore, one important consequence of this approximation is that the base temperature is not a constant, as often seems to be assumed, but reflects the linear fit for the temperature range under which the experimental data were collected. Base temperature is expected to increase as the range of temperatures being included in the linear approximation increases. For more information refer to Sinclair (1994).

A 3-segment linear function can be used to describe the curvilinear response curve. Four parameters (base temperature, lower optimum temperature, upper optimum temperature, and ceiling temperature) define the endpoints of each segment (Soltani *et al.*, 2006a, b). In the 3-segment function, rate of development is zero at temperatures lower than base temperature (TBD, °C) (Fig. 6.2a). Between base and lower optimum temperatures, the rate increases linearly from zero to its maximum value. Lower (TP1D, °C) and upper (TP2D, °C) optimum temperatures define a plateau where the rate of development has reached a maximum and can be assumed constant for this temperature range. The rate of development again decreases linearly above the upper optimum temperature. The decrease in development reaches zero at a ceiling temperature (TCD, °C). At temperatures greater than TCD, development rate is zero.

The output of the 3-segment linear function is a scalar factor between 0 and 1, *tempfun*. The function can be written in equations as:

$$
\begin{aligned}
tempfun &= 0 \quad \textit{if} \quad \text{TMP} \leq \text{TBD} \\
&= (\text{TMP} - \text{TBD}) / (\text{TP1D} - \text{TBD}) \quad \textit{if} \quad \text{TBD} < \text{TMP} < \text{TP1D} \\
&= 1 \quad \textit{if} \quad \text{TP1D} \leq \text{TMP} \leq \text{TP2D} \\
&= (\text{TCD} - \text{TMP}) / (\text{TCD} - \text{TP2D}) \quad \textit{if} \quad \text{TP2D} < \text{TMP} < \text{TCD} \\
&= 0 \quad \textit{if} \quad \text{TMP} \geq \text{TCD}
\end{aligned}
\tag{6.1}
$$

A special case exists when the lower and upper optimum temperatures are the same, so that the 3-piece linear-segment function becomes a 2-piece linear-segment function (see Fig. 6.2b). If the value of the upper optimum temperature is very high and exceeds any temperature expected in the simulated

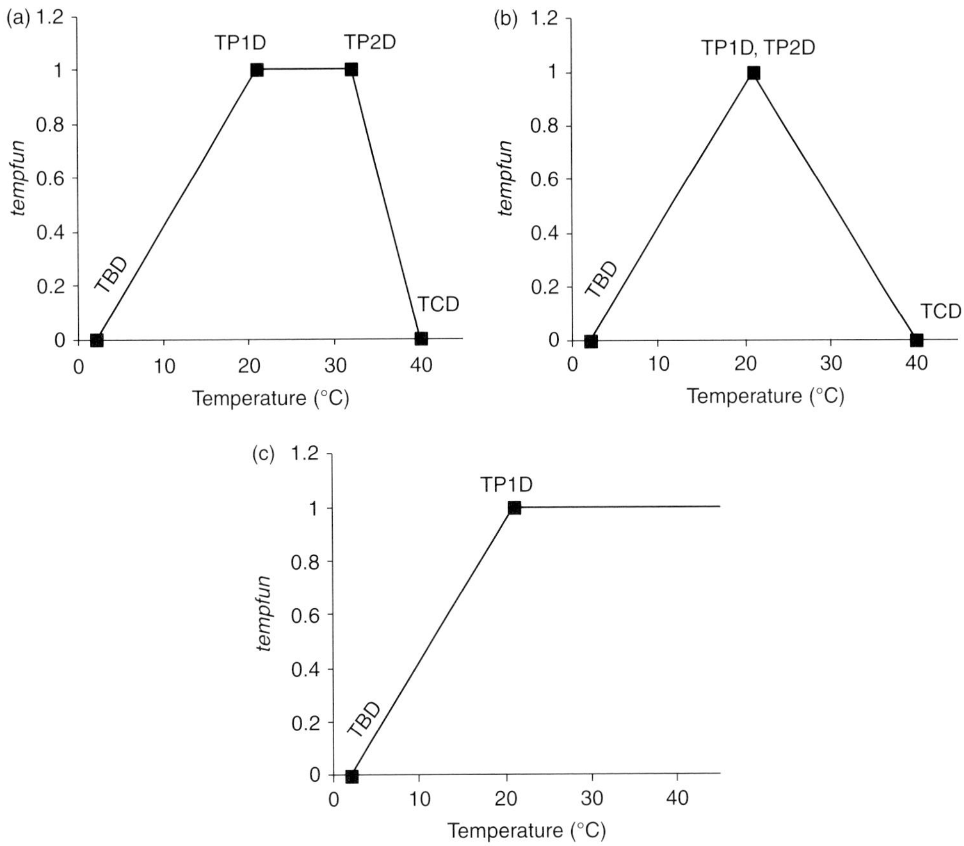

Fig. 6.2. Response of relative development rate, *tempfun*, to temperature as described by a 3-piece segmented function (a). The 3-piece segmented function converts to a 2-piece segmented function when lower and upper optimum temperatures are the same (b), and it converts to a linear-plateau function when upper optimum temperature is very high (c). Markers from left to right are base temperature (TBD, 2°C), lower optimum temperature (TP1D, 21°C), upper optimum temperature (TP2D, 21 or 32°C) and ceiling temperature (TCD, 40°C).

environment, the function is a 2-piece function with a linear increase up to a plateau region at high temperatures (see Fig. 6.2c).

Base, lower optimum, upper optimum, and ceiling temperature are called cardinal temperatures. Cardinal temperatures for a number of major crops are presented in Table 6.3. For many purposes the cardinal temperatures in Table 6.3 can be used because these values tend to be fairly stable within a species. Of course, if experimental evidence shows significant differences among genotypes (Sinclair *et al.*, 1991; Grimm *et al.*, 1993, 1994; Robertson *et al.*, 2002b; Soltani *et al.*, 2006a, b), these differences need to be considered in the model. Cardinal temperatures may also vary depending on developmental stage as found by Piper *et al.* (1996) for soybean. However, constant cardinal

Table 6.3. Estimates of cardinal temperatures for phenological development in important crops. The estimates are synthesized from different crop models (Hammer *et al.*, 1995; Jones *et al.*, 2003; Keating *et al.*, 2003; Stockle *et al.*, 2003; Yin and van Laar, 2005; Soltani *et al.*, 2006a, b).

Crop	TBD	TP1D	TP2D	TCD
Wheat	0	25	28	40
Barley	0	25	28	40
Rice	8	30	37	45
Maize	8	30	37	45
Sorghum	8	30	37	45
Soybean	8	30	35	45
Peanut	8	30	35	45
Canola	0	25	28	40
Sunflower	8	30	34	45
Dry bean	8	30	35	45
Chickpea	0	25	30	40

TBD: base temperature for development (°C)
TP1D: lower optimum temperature for development (°C)
TP2D: upper optimum temperature for development (°C)
TCD: ceiling temperature for development (°C)

temperature during whole crop life cycle is assumed in this book, although it is not difficult to incorporate into the model variation in cardinal temperatures.

With *tempfun* calculated from daily mean temperature, temperature units experienced by a crop on a given day (DTU, °C) can be obtained. First, DTU is computed assuming the temperature of the day is optimal for phenological development (i.e. DTU = TP1D – TBD), then it is corrected for actual temperature of the day using *tempfun*:

$$DTU = (TP1D - TBD) \times tempfun \tag{6.2}$$

Figure 6.3 illustrates calculations of DTU for the crop with cardinal temperatures presented in Fig. 6.2. Once the daily value of DTU has been determined, cumulative temperature unit during a particular phenological stage up to the current day (CTU_i) is obtained by adding DTU to CTU of the previous day (CTU_{i-1}, °C):

$$CTU_i = CTU_{i-1} + DTU \tag{6.3}$$

Figure 6.4 presents the flowchart of the method described above to calculate temperature unit prediction of different phenological stages.

To predict the transition from one phenological stage to the next stage (say to Stage *b* from Stage *a*), cardinal temperatures and the cumulative temperature unit required to complete Stage *a* (CTU_{ab}, °C) must be known. Once the value of CTU_i exceeds CTU_{ab}, Stage *a* ends and Stage *b* begins.

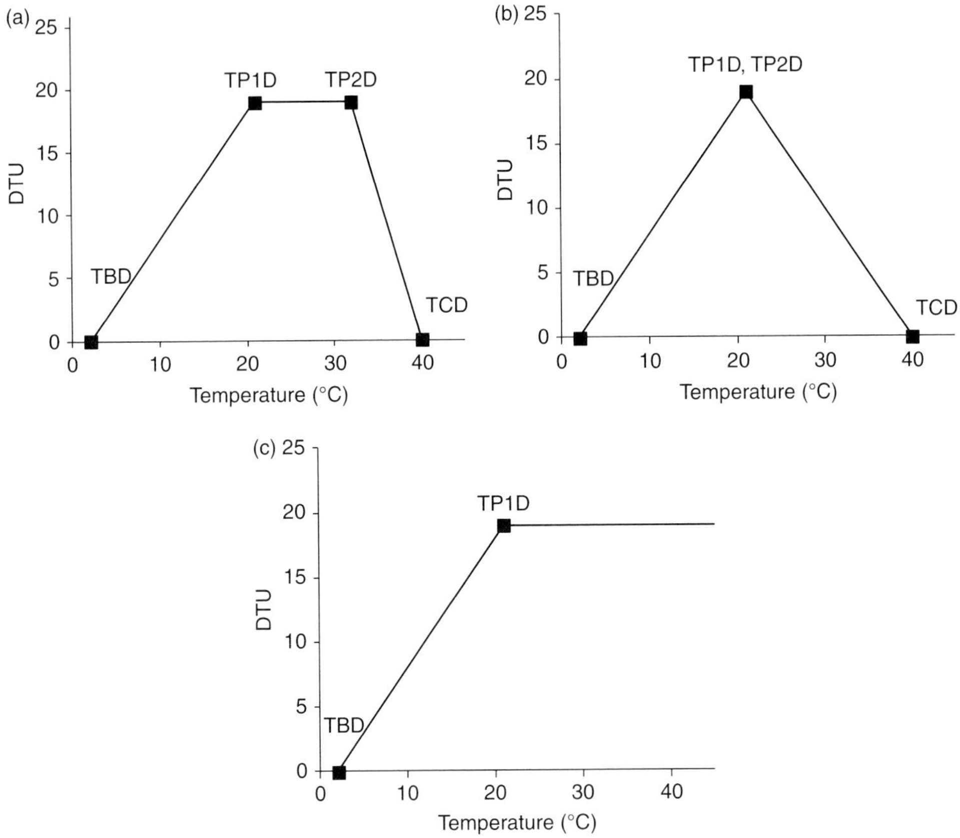

Fig. 6.3. Daily temperature unit (DTU, °C per day) as a function of mean daily temperature. This example indicates how temperature unit is obtained from *tempfun* for the crop with cardinal temperatures shown in Fig. 6.2. Markers from left to right are base temperature (TBD, 2°C), lower optimum temperature (TP1D, 21°C), upper optimum temperature (TP2D, 21 or 32°C) and ceiling temperature (TCD, 40°C).

The *minimum* number of calendar days required for a phenological stage (CBD_{ab}, days) can be calculated directly from CTU_{ab} and the maximum DTU per day (= TP1D – TBD). This conversion is done by dividing CTU_{ab} by maximum DTU per day. That is,

$$CBD_{ab} = CTU_{ab} / (TP1D - TBD) \tag{6.4}$$

For example, CBD_{ab} can be calculated for the emergence of a crop if cumulative temperature unit from sowing to emergence (CTU_{emerg}) is known. Assuming a base temperature of 0°C and an (lower) optimum temperature of 20°C, if CTU_{emerg} is 100°C, the value of CBD_{emerg} is equal to 5 (= 100 / 20) days. That is, it will take 5 days for the crop to emerge under optimum temperature.

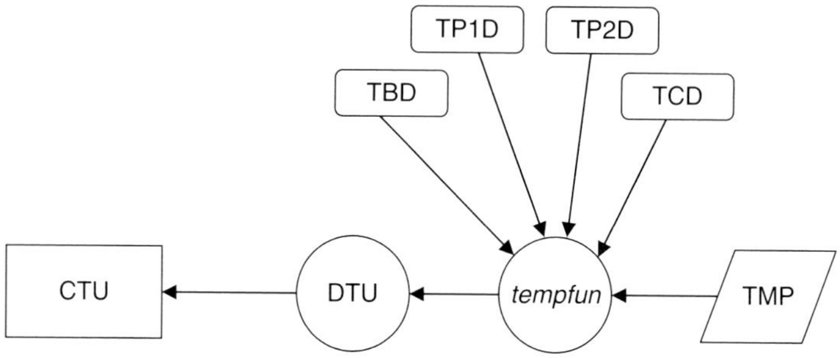

Fig. 6.4. Relational diagram of phenology submodel. TMP is the mean daily temperature (°C), *tempfun* the temperature function, DTU the daily temperature unit (°C), CTU the cumulative temperature unit (°C), TBD the base temperature for development (°C), TP1D the lower optimum temperature for development (°C), TP2D the upper optimal temperature for development (°C), and TCD the ceiling temperature for development (°C). Shapes are defined in Table 1.2.

Box 6.1 summarizes the method to quantify phenological development.

Box 6.1. Summary of computation in phenological development submodel from sowing to harvest.

- Temperature unit requirements of different phenological stages are defined.
- Daily mean temperature is used to find a scalar factor (0–1) that accounts for the effect of temperature on development rate.
- Daily temperature unit is calculated and corrected for daily mean temperature.
- Cumulative temperature unit is obtained.
- A phenological stage is predicted to occur if cumulative temperature unit has just reached or passed temperature unit requirement of that stage.

Parameter Estimation

Cardinal temperatures

The simplest way experimentally to determine values of cardinal temperatures is to examine germination or emergence of the desired species or cultivars under different temperatures under laboratory conditions (e.g. Ghaderi *et al.*, 2008) or in the field (e.g. Jame and Cutforth, 2004). Plants in these experiments should not encounter other limitations like drought. Under field conditions, different sowing dates can be used to create different temperature regimes. For example, Soltani *et al.* (2006a) used a serially sown field experiment with 12 sowing dates (one for each month) to obtain cardinal temperatures for chickpea emergence. From these experiments, the observed number of calendar days for emergence (DTE) is recorded for each experiment or treatment. The daily rate of progress toward emergence is, therefore, equal to 1/DTE.

The daily rate of developmental progress in emergence of chickpea (1/DTE) is graphed in Fig. 6.5 against the mean temperature for the individual experiments/treatments. Assuming the data fit a 3-segment linear function, the four cardinal

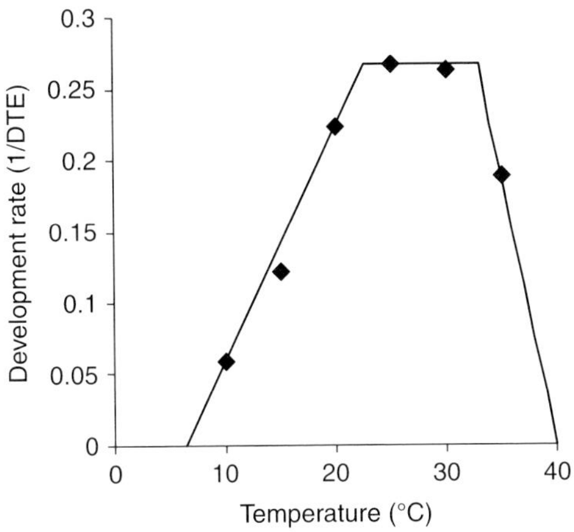

Fig. 6.5. Example fit of a 3-piece segmented function to data of development rate (sowing to emergence) versus mean temperature in chickpea sown at 3cm depth (A. Soltani and M. Yousefi, unpublished data).

temperatures can be resolved by regression analysis of the data. More information of such an analysis is presented by Soltani *et al.* (2006a).

Sometimes, cardinal temperatures are derived from observations of leaf appearance rate in response to temperature. In this case, the number of leaves or nodes on the main stem is observed periodically for various temperature regimes. The slope of a linear relationship between leaf number and calendar days for each temperature regime is a leaf appearance rate. Graphing leaf appearance rate against the mean temperature of each regime gives a description of response to temperature. Cardinal temperatures can be obtained by fitting regressions to the resulting graph. Measurement of leaf appearance often seems to provide good estimates of cardinal temperatures but such experiments require sophisticated facilities such as growth chambers. An example of examining leaf appearance rate is presented by Villalobos and Ritchie (1992).

As indicated previously, cardinal temperatures within a species tend to be fairly constant across all development stages. The values given in Table 6.3 can generally be used with a good degree of reliability.

Temperature unit requirements

As mentioned earlier and as will be discussed in Chapters 9, 10 and 11, the following phenological events often need to be simulated as part of crop models:

- emergence;
- termination leaf growth on main stem (TLM);
- beginning seed growth (BSG);
- termination seed growth (TSG); and
- harvest maturity.

Experimental observations at some point are required to define the duration between events in terms of cumulative temperature units (CTU_{ab}). The first step is to determine the cardinal temperatures that are necessary in calculating CTU_{ab}. Here we have assumed cardinal temperatures are stable across phenological stages so these can be determined as described earlier. Having determined the cardinal temperatures, Eqn 6.2 is used to determine the daily temperature unit on each day (DTU, °C). The sum of the DTU between the beginning and end of a specific development stage (Eqn 6.3) is CTU_{ab}. Assuming a small variation in CTU_{ab} among experiments/treatments, the average of the CTU_{ab} is used as the mean value to describe phenological development for this particular stage. A program, *tu_calc.xls*, on the book's website helps in the calculations.

The exact temperature unit of termination of leaf growth on the main stem (TLM) can be found from a plot of main stem leaf (node) number versus temperature unit (Fig. 6.6). This relationship can be simplified using two intersecting lines, a sloping line for the linear increase in leaf (node) number and a horizontal line, which determines maximum leaf (node) number on the main stem. The inflection time between the two intersecting lines will be the temperature unit of TLM. For more information refer to Soltani *et al.* (2006c).

The exact point for the beginning of seed growth (BSG) and termination of seed growth (TSG) can be found from a plot of harvest index on the *y* axis and time or temperature unit on the *x* axis (Fig. 6.7). Again, this relationship can be simplified using two intersecting lines, a sloping line for the linear increase in harvest index and a horizontal line, which reflects the final harvest index. BSG is the time or temperature unit when the sloping line intersects the *x*-axis and TSG is the time or temperature unit when the sloping line of harvest index increase intersects the horizontal line of the final harvest index. For more information refer to Soltani *et al.* (2004a).

It is possible to relate TLM, BSG, and TSG obtained from the above analysis to standard observable phenological keys. For example, TLM coincides with ligule appearance of the flag leaf in wheat and R3 stage in soybean and

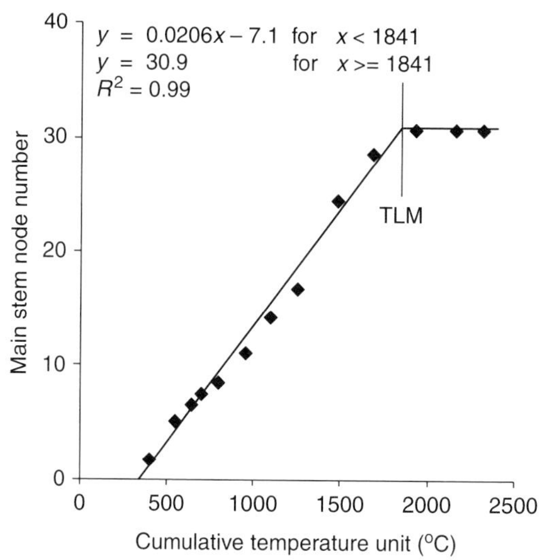

Fig. 6.6. A scatter plot of main stem node number versus cumulative temperature unit in chickpea (Soltani *et al.*, 2006c). This plot can be used to find phenological stage of termination leaf growth on main stem (vertical line) in terms of temperature unit.

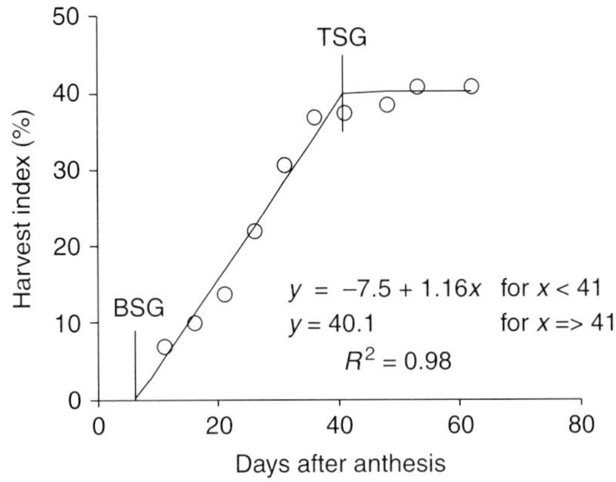

Fig. 6.7. A scatter plot of harvest index versus days after anthesis in wheat (Soltani *et al.*, 2004a). This plot can be used to find phenological stages of beginning seed growth and termination seed growth (vertical lines) in terms of days after anthesis.

chickpea (Tables 6.1 and 6.2). BSG occurs about 120°C temperature units after anthesis in wheat and R5 stage in chickpea. Finding these readily observed indices may minimize the intensive and time-consuming observations required to determine TLM, BSG, and TSG.

Some estimates of temperature unit requirements (CTU_{ab}) are given in Table 6.4. However, it should be remembered that while cardinal temperatures are relatively constant within a species, the values of temperature unit might

Table 6.4. Rough estimates of temperature unit requirements for different phenological phases in important crops in north east of Iran (A. Soltani, unpublished data).

Crop	tuSOWEMR	tuEMRTLM	tuTLMBSG	tuBSGTSG	tuTSGMAT
Wheat	140	724	446	588	285
Barley	100	700	450	550	250
Rice	70	980	320	630	100
Maize	100	1025	205	850	70
Sorghum	100	1000	350	550	200
Soybean	70	950	190	700	240
Peanut	100	800	1	750	1
Canola	100	890	89	700	200
Sunflower	100	900	100	500	400
Dry bean	88	527	132	616	44
Chickpea	100	1590	100	450	100

tuSOWEMR: temperature unit from sowing to emergence (°C)
tuEMRTLM: temperature unit from emergence to termination leaf growth on main stem (°C)
tuTLMBSG: temperature unit from termination leaf growth on main stem to beginning seed growth (°C)
tuBSGTSG: temperature unit from beginning to termination seed growth (°C)
tuTSGMAT: temperature unit from termination seed growth to harvest maturity (°C)

differ greatly depending on species, cultivar, and location. Therefore, temperature unit values of Table 6.4 are given only as a guide and should not be considered constants.

Programming

A flowchart of the method described above to calculate temperature units for predictions of phenological stages is shown in Fig. 6.4, and Box 6.2 presents a submodel to predict phenological development based on temperature. The submodel is written in Visual Basic for application (VBA), and uses Excel's sheets for input and output. There are many textbooks and web resources for programming by VBA. The parameter estimates in this submodel are those for the wheat cultivar "Tajan". For a description of the variable names in this

Box 6.2. Program of phenology submodel as written in Visual Basic for Application in Excel. For the name of variables refer to the text or Appendix III.

```
Phenology:
'------------------------------ Parameters and Initials
  If iniPheno = 0 Then
  TBD = Sheet5.[b7]
  TP1D = Sheet5.[b8]
  TP2D = Sheet5.[b9]
  TCD = Sheet5.[b10]
  tuSOWEMR = Sheet5.[b11]
  tuEMRTLM = Sheet5.[b12]
  tuTLMBSG = Sheet5.[b13]
  tuBSGTSG = Sheet5.[b14]
  tuTSGMAT = Sheet5.[b15]

  tuEMR = tuSOWEMR
  tuTLM = tuEMR + tuEMRTLM
  tuBSG = tuTLM + tuTLMBSG
  tuTSG = tuBSG + tuBSGTSG
  tuMAT = tuTSG + tuTSGMAT

  DAP = 0:   CTU = 0:   iniPheno = 1
  End If
'------------------------------Thermal time calculation
  If TMP <= TBD Or TMP >= TCD Then
    tempfun = 0
  ElseIf TMP > TBD And TMP < TP1D Then
    tempfun = (TMP - TBD) / (TP1D - TBD)
  ElseIf TMP > TP2D And TMP < TCD Then
    tempfun = (TCD - TMP) / (TCD - TP2D)
  ElseIf TMP >= TP1D And TMP <= TP2D Then
    tempfun = 1
  End If
```

Continued

Box 6.2. Continued.

```
    DTU = (TP1D – TBD) * tempfun
    CTU = CTU + DTU
    DAP = DAP + 1

    If CTU < tuEMR Then DTEMR = DAP + 1 'Saving days to EMR
    If CTU < tuTLM Then DTTLM = DAP + 1 'Saving days to TLM
    If CTU < tuBSG Then DTBSG = DAP + 1 'Saving days to BSG
    If CTU < tuTSG Then DTTSG = DAP + 1 'Saving days to TSG
    If CTU < tuMAT Then DTMAT = DAP + 1 'Saving days to MAT

    If CTU > tuMAT Then MAT = 1
Return
```

submodel refer to Appendix III. In Chapter 12, this submodel is used to predict phenological development in a simulation model of potential production. A similar program (*tu_calc.xls*) to this submodel is used to calculate temperature unit requirements between stages (please see book's website).

Additional Notes

Using temperature unit when the crop is sensitive to photoperiod

As mentioned earlier, the temperature unit method to predict crop phenology is generally valid if the crop is not sensitive to photoperiod. However, the method can be used for sensitive crops and/or phenophases with the following considerations.

1. The method can be used to predict phenological development in the same latitude and in a narrow range of sowing dates that have been used to calculate temperature unit requirements. Change in latitude and sowing date result in the crop being exposed to different photoperiods, and if the crop is sensitive to photoperiod, temperature unit requirements will change. For example, it may take 1000°C from sowing to flowering of a soybean cultivar at 37°N, but the same cultivar might need 1200°C from sowing to flowering at 40°N. This is due to longer photoperiods of a higher latitude location that retard development rate toward flowering in soybean, which is a short-day plant. Therefore, calculated temperature unit at a location of 37°N is not applicable to a location with 40°N and vice versa.

2. It is still possible to use temperature unit by incorporating a relationship describing temperature unit requirement as a function of sowing date. Changing sowing date results in change in photoperiods that the crop experiences. One example of such a relationship between sowing date and temperature unit requirement is given in Fig. 6.8 for chickpea in Gorgan, in the northeast of Iran. In the figure, temperature unit requirement from emergence to flowering has been defined as a function of sowing date as days after 1 September. This relationship is valid only in Gorgan and cannot be used in other locations because different locations have different photoperiod regimes.

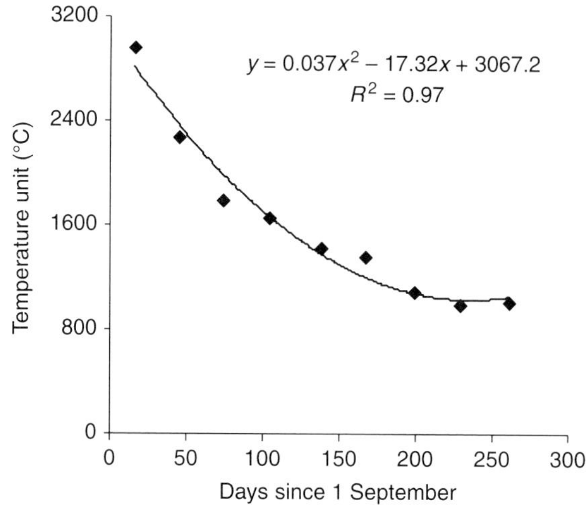

Fig. 6.8. Temperature unit from sowing to flowering as a function of sowing date in chickpea in Gorgan, northeast Iran (A. Soltani, unpublished data).

Other temperature functions

In this chapter a simple linear function was used to account for the effect of temperature on development rate. It seems this function is adequate for many crops and cultivars. However, if necessary, other response functions can be used. For examples of such other functions refer to Soltani *et al.* (2006a, b), who indicated a 3-piece linear-segment function was superior in comparison to some other curvilinear functions such as a beta function.

Time-step in calculation of thermal time

The time-step in modeling is always an important consideration, including the model for phenology development. For phenological development is it satisfactory to use daily time steps considering that temperature is dynamic through the daily cycle? For example, suppose we have a crop with base temperature of 5°C and average daily temperature is also 5°C. The relationships presented above in this chapter predict that the daily temperature unit experienced by the crop is 0°C. However, temperature is not constant throughout the day and night. It is likely that temperature will be higher than 5°C at least in a few hours during the day, so some progress in development is expected.

Due to temperature variation through the day and the nature of the plant response to temperature, it may be preferable to use hourly time steps in the calculation of temperature instead of daily time steps. The hourly approach requires, of course, input temperature data on an hourly basis, which are not often reported. Methods have been developed to predict hourly temperature data from minimum and maximum daily temperature data (Jones and Kiniry, 1986; Goudriaan and van Laar, 1994; Yin *et al.*, 1997). However, there are reports that hourly time steps do not improve prediction of phenological development (Purcell, 2003; Soltani *et al.*, 2009).

Soltani *et al.* (2009) indicated that the absolute difference between calculated temperature units using different time steps is greater around cardinal temperatures, and there are not significant differences between them in considerable ranges of sub- and supra-optimal temperatures. At temperatures around base and ceiling temperature, using hourly time steps resulted in a greater temperature unit, while at temperatures around optimal temperatures it led to a lower temperature unit. The larger the temperature amplitude, the larger was the difference in estimated temperature unit between length of time steps (Fig. 6.9).

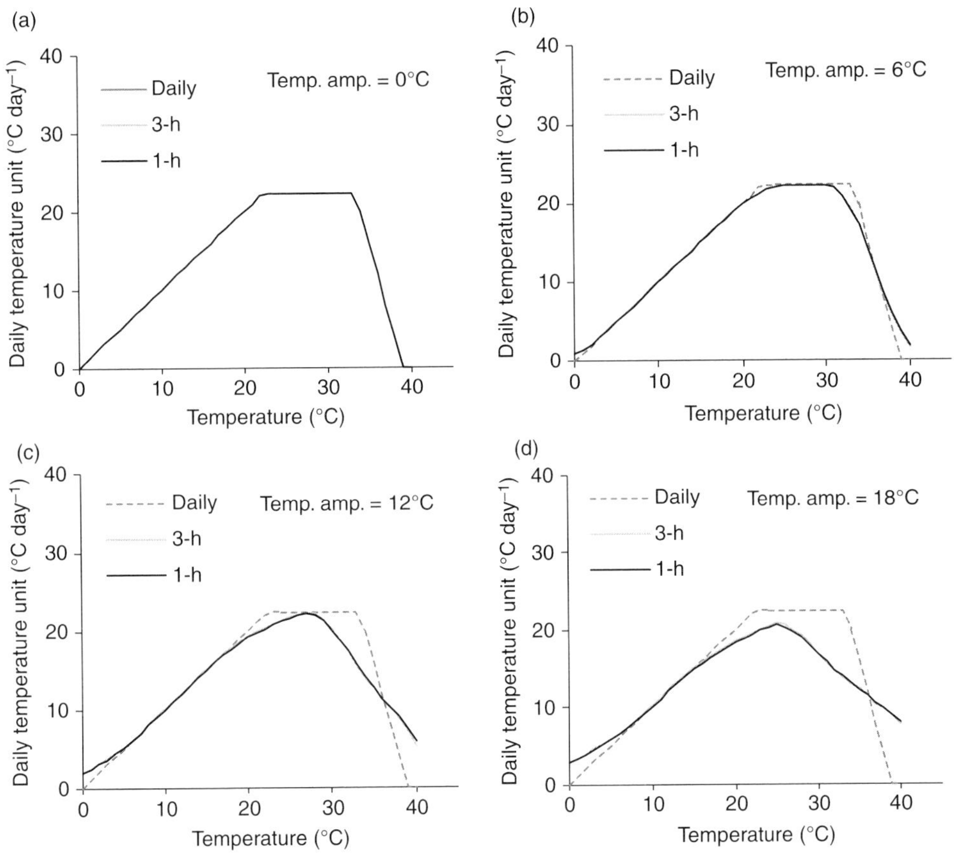

Fig. 6.9. Calculated daily temperature unit based on a 3-piece segmented function versus mean temperature at daily temperature amplitudes (= Tmax – Tmin) of 0, 6, 12, and 18°C. Time steps used are 1, 3, and 24 h (daily). In all cases, the results for the 3-h time step virtually coincide with the daily time step.

Exercises

1. Suppose there is a crop with cardinal temperatures of 0, 25, 25, and 35°C for base, lower optimum, upper optimum, and ceiling temperatures, respectively. Try to complete temperature unit calculations for the crop in the table below. Assume this crop needs 5 biological days from sowing to emergence, which is equal to temperature unit of 125°C. Now, calculate how much temperature unit has been accumulated until Day 6. How many biological days have been accumulated until Day 6?

Sample calculation of daily temperature unit and biological day (*tempfun*) during 6 consecutive days for a crop with 0, 25, 25, and 35°C for base, lower optimum, upper optimum, and ceiling temperatures, respectively. Fill in the empty spaces. The sum of *tempfun* will give cumulative biological days.

Day	TMIN	TMAX	TMP	*tempfun*	(TP1D–TBD)	DTU
1	–5	5	?	?	?	0
2	5	15	?	?	25	?
3	15	25	20	0.8	25	20.0
4	25	35	?	0.5	?	?
5	30	40	?	?	?	?
6	35	45	40	?	?	0
Sum	–	–	–	?	–	?

2. Using cardinal temperatures presented in Table 6.3, try to classify the crops into two categories. Does this classification fit any other classification that you are already familiar with?

3. Calculate temperature unit requirements of different phenological stages for crops in your region. Cardinal temperatures given in Table 6.3 and Excel program *tu_calc.xls* can be used for this purpose.

7 Phenology – Temperature and Photoperiod

In Chapter 6, a model of phenological development based solely on response to temperature was presented to predict crop development through various phenological stages. In this chapter, modeling of phenological development will be extended to include the influence of photoperiod. By including the influence of photoperiod, the phenology model will be much more general and crop development can usually be simulated across a range of latitudes and sowing dates.

Background

Studies by Garner and Allard in the 1920s resulted in identification of photoperiod as a controlling environmental factor in flowering (a key factor in phenological development). Later studies demonstrated that the length of night, rather than length of day, was actually responsible for control of plant development response (Gardner *et al.*, 1985). Since length of the photoperiod on any particular day depends on latitude and day of the year (Sinclair and Weiss, 2010), these variables are important inputs for the photoperiod submodel.

Most often the phenological development stages sensitive to photoperiod are during the period preceding anthesis. Anthesis date is when the sexual organs are mature, leading to pollination and embryo fertilization. This development stage is commonly referred to as "flowering". Three main categories of photoperiodic response are recognized in crop plants (Roberts and Summerfield, 1987): photoperiod-insensitive or day-neutral plants, short-day plants, and long-day plants. Photoreceptors called phytochromes exist in plants and slow shifts in the state of phytochrome during the dark period provide the clock for measuring photoperiod. For more physiological information refer to Taiz and Zeiger (2010).

In short-day plants, development rate towards flowering is accelerated by daylength shorter than a certain photoperiod, which is called the "critical" photoperiod. In long-day plants, however, flowering is accelerated by photoperiod longer than a critical photoperiod. In both cases, the critical photoperiod varies among and within species.

In the complete crop model, predictions are needed for phenological stages of emergence, termination leaf growth on main stem (TLM), beginning seed growth (BSG), termination seed growth (TSG), and harvest maturity (MAT). However, more detailed descriptions of plant development may be needed to estimate these events. In wheat, for example, phenological development is sensitive to photoperiod from emergence to formation of terminal spikelet or just before stem elongation (Ritchie, 1991), which occurs well before TLM (or flag leaf ligule visible in wheat). Therefore, for wheat this phase, i.e. terminal spikelet or stem elongation, should be included in the phenology model. Similarly, in a crop like chickpea, phenological development is sensitive to photoperiod from emergence until flowering, which occurs before TLM (first-pod or R3 in this crop, Soltani *et al.*, 2006b). Therefore, for this crop prediction of flowering needs to be added in the model although it is not required for simulation of other crop processes. Table 7.1 indicates phenological stages of wheat and chickpea that are predicted by the phenology model including photoperiod effect. Other phenological stages can be added for prediction if required. Box 7.1 includes a summary of the method developed in this chapter.

Table 7.1. Phenological stages in wheat and chickpea, their response to temperature and photoperiod (Ritchie, 1991 for wheat and Soltani *et al.*, 2006b for chickpea) and later need of them for simulating other crop processes. TLM is termination leaf growth on main stem, BSG is beginning seed growth, TSG is termination seed growth, and MAT is maturity.

Phenological stages	Response to temperature and photoperiod	Later need for simulation of other processes?
Wheat		
Sowing–emergence	Temp	Yes
Emergence–stem elongation	Temp + photoperiod	No
Stem elongation–TLM	Temp	Yes
TLM–BSG	Temp	Yes
BSG–TSG	Temp	Yes
TSG–MAT	Temp	Yes
Chickpea		
Sowing–emergence	Temp	Yes
Emergence–flowering	Temp + photoperiod	No
Flowering–TLM	Temp	Yes
TLM–BSG	Temp	Yes
BSG–TSG	Temp	Yes
TSG–MAT	Temp	Yes

Box 7.1. Summary of computation in phenological development submodel from sowing to maturity.

- Biological day requirement of different phenological stages are defined.
- Biological days when response to photoperiod is started and ended are defined.
- Daily mean temperature is used to find a temperature function (0–1) that accounts for the effect of temperature on development rate.
- Daily photoperiod is used to find a photoperiod function (0–1) that accounts for the effect of photoperiod on development rate if the stage is sensitive to photoperiod.
- Biological day is obtained by multiplying temperature and photoperiod functions.
- Cumulative biological day is obtained.
- A phenological stage is predicted to occur if cumulative biological day has just reached or passed biological day requirement of that stage.

Basics

An important advance in the analysis of crop phenological stages to account for the influence of photoperiod was the use of development rate, or the inverse of duration (Chapter 6). For example, analysis of flowering in different crops can be defined as the developmental rate from emergence to flowering by calculating the inverse of the time between these two phenological stages (Sinclair *et al.*, 1991, 2005b; Ellis *et al.*, 1994; Lawn and James, 2011). Based on this definition, a cultivar with a long duration (D_{ab}, day), for example from emergence to flowering, would have a small development rate ($R = 1/D_{ab}$, day^{-1}) and vice versa.

A linear additive model has been often used to describe rate of development (R, day^{-1}) as a function of average temperature (T_{ave}, °C) and photoperiod (P_{ave}, h) for the entire phenological interval being considered:

$$R = a + b\,T_{ave} + c\,P_{ave} \tag{7.1}$$

where a, b, and c are empirical coefficients.

Some researchers have challenged the use of the additive, linear model, which is based on average temperature and photoperiod over an entire phenological stage rather than one based on daily values for these variables (Yin *et al.*, 1997; Carberry *et al.*, 2001).

Alternatively, a multiplicative model has been used to describe the effects of temperature and photoperiod on crop development rate (e.g. Sinclair *et al.*, 1991, 2005b; Grimm *et al.*, 1993). In this model, development rate is predicted as a function of maximum development rate (R_{max}, day^{-1}) multiplied by scaling factors for daily temperature (*tempfun*) and daily photoperiod (*ppfun*):

$$R = R_{max} \times tempfun \times ppfun \tag{7.2}$$

R_{max} defines the maximum development rate during a specific stage in a specific cultivar, which would be observable under optimal temperature and photoperiod.

The inverse of R_{max} defines the minimum number of days between two stages under optimum conditions (D_{min}, day). Therefore, for a species/cultivar with D_{min} equal to 25 days, the value of R_{max} is 1/25 or 0.04 day^{-1}. Equation 7.2 can be readily rewritten as:

$$R = tempfun \times ppfun / D_{min} \tag{7.3}$$

Having response functions for temperature and photoperiod and knowing the values of D_{min} (or R_{max}) for a specific phenological interval (say Stage *a* to Stage *b*), phenological development can be calculated for each day using Eqn 7.2 or 7.3. The development rates for each day are summed following Stage *a* to obtain cumulative development. When ΣR reaches or just exceeds a value of 1.0, this is the time when Stage *b* has been reached.

Alternatively, the biological day can be calculated for each day. The two approaches only differ in their approach to accounting for the influence of *tempfun* and *ppfun*. In the biological day approach, on each day a fraction (BD) is calculated to reflect the inhibition of potential development because of restricting temperature and photoperiod functions. In this method, the two functions of temperature and photoperiod are multiplied to calculate the progress in development on each day discounted for restricting temperature and photoperiod. If both temperature and photoperiod allow maximum development rate, then the BD on that day equals 1.0.

$$BD = tempfun \times ppfun \tag{7.4}$$

Cumulative BD in progressing through phenological interval (CBD_i, day) is then obtained by adding value of BD for the current day to cumulative BD on the previous day (CBD_{i-1}).

$$CBD_i = CBD_{i-1} + BD \tag{7.5}$$

Stage *b* is predicted to occur on the day when the value of CBD_i reaches or surpasses CBD_{ab} (or D_{min} for that specific phenophase). In this chapter, Eqns 7.4 and 7.5 are used to predict different phenological stages. During the stages when the crop is not sensitive to photoperiod, *ppfun* is equal to 1.0 and BD responds only to *tempfun*. Thus, this generalized model can be readily used to simulate only the influence of temperature on development rate. For example, in chickpea *ppfun* = 1 from sowing to emergence and also from flowering to maturity (Soltani *et al.*, 2006b). However, soybean responds to photoperiod from emergence to physiological maturity (Grimm *et al.*, 1994).

To model phenological development between two specified phenological events in response to temperature and photoperiod, it is necessary that response functions to temperature and photoperiod be available. Also, a value needs to be inputted for maximum development rate (R_{max}) or minimum number of days for phenological interval under consideration (D_{min}). Options for describing the temperature function (*tempfun*) were discussed in Chapter 6. The most generalized temperature function is the 3-segment linear function, as introduced in Chapter 6. If required, this function can be replaced with another suitable function.

Photoperiod Function

First, it should be noted that photoperiod used in plant science is different from the usual description of daylength. In many cases, daylength refers to the time each day from the moment the upper limb of the sun's disk appears above the horizon during sunrise to the moment when the upper limb disappears below the horizon during sunset. However, due to reflection and refraction of sunlight by the atmosphere, there is actually sufficient radiation energy even when the sun is slightly below the horizon to influence phytochrome in many plants (Summerfield and Roberts, 1987). Therefore, the daylength to trigger plant photoperiod responses is defined to include civil twilight when the sun is still below the horizon.

In calculations related to phenological development, photoperiod is generally the time period between the time when the sun is −6° below the horizon at sunrise and when it is below the horizon at −6° at sunset. While sun angle of −6° is most commonly used, other angles such as −2 and −4° have also been reported and used (Summerfield and Roberts, 1987). The exact sun angle to be used depends on the light irradiance that affects phytochome in a particular crop species resulting in altered phenological development. Keisling (1982) has presented a relationship between the critical light irradiance and sun angle below horizon (Fig. 7.1). Box 7.2 includes a simple program to calculate photoperiod for any latitude (negative for south latitudes) and sun angle below horizon during all days of the year, which is based on Keisling (1982).

To quantify the response of development rate to photoperiod, different functions can be used. Here, a 2-segment linear function is used that is applicable for both long-day and short-day plants (Fig. 7.2). Based on this function, development rate of a long-day plant is at its maximum under long photoperiods (Fig. 7.2a). Photoperiod (PP, h) at which development rate starts to decline is called the critical photoperiod (CPP, h). With shorter photoperiod below the CPP, development rate decreases linearly and eventually reaches a minimum development rate characterized by a minimum photoperiod, or a "base" photoperiod. The slope of this linear decrease is

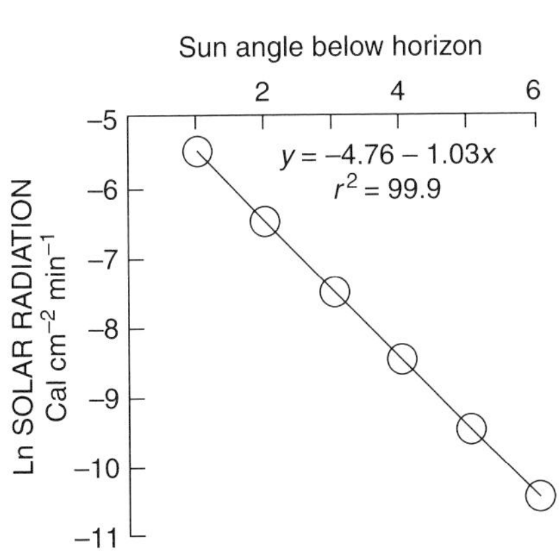

Fig. 7.1. Relationship between the angle of the sun below the horizon and the natural logarithm of the light intensity (Keisling, 1982).

Box 7.2. A program to calculate photoperiod based on Keisling (1982).

```
Pi = 3.141592654:          RDN = Pi / 180
ALPHA = 90 + SABH
SMA3 = 0.9856 * DOY – 3.251
LANDA = SMA3 + 1.916 * Sin (SMA3 * RDN) + 0.02 * Sin (2 * SMA3 * RDN)+282.565
DEC = 0.39779 * Sin (LANDA * RDN)
DEC = Atn (DEC / Sqr (1 – DEC ^ 2))
DEC = DEC / RDN
TALSOC = 1 / Cos (lat * RDN)
CEDSOC = 1 / Cos (DEC * RDN)
SOCRA = (Cos (ALPHA * RDN) * TALSOC * CEDSOC) – (Tan(lat * RDN) * Tan
  (DEC * RDN))
pp = Pi / 2 – (Atn (SOCRA / Sqr(1 – SOCRA ^ 2)))
pp = pp / RDN
pp = 2 / 15 * pp
```

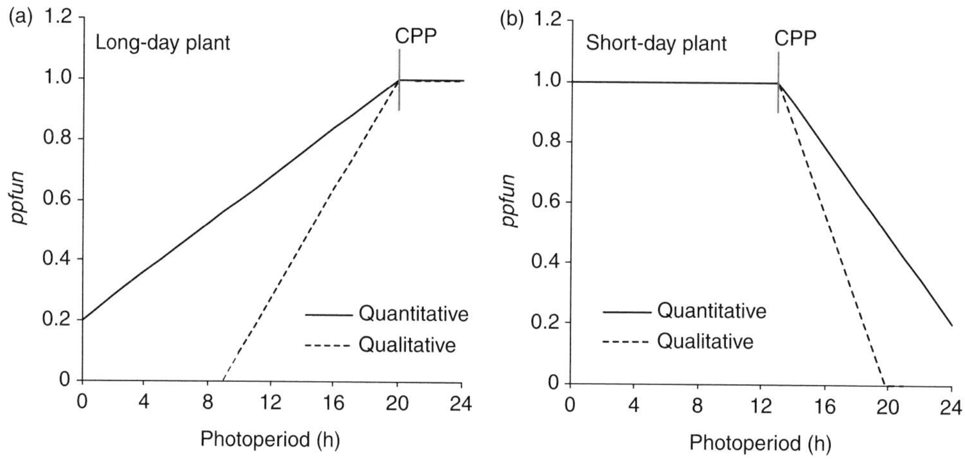

Fig. 7.2. Response of development rate to photoperiod in long-day (a) and short-day (b) plants. Quantitative and qualitative responses are indicated.

characterized using a photoperiod sensitivity coefficient (*ppsen*). In qualitative or obligatory, long-day plants, minimum development rate at and below the base photoperiod will be zero. However, in quantitative or facultative, long-day plants, minimum development rate is not zero. Mathematically, this function can be represented as:

$$
\begin{aligned}
ppfun &= 1 - ppsen \times (\mathrm{CPP} - \mathrm{PP}) && \text{if} \quad \mathrm{PP} < \mathrm{CPP} \\
ppfun &= 1 && \text{if} \quad \mathrm{PP} \geq \mathrm{CPP}
\end{aligned}
\tag{7.6}
$$

Note that if *ppfun* from Eqn 7.6 is lower than zero (negative), a zero is used because phenological development is only a forward process and cannot be negative. A similar 2-segment linear function can be used in short-day plants. In these plants, critical photoperiod is a photoperiod above which development rate starts to decline but at photoperiods shorter than that development rate is at its maximum. In the same way, base photoperiod is a photoperiod at which development rate reaches its minimum. Again, this minimum in qualitative short-day plants is equal to zero (Fig. 7.2b):

$$\begin{aligned} ppfun &= 1 - ppsen \times (\mathrm{PP} - \mathrm{CPP}) \quad &if \quad \mathrm{PP} > \mathrm{CPP} \\ ppfun &= 1 \quad &if \quad \mathrm{PP} \leq \mathrm{CPP} \end{aligned} \tag{7.7}$$

The value of *ppfun* is limited between 0 and 1. In day-neutral plants, *ppfun* is always equal to 1, which means photoperiod does not affect development rate in these plants. Also, in some crops there are phenological stages that are not sensitive to photoperiod and *ppfun* for these stages is set equal to 1.

For calculation of *ppfun*, initiation and termination of the phenological stages that are sensitive to photoperiod are required. Some crops, mainly warm-season cereals such as rice and maize, have a basic vegetative period during which they do not respond to photoperiod (Ellis *et al.*, 1994). Many other crops respond to photoperiod right after emergence (e.g. Grimm *et al.*, 1993; Robertson *et al.*, 2002b; Soltani *et al.*, 2006b).

Simple linear functions have been presented above to describe the effect of photoperiod on development rate. Other, somewhat more complex functions defining *ppfun* have been used for this purpose by other researchers (e.g. Sinclair *et al.*, 2005b). Figure 7.3 represents a quadratic function which is used in the DSSAT-CERES model (Jones *et al.*, 2003) and a negative exponential model (Grimm *et al.*, 1993).

Figure 7.4 graphically indicates calculations that are performed by the phenology model of this chapter.

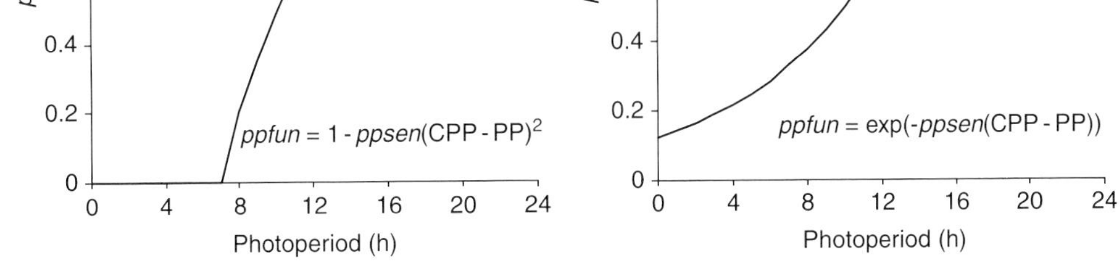

Fig. 7.3. Response of development rate to photoperiod described by quadratic (a) and negative exponential (b) functions. For quadratic function sensitivity coefficient (ppsen) is 0.008 and critical photoperiod (CPP) is 18h. For negative exponential function sensitivity coefficient (ppsen) is 0.14 and critical photoperiod (CPP) is 15h.

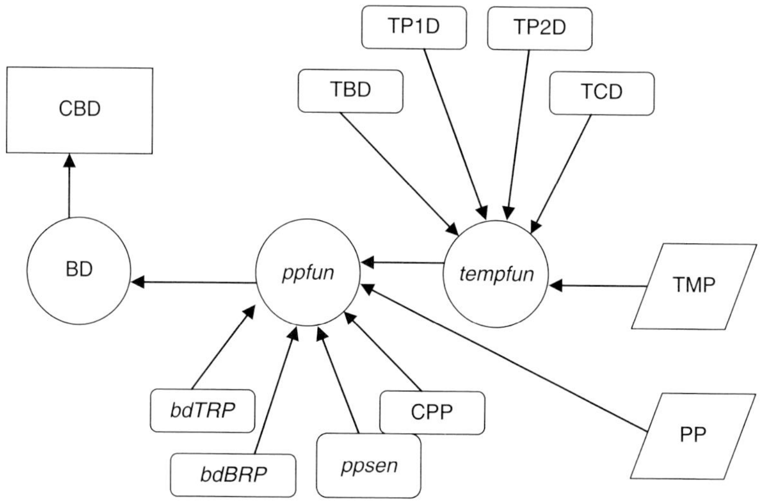

Fig. 7.4. Relational diagram of phenology model. TMP is the mean daily temperature (°C), PP the photoperiod (h), *tempfun* the temperature function, *ppfun* the photoperiod function, BD the biological day per calendar day, CBD the cumulative biological day, TBD the base temperature for development (°C), TP1D the lower optimum temperature for development (°C), TP2D the upper optimal temperature for development (°C), TCD the ceiling temperature for development (°C), CPP the critical photoperiod (h), *ppsen* the photoperiod sensitivity coefficient, *bdBRP* the biological day when response to photoperiod begins, and *bdTRP* the biological day when response to photoperiod ends. Shapes are defined in Table 1.2.

Parameter Estimation

Required parameter estimates to model phenological development as a function of both temperature and photoperiod are:

- cardinal temperatures for plant development (see Chapter 6);
- critical photoperiod (CPP) and photoperiod sensitivity coefficient (*ppsen*);
- biological day requirements when plant sensitivity to photoperiod begins and ends; and
- biological day requirements of different phenological phases.

Parameters related to response of development rate to photoperiod can be obtained from growth-chamber and/or field experiments. In a growth chamber, plants are exposed to different photoperiods while the temperature is optimal and not limiting for development rate. (If the crop has a vernalization response, the plants need to be vernalized.) For example, Ritchie (1991) performed a growth cabinet experiment to observe response of development rate of wheat seedlings placed in growth chambers under various photoperiod treatments. The development rate for the 19-h treatment was assumed to be adequate for the maximum development rate and all other shorter-photoperiod treatments were referenced to it. A quadratic function (similar to Fig. 7.3a) was used to describe response of development rate to photoperiod.

Often, critical photoperiod can be relatively constant among cultivars within a species and the photoperiod sensitivity coefficient characterizes cultivar differences. For example, Major and Kiniry (1991) reviewed published data and concluded critical photoperiod was about 12.5 h for maize, 12 h for sorghum, and 17.7 h for a number of long-day crops including wheat, barley, oat, rye, flax, and rapeseed. However, they also found that critical photoperiod varied from less than 10 h to more than 13 h in rice and from 12 h to more than 14 h in mungbean. Ritchie (1991) showed that critical photoperiod of 19 h was applicable in several spring and winter wheat cultivars. Soltani *et al.* (2006b) indicated that for several chickpea genotypes the critical photoperiod was constant (21 h). However, it has been shown that critical photoperiod in soybean is not constant and is different depending on cultivar maturity group (Grimm *et al.*, 1993; Sinclair *et al.*, 2005b).

It should be noted that the value of critical photoperiod also depends on the response function used to calculate *ppfun*. For instance, a critical photoperiod of 21 h in chickpea with a quadratic function decreases to 18 h when a 2-piece segmented function is used (Soltani *et al.*, 2006b; A. Soltani, unpublished data).

The phenological stage when the plants are responsive to photoperiod can be obtained from an experiment using a reciprocal transfer between photoperiod treatments. In this study, plants are transferred from long- to short-day regimes and vice versa at various times between sowing and flowering. Ellis *et al.* (1992) devised an approach to estimate these times from reciprocal transfer experiments and the approach has been used in different crops. In their approach, time from sowing to flowering is divided into three phases, i.e. pre-inductive phase (i.e. basic vegetative phase), inductive phase, and post-inductive phase, and length of each of these phases is determined. During pre- and post-inductive phases plants do not respond to photoperiod. Robertson *et al.* (2002b) used this approach in seven canola genotypes and found that there is no obvious pre-inductive phase in this crop and plants respond to photoperiod immediately after emergence. The canola genotypes became insensitive to photoperiod after visible-bud stage. For more information about details of the approach and its application refer to Robertson *et al.* (2002b) and Alcalde and Larraín (2006).

Another approach to detect photoperiod-sensitive stages is based on field experiments. If raw data on temperature unit requirements of a specific developmental phase appear to vary with location and/or sowing data (i.e. differing photoperiod conditions), the results indicate a sensitivity of that stage to photoperiod. For example, Kirby *et al.* (1999) used such an approach to determine photoperiod-sensitive developmental stages in several winter and spring wheat varieties at two sites with several sowing dates and years.

Controlled-environment equipment is usually not readily available to determine photoperiod parameters and biological day requirement due to the extensive requirement for such facilities. Therefore, some programs and software have been developed to estimate photoperiod function parameters (along with temperature function parameters and biological day requirements, D_{min}) from field data of observed phenological events and daily photoperiod and temperature data. These programs are mainly based on iterative optimization methods such as SIMPLEX. They start with some initial values for the parameters provided by the user and will find optimized final values for the parameters. For examples

Table 7.2. Parameter estimates of photoperiod function for wheat (based on Jones *et al.*, 2003; Ahmadi, 2007), soybean (based on Grimm *et al.*, 1993), and chickpea (based on Soltani *et al.*, 2006b). CPP is critical photoperiod (h) and *ppsen* is sensitivity coefficient to photoperiod.

Crop	Cultivar	CPP	*ppsen*
Wheat	Spring high latitude	17	0.10
	Spring low latitude	17	0.11
	Winter Europe	16	0.09
	Winter USA	16	0.09
	Kohdasht (Spring Iran)	14	0.10
	Tajan (Spring Iran)	14	0.17
	Zagros (Spring Iran)	14	0.11
Soybean	Maturity Group 000	14.6	0.129
	Maturity Group 00	14.4	0.148
	Maturity Group 0	14.1	0.171
	Maturity Group 1	13.8	0.203
	Maturity Group 2	13.6	0.249
	Maturity Group 3	13.4	0.285
	Maturity Group 4	13.1	0.294
	Maturity Group 5	12.8	0.303
	Maturity Group 6	12.8	0.311
	Maturity Group 7	12.3	0.320
	Maturity Group 8	12.1	0.330
	Maturity Group 9	11.9	0.340
	Maturity Group 10	11.8	0.349
Chickpea	Beauvanij	17.5	0.13
	Arman	20.0	0.09
	Hashem	19.0	0.11
	Jam	18.0	0.10

of using these programs refer to Sinclair *et al.* (1991, 2006b), Piper *et al.* (1996), and Robertson *et al.* (2002b).

DEVEL is another example of such a program to determine response functions from field data for temperature and photoperiod. DEVEL can be obtained from its developers (Holzworth and Hammer, 1996). Soltani *et al.* (2006b) modified DEVEL to include more functions for response of development rate to temperature and photoperiod. Table 7.2 provides parameter estimates of photoperiod function in wheat, soybean, and chickpea.

Programming

Figure 7.4 presents a diagram of the relationship of the variables required to simulate phenological development in response to photoperiod and temperature. A program that simulates phenological development is included in Box 7.3. A sample of parameter estimates of this model for soybean (cv. Sahar, MG III) and chickpea (cv. Hashem) are indicated in Table 7.3. For a complete summary of the variable names in this submodel refer to Appendix III.

Box 7.3. Program of the phenology model developed in this chapter. This program can be used in a standalone mode to predict phenological development or can be incorporated as a submodel in a crop simulation model. For a list of variables refer to Appendix III.

```
Sub Pheno_bd()
  GoSub ManagInputs
  GoSub InitialsHeaders
  GoSub FindSowingDate
  Do Until MAT = 1
    GoSub Weather
    GoSub PhenologyBD
    GoSub DailyPrintOut
  Loop
  GoSub SummaryPrintOut
  Exit Sub

ManagInputs:
  lat = Sheet1.[b7]
  pyear = Sheet1.[b8]
  pdoy = Sheet1.[b9]
Return

InitialsHeaders:
  MAT = 0
  iniPheno = 0
  Sheet4.Cells(2, 1) = "Year"
  Sheet4.Cells(2, 2) = "DOY"
  Sheet4.Cells(2, 3) = "DAP"
  Sheet4.Cells(2, 4) = "TMP"
  Sheet4.Cells(2, 5) = "tempfun"
  Sheet4.Cells(2, 6) = "pp"
  Sheet4.Cells(2, 7) = "ppfun"
  Sheet4.Cells(2, 8) = "bd"
  Sheet4.Cells(2, 9) = "CBD"
Return

FindSowingDate:
  Row = 10
  Do
    Row = Row + 1
    Yr = Sheet2.Range("A" & Row)
    DOY = Sheet2.Range("B" & Row)
    SRAD = Sheet2.Range("C" & Row)
    TMAX = Sheet2.Range("D" & Row)
    TMIN = Sheet2.Range("E" & Row)
    RAIN = Sheet2.Range("F" & Row)
  Loop Until Yr = pyear And DOY = pdoy
Return

Weather:
  Row = Row + 1
```

Continued

Box 7.3. Continued.

```
  Yr = Sheet2.Range("A" & Row)
  DOY = Sheet2.Range("B" & Row)
  SRAD = Sheet2.Range("C" & Row)
  TMAX = Sheet2.Range("D" & Row)
  TMIN = Sheet2.Range("E" & Row)
  RAIN = Sheet2.Range("F" & Row)

  TMP = (TMAX + TMIN) / 2
Return

PhenologyBD:
  If iniPheno = 0 Then
    TBD = Sheet5.[b12]
    TP1D = Sheet5.[b13]
    TP2D = Sheet5.[b14]
    TCD = Sheet5.[b15]

    ppres = Sheet5.[b17]
    bdBRP = Sheet5.[b18]
    bdTRP = Sheet5.[b19]
    SABH = Sheet5.[b20]
    cpp = Sheet5.[b21]
    ppsen = Sheet5.[b22]

    bdSOWEMR = Sheet5.[b24]
    bdEMRTLM = Sheet5.[b25]
    bdTLMBSG = Sheet5.[b26]
    bdBSGTSG = Sheet5.[b27]
    bdTSGMAT = Sheet5.[b28]

    bdEMR = bdSOWEMR
    bdTLM = bdEMR + bdEMRTLM
    bdBSG = bdTLM + bdTLMBSG
    bdTSG = bdBSG + bdBSGTSG
    bdMAT = bdTSG + bdTSGMAT

    CBD = 0:     DAP = 0:
    iniPheno = 1
  End If

'--------------------------- Temperature function
  If TMP <= TBD Or TMP >= TCD Then
    tempfun = 0
  ElseIf TMP > TBD And TMP < TP1D Then
    tempfun = (TMP – TBD) / (TP1D – TBD)
  ElseIf TMP > TP2D And TMP < TCD Then
    tempfun = ((TCD – TMP) / (TCD – TP2))
  ElseIf TMP >= TP1D And TMP <= TP2D Then
    tempfun = 1
  End If
```

Continued

Box 7.3. Continued.

```
'--------------------------- Photoperiod function
  Pi = 3.141592654:        RDN = Pi / 180
  ALPHA = 90 + SABH
  SMA3 = 0.9856 * DOY – 3.251
  LANDA = SMA3 + 1.916 * Sin(SMA3 * RDN) + 0.02 * Sin(2 * SMA3 * RDN) + 282.565
  DEC = 0.39779 * Sin(LANDA * RDN)
  DEC = Atn(DEC / Sqr(1 – DEC ^ 2))
  DEC = DEC / RDN
  TALSOC = 1 / Cos(lat * RDN)
  CEDSOC = 1 / Cos(DEC * RDN)
  SOCRA = (Cos(ALPHA * RDN) * TALSOC * CEDSOC) – (Tan(lat * RDN) *
    Tan(DEC * RDN))
  pp = Pi / 2 – (Atn(SOCRA / Sqr(1 – SOCRA ^ 2)))
  pp = pp / RDN
  pp = 2 / 15 * pp

  If CBD >= bdBRP And CBD <= bdTRP Then
    If ppres = 1 Then '<--- Long-day plant
      If pp < cpp Then
        ppfun = 1 – ppsen * (cpp – pp)
      Else
        ppfun = 1
      End If
    ElseIf ppres = 2 Then '<--- Short-day plant
      If pp > cpp Then
        ppfun = 1 – ppsen * (pp – cpp)
      Else
        ppfun = 1
      End If
    End If
  Else
    ppfun = 1
  End If
'--------------------------- Biological day
  DTU = (TP1D – TBD) * tempfun 'daily temperature unit
  bd = tempfun * ppfun
  CBD = CBD + bd
  DAP = DAP + 1
'--------------------------- Days to stages
  If CBD < bdEMR Then dtEMR = DAP + 1
  If CBD < bdBRP Then dtBRP = DAP + 1
  If CBD < bdTRP Then dtTRP = DAP + 1
  If CBD < bdTLM Then dtTLM = DAP + 1
  If CBD < bdBSG Then dtBSG = DAP + 1
  If CBD < bdTSG Then dtTSG = DAP + 1
  If CBD < bdMAT Then dtMAT = DAP + 1
```

Continued

Box 7.3. Continued.

```
  If CBD > bdMAT Then MAT = 1
Return

DailyPrintOut:
  Sheet4.Cells(DAP + 2, 1) = Yr
  Sheet4.Cells(DAP + 2, 2) = DOY
  Sheet4.Cells(DAP + 2, 3) = DAP
  Sheet4.Cells(DAP + 2, 4) = TMP
  Sheet4.Cells(DAP + 2, 5) = tempfun
  Sheet4.Cells(DAP + 2, 6) = pp
  Sheet4.Cells(DAP + 2, 7) = ppfun
  Sheet4.Cells(DAP + 2, 8) = bd
  Sheet4.Cells(DAP + 2, 9) = CBD
Return

SummaryPrintOut:
  Sheet1.[g7] = dtEMR
  Sheet1.[g8] = dtBRP
  Sheet1.[g9] = dtTRP
  Sheet1.[g10] = dtTLM
  Sheet1.[g11] = dtBSG
  Sheet1.[g12] = dtTSG
  Sheet1.[g13] = dtMAT
Return

End Sub
```

Table 7.3. Parameter estimates of phenology model for soybean (cv. Sahar, MG III) and chickpea (cv. Hashem) (A. Soltani, unpublished data). BD is biological day, TLM the termination of leaf growth on main stem, BSG the beginning seed growth, TSG the termination seed growth, and MAT the harvest maturity.

Parameter	Soybean	Chickpea
Base temperature for development (°C)	7	2
Lower optimum temperature for development (°C)	27	21
Upper optimum temperature for development (°C)	34	30
Ceiling temperature (°C)	45	40
Beginning response to photoperiod (BD)	4	5
Termination response to photoperiod (BD)	67	36
Sun angle below horizon (°)	6	4
Critical photoperiod	13.4	18
Photoperiod sensitivity coefficient	0.285	0.12
BD from sowing to emergence	4	5
BD from emergence to TLM	25	36
BD from TLM to BSG	8	5
BD from BSG to TSG	34	24
BD from TSG to MAT	12	5

Exercises

1. There are two wheat cultivars; cultivar *A* has a critical photoperiod of 16 h and a photoperiod sensitivity coefficient of 0.2 h^{-1}. The same figures are 16 h and 0.1 h^{-1} for cultivar *B*. Prepare a graph of *ppfun* versus photoperiod for these two cultivars (both cultivars in one graph). Use Eqn 7.6 to calculate *ppfun* for different values of photoperiod. Explain the graph.
2. Prepare a similar graph of Exercise 1 for a wheat cultivar with critical photoperiod of 16 h and a photoperiod sensitivity coefficient of 0.2 h^{-1} plus a soybean cultivar with critical photoperiod of 13 h and a photoperiod sensitivity coefficient of 0.3 h^{-1}. Explain the graph.
3. Table 7.2 includes critical photoperiod and photoperiod sensitivity coefficient for different maturity groups of soybean. Prepare a graph with maturity group as *x*-axis and critical photoperiod as *y*-axis (ignore maturity groups 00 and 000). How is critical photoperiod related to maturity group? Explain it.
4. The same table (Table 7.2) includes photoperiod sensitivity coefficient for different maturity groups of soybean. Draw a similar graph with photoperiod sensitivity coefficient as *y*-axis and maturity group as *x*-axis. Explain how these two are related.
5. Below is a table which includes temperature and photoperiod and *tempfun* and *ppfun* during 10 hypothetical consecutive days. Complete the table. What is the cumulative biological day in day 10? (For more information about calculation of *tempfun* refer to Chapter 6.) Assume cardinal temperatures are 0°C for base, 25°C for lower optimum, 28°C for upper optimum, and 40°C for ceiling temperature. Consider a critical photoperiod of 16 h and photoperiod sensitivity coefficient of 0.2 h^{-1}.

Day	TMP	PP	*tempfun*	*ppfun*	BD	CBD
1	10	12		0.2		0.08
2	20	10	0.80	0		
3	20	14				
4	28	18		1	1.00	
5	30	16	0.83			
6	10	14	0.40			2.63
7	20	18				
8	28	16	1.00			
9	30	14		0.6	0.50	4.93
10	30	12			0.17	

6. Box 7.2 presents a program to obtain daily photoperiod for any location of interest. The program can be found on the book's website, too. Using the program, calculate daily photoperiod for all days of the year (1–365) for two locations with latitudes of 30° and 40°. Then, complete your calculations by obtaining *ppfun* for wheat and soybean cultivars of Exercise 2 for every day of the year. Examine how different days are suitable for the development of the cultivars with respect to photoperiod effect.

7. Obtain parameter estimates for photoperiod impact on development rate for a crop cultivar of your interest at a location of your interest. For this exercise you will need a lot of field data and appropriate software (e.g. DEVEL).

8 Phenology – Vernalization

Development in some crop plants that originate from temperate climates requires exposure to cold temperature, i.e. vernalization. In these plants, development rate before flowering is influenced by the amount of cold period they have experienced. If these plants do not experience a sufficient cold period, development rate is reduced if not completely inhibited (Robertson *et al.*, 2002b).

Vernalization is important with respect to adaptation of crops to their natural environments and is usually seen in winter crops like wheat and rapeseed. Vernalization requirement delays development of flowering in these crops until after they experience a defined duration of cold temperatures. Without vernalization, these species might initiate reproductive development during cold and freezing temperatures in winter or early spring that is lethal for flowering organs (Roberts and Summerfield, 1987). Crops sensitive to vernalization become responsive after water absorption by seeds. Vernalization generally only impacts flowering and there is no further impact in wheat beyond formation of terminal spikelet (Ritchie, 1991) or anthesis (Wang and Engel, 1998).

In Chapter 7, a model of phenological development was presented, which included the effect of both temperature and photoperiod on phenological development. This chapter extends the phenology model of Chapter 7 to include the vernalization effect as well. Similar to the previous phenology model, this model will be a general one and able to predict phenological development of crops that respond to vernalization across environments and locations with a constant set of parameters.

Vernalization Model

In Chapter 7, progress toward a phenological stage to a second stage (say Stage *a* to Stage *b*) was quantified by calculating biological day. Starting at the beginning of Stage *a*, biological day per each day (BD) was defined in Eqn 7.4 by multiplying the value of the temperature (*tempfun*) and photoperiod (*ppfun*) functions for that day, i.e.:

$$BD = tempfun \times ppfun \quad (8.1)$$

Completion of a stage was predicted to occur when cumulative BD exceeds the minimum biological day requirement for that stage (CBD_{ab}).

The effect of vernalization can be included in Eqn 8.1 by including a function that accounts for vernalization (*verfun*).

$$BD = tempfun \times ppfun \times verfun \quad (8.2)$$

Like *tempfun* and *ppfun*, the vernalization function has a value between 0 and 1. During phenological stages when the crop is not sensitive to vernalization, the value of *verfun* is fixed at 1. For example, development rate is not sensitive to vernalization in wheat from sowing to emergence and then from terminal spikelet to maturity (Ritchie, 1991; Streck *et al.*, 2003b). Thus, *verfun* = 1 is used during these phases. In this chapter, Eqn 8.2 instead of Eqn 8.1 is used to calculate biological day and prediction of different phenological stages.

Vernalization function

Vernalization results from cumulative exposure to low temperature. As a result, the calculation of *verfun* is somewhat complex. First, vernalization day (VERDAY, day) needs to be calculated for each calendar day to reflect the contribution of each day to vernalization. Cumulative vernalization days are needed to assess *verfun* on each day during the stage sensitive to vernalization.

The calculation of vernalization day is based on the temperature to which the plants are subjected each day. Vernalization typically takes place at temperatures between −5 and 16°C with maximum effect between 0 and 8°C (Roberts and Summerfield, 1987). Effective temperatures for induction of vernalization may vary from one crop to another. Figure 8.1 represents effectiveness of different temperatures for vernalization in wheat. A 3-segment linear function is used to obtain vernalization day experienced by the crop each calendar day (VERDAY). Thus, four cardinal temperatures are defined, i.e. base (TBV), lower optimum (TP1V), upper optimum (TP2V), and ceiling (TCV) temperatures for vernalization:

$$\begin{aligned} VERDAY &= 0 \quad \textit{if} \quad TMP \le TBV \\ &= (TMP - TBV) / (TP1V - TBV) \quad \textit{if} \quad TBV < TMP < TP1V \\ &= 1 \quad \textit{if} \quad TP1V \le TMP \le TP2V \\ &= (TCV - TMP) / (TCV - TP2V) \quad \textit{if} \quad TP2V < TMP < TCV \\ &= 0 \quad \textit{if} \quad TMP \ge TCV \end{aligned} \quad (8.3)$$

As defined in Eqn 8.3, a value for VERDAY of 1.0 on a day is calculated when the plant is exposed to the optimum temperature for vernalization. The optimum temperature for vernalization in wheat is between 0 and 8°C (Ritchie, 1991).

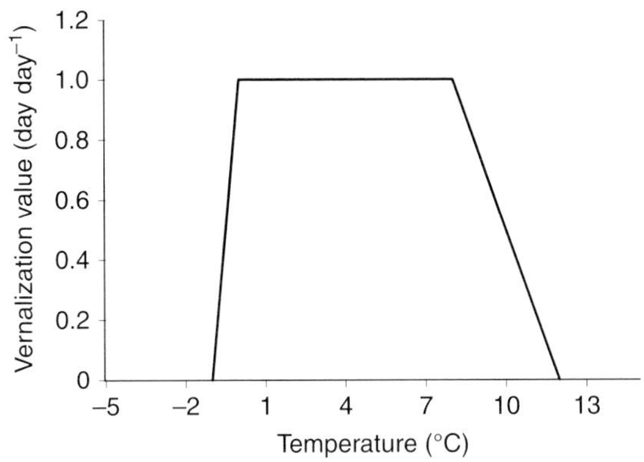

Fig. 8.1. Effectiveness of different temperatures in induction of vernalization in wheat.

Temperatures lower than –1°C or higher than 12°C do not contribute to vernalization, and the value of VERDAY is 0. For temperatures in the range of 0 to –1°C and from 8 to 12°C, the effectiveness in vernalization is decreased and the value VERDAY is between 0 and 1.

In the APSIM model, vernalization can occur at temperatures between 0 and 15°C with maximum effectiveness at 2°C for both wheat and canola (Robertson *et al.*, 2002b; Keating *et al.*, 2003). In a review of wheat literature, Porter and Gawith (1999) reported –1.3°C for base, 3.8°C for lower optimum, 6.0°C for upper optimum, and 15.7°C for ceiling temperature in wheat response to vernalization. According to Fig. 8.1 and Eqn 8.3, placing a wheat seedling at 7°C for 5 days results in 5 VERDAY. However, placing this seedling at 10 or –0.5°C for 5 days results in only 2.5 VERDAY.

Once the daily value of VERDAY has been calculated, the cumulative vernalization days during the phenological stage of interest (CUMVER) is calculated by adding VERDAY to the previous day's $CUMVER_{i-1}$.

$$CUMVER_i = CUMVER_{i-1} + VERDAY \quad (8.4)$$

High temperatures during a sensitive period of vernalization can cause de-vernalization during the early stages of vernalization (Ritchie, 1991). If a crop had already experienced 10 days of vernalization ($CUMVER_i > 10$), occurrence of high temperatures will not result in de-vernalization. However, if cumulative vernalization day is lower than 10 days and maximum temperature (TMAX, °C) is higher than 30°C, then cumulative vernalization day is reduced by 0.5 day per each degree celsius greater than 30°C:

$$CUMVER_i = CUMVER_{i-1} - 0.5\,(TMAX - 30) \quad (8.5)$$

Now, the information is available to calculate the vernalization function, *verfun*. The value of *verfun* is dependent on three factors: (i) cumulative vernalization day experienced by the crop ($CUMVER_i$) up to that day; (ii) total amount of vernalization days needed to saturate the vernalization response (VDSAT);

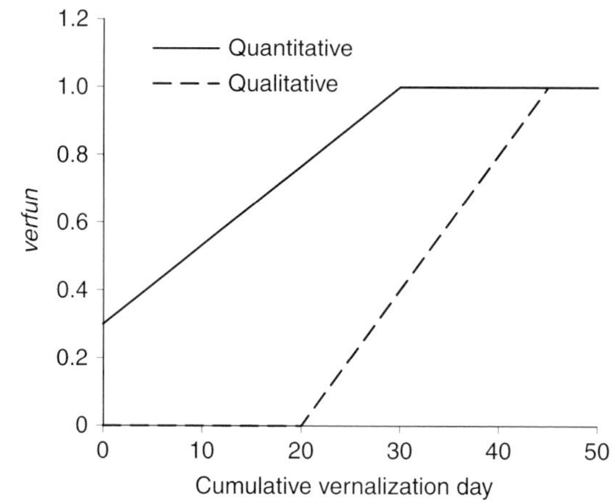

Fig. 8.2. Vernalization function (*verfun*) as influenced by cumulative vernalization day experienced by the crop. Quantitative and qualitative responses are indicated.

and (iii) a sensitivity coefficient of development rate response to vernalization (*vsen*) (Fig. 8.2):

$$verfun = 1 - vsen\,(\mathrm{VDSAT} - \mathrm{CUMVER_i}) \quad if \quad \mathrm{CUMVER_i} < \mathrm{VDSAT}$$
$$verfun = 1 \quad if \quad \mathrm{CUMVER_i} \geq \mathrm{VDSAT} \qquad (8.6)$$

The value of VDSAT can vary from a few days to about 60 days, or longer, depending on species and genotype (Roberts and Summerfield, 1987). Saturation of vernalization in many wheat cultivars has been reported to be 50 days, i.e. VDSAT = 50 (Ritchie, 1991). However, Baloch *et al.* (2003) reported higher values of 60 to 70 days for Pacific Northwest wheats. The value of VDSAT for canola has been reported to be 30 days (Robertson *et al.*, 2002b). Table 8.1 includes parameter estimates for vernalization function in wheat and rapeseed.

As indicated in Fig. 8.2, some genotypes/crops need to experience a certain number of cold days before *verfun* is greater than zero. These plants are identified as being qualitative in their response to the vernalization requirement. In some others, experiencing a cold period accelerates development rate toward flowering, but even without any vernalization the value of *verfun* is greater than zero. These plants are identified as quantitative with respect to vernalization.

Other functions also exist to obtain *verfun*. For example, Streck *et al.* (2003a) developed a non-linear vernalization response function for winter wheat (Fig. 8.3).

Crown temperature

In crop plants that are sensitive to vernalization, the growing point is often located below the soil surface during the vernalization period. Therefore, soil surface or crown temperature (T_{cr}) instead of air temperature should be used in calculation of vernalization. Soil surface temperature is assumed to be similar to air temperature but snow cover causes divergence between these two

Table 8.1. Parameter estimates of vernalization function for wheat (based on Ritchie, 1991; Jones *et al.*, 2003; Mirdavardoost, 2009) and canola (Nikoobin, 2009). *vsen* is sensitivity coefficient to vernalization and VDSAT is saturated vernalization day.

Crop	Cultivar	*vsen*	VDSAT
Wheat	Spring high latitude	0.003	50
	Winter Europe	0.033	50
	Winter US	0.033	50
	Winter-spring	0.011	50
	Agent	0.005	50
	Lancota	0.014	50
	Centurk	0.026	50
	Sage	0.027	50
	Sturdy	0.03	50
	Triumph	0.031	50
	Bezastaya	0.031	50
	Pawnee	0.04	50
	Kohdasht (Spring – Iran)	0.0015	40
	Tajan (Spring – Iran)	0.007	13
	Zagros (Spring – Iran)	0.009	23
Canola	Option 500 (Iran)	0.0012	50
	Hayula 60 (Iran)	0.0027	50
	Hayula 308 (Iran)	0.0045	30

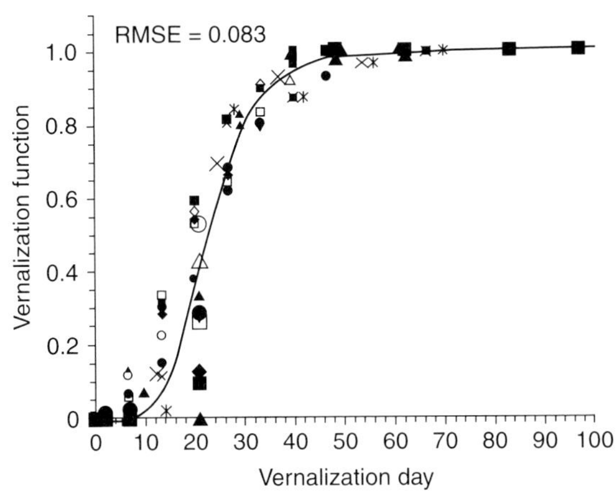

Fig. 8.3. The vernalization response function of 12 winter wheat cultivars from 19 trials described by a non-linear function (Streck *et al.*, 2003a). Symbols indicate different wheat cultivars.

temperatures. The interaction between soil temperature and air temperature is complex but a simple approximation has been developed (Ritchie, 1991). Data obtained by Aase and Siddoway (1979) on measured soil temperature (T_{cr}) at 0.03 m depth, air temperature (T_a) at 1.5 m height, and snow depth (D_s, m) were used to generate the following equation:

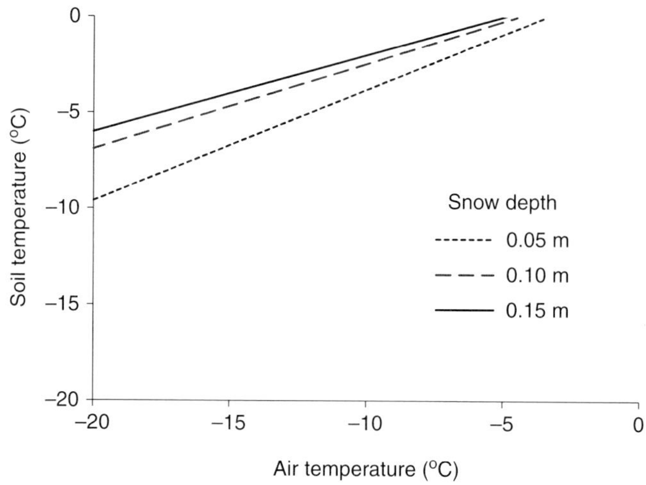

Fig. 8.4. Crown temperature as influenced by air temperature and snow depth (drawn using Eqn 8.7).

$$T_{cr} = 2 + T_a (0.4 + 18 (D_s - 0.15)^2) \quad (8.7)$$

In Eqn 8.7, if snow depth is greater than 0.15 m, a snow depth of 0.15 m is used because snow depth more than 0.15 m adds little additional insulation effect. The equation is also appropriate for air temperature less than 0°C. Daily minimum and maximum crown temperatures are estimated from Eqn 8.7 using daily minimum and maximum air temperatures as T_a. Mean daily crown temperature is then computed by averaging daily minimum and maximum crown temperatures. Figure 8.4 represents the relationship between crown and air temperature as predicted by Eqn 8.7 for different snow depths.

Meteorological data do not often report snow depth. Estimates of snow depth can be calculated from daily temperature and precipitation data (Ritchie, 1991). Precipitation is assumed to be in the form of snow if the maximum temperature is lower or equal to 1°C. The increment of daily snow accumulation is assumed to be 10 mm for each 1 mm of precipitation. Snow melt is assumed to occur when the daily value of maximum air temperature is higher than 1°C. The snow melt increment is assumed to be 10 mm per degree above 0°C plus 4 mm per 1 mm rainfall. For instance, on a day with a maximum temperature of 10°C and no rain, the quantity of melted snow would be 100 mm. Similarly, with a 10°C maximum temperature and 20 mm of rainfall, the snow melt would be 180 mm.

Parameter Estimation

Parameters related to response of development rate to vernalization can be estimated using field and/or controlled-environment experiments.

Ritchie (1991) presented a procedure to obtain vernalization parameters. Several wheat cultivars were sown and kept at 15°C until emergence. The plants were then placed in a growth chamber at 2°C for the duration of a vernalization period that ranged from 0 to 50 days. After this period of vernalization, the plants were grown for a week at 15°C to prevent de-vernalization and then were grown until ear emergence at 20°C. The photoperiod was 18h, which was considered non-limiting on development. Date of appearance of terminal spikelet was recorded. Relative development rate was calculated as the ratio of development rate toward terminal spikelet in each treatment to development rate at 50-day vernalization treatment, which was assumed sufficient to saturate the vernalization requirement. By fitting Eqn 8.6 with VDSAT fixed at 50 days, a vernalization sensitivity coefficient was estimated for different cultivars. Robertson *et al.* (2002b) used a similar method to quantify vernalization response in canola (rapeseed). For additional examples of using field experiments refer to Streck *et al.* (2003a).

Table 8.1 includes estimates of vernalization parameters in wheat and canola.

Programming

A program to predict phenological development as discussed in this chapter is included in Box 8.1. A sample of parameter estimates of this model for wheat (cv. Tajan) is given in Table 8.2. For a description of the variable names in this submodel refer to Appendix III.

Box 8.1. Program of the phenology model developed in this chapter. This program can be used as standalone to predict phenological development or can be incorporated as a submodel in a crop simulation model. For a list of variables refer to Appendix III.

```
Sub Pheno_vz()

  GoSub ManagInputs
  GoSub InitialsHeaders
  GoSub FindSowingDate
  Do Until MAT = 1
    GoSub Weather
    GoSub PhenologyBD
    GoSub DailyPrintOut
  Loop
  GoSub SummaryPrintOut
  Exit Sub

ManagInputs:
  lat = Sheet1.[b7]
  pyear = Sheet1.[b8]
  pdoy = Sheet1.[b9]
Return

InitialsHeaders:
  MAT = 0
  iniPheno = 0
```

Continued

Box 8.1. Continued.

```
  Sheet4.Cells (2, 1) = "Year"
  Sheet4.Cells (2, 2) = "DOY"
  Sheet4.Cells (2, 3) = "DAP"
  Sheet4.Cells (2, 4) = "TMP"
  Sheet4.Cells (2, 5) = "tempfun"
  Sheet4.Cells (2, 6) = "CUMVER"
  Sheet4.Cells (2, 7) = "verfun"
  Sheet4.Cells (2, 8) = "pp"
  Sheet4.Cells (2, 9) = "ppfun"
  Sheet4.Cells (2, 10) = "bd"
  Sheet4.Cells (2, 11) = "CBD"
  Sheet4.Cells (2, 12) = "SNOW"
Return

FindSowingDate:
  Row = 10
  Do
    Row = Row + 1
    Yr = Sheet2.Range ("A" & Row)
    DOY = Sheet2.Range ("B" & Row)
    SRAD = Sheet2.Range ("C" & Row)
    TMAX = Sheet2.Range ("D" & Row)
    TMIN = Sheet2.Range ("E" & Row)
    RAIN = Sheet2.Range ("F" & Row)
  Loop Until Yr = pyear And DOY = pdoy
Return

Weather:
    Row = Row + 1

    Yr = Sheet2.Range ("A" & Row)
    DOY = Sheet2.Range ("B" & Row)
    SRAD = Sheet2.Range ("C" & Row)
    TMAX = Sheet2.Range ("D" & Row)
    TMIN = Sheet2.Range ("E" & Row)
    RAIN = Sheet2.Range ("F" & Row)

    TMP = (TMAX + TMIN) / 2
Return

PhenologyBD:
  If iniPheno = 0 Then
    TBD = Sheet5. [b12]
    TP1D = Sheet5. [b13]
    TP2D = Sheet5. [b14]
    TCD = Sheet5. [b15]

    vzres = Sheet5. [b17]
    bdBRV = Sheet5. [b18]
    bdTRV = Sheet5. [b19]
    TBVER = Sheet5. [b20]
```

Continued

Box 8.1. Continued.

```
    TP1VER = Sheet5. [b21]
    TP2VER = Sheet5. [b22]
    TCVER = Sheet5. [b23]
    VDSAT = Sheet5. [b24]
    vsen = Sheet5. [b25]

    ppres = Sheet5. [b27]
    bdBRP = Sheet5. [b28]
    bdTRP = Sheet5. [b29]
    SABH = Sheet5. [b30]
    cpp = Sheet5. [b31]
    ppsen = Sheet5. [b32]

    bdSOWEMR = Sheet5. [b34]
    bdEMRTLM = Sheet5. [b35]
    bdTLMBSG = Sheet5. [b36]
    bdBSGTSG = Sheet5. [b37]
    bdTSGMAT = Sheet5. [b38]

    bdEMR = bdSOWEMR
    bdTLM = bdEMR + bdEMRTLM
    bdBSG = bdTLM + bdTLMBSG
    bdTSG = bdBSG + bdBSGTSG
    bdMAT = bdTSG + bdTSGMAT

    SNOW = 0:   CUMVER = 0:   CBD = 0:   DAP = 0:
    iniPheno = 1
  End If
'---------------------------- Vernalization function
  If vzres = 1 Then
    '- Snow Cover and Melt
    SNOMLT = 0
    If TMAX <= 1 Then
      SNOW = SNOW + RAIN
      RAIN = 0
    ElseIf SNOW > 0 Then
      SNOMLT = TMAX + RAIN * 0.4
      If SNOMLT > SNOW Then SNOMLT = SNOW
      SNOW = SNOW – SNOMLT:   RAIN = RAIN + SNOMLT
    End If

  '- Crown Temperature
  TMINCR = TMIN
  TMAXCR = TMAX
  XS = SNOW
  If XS > 15 Then XS = 15
  If TMIN < 0 And XS > 0 Then TMINCR = 2 + TMIN * (0.4 + 0.0018 * (XS – 15) ^ 2)
  If TMAX < 0 And XS > 0 Then TMAXCR = 2 + TMAX * (0.4 + 0.0018 * (XS – 15) ^ 2)
  TMPCR = (TMAXCR + TMINCR) / 2
```

Continued

Box 8.1. Continued.

```
'- Vernalization day per calendar day
VERDAY = 0
If TMPCR <= TBVER Or TMPCR >= TCVER Then
    VERDAY = 0
  ElseIf TMPCR > TBVER And TMPCR < TP1VER Then
    VERDAY = (TMPCR – TBVER) / (TP1VER – TBVER)
  ElseIf TMPCR > TP2VER And TMPCR < TCVER Then
    VERDAY = ((TCVER – TMPCR) / (TCVER – TP2VER))
  ElseIf TMPCR >= TP1VER And TMPCR <= TP2VER Then
    VERDAY = 1
  End If

  '- Cumulative vernalization day
  CUMVER = CUMVER + VERDAY
  If CUMVER < 10 And TMAX > 30 Then CUMVER = CUMVER – 0.5 * (TMAX – 30)
  If CUMVER < 0 Then CUMVER = 0

  If CBD >= bdBRV And CBD <= bdTRV Then
      verfun = 1 – vsen * (VDSAT – CUMVER)
      If verfun < 0 Then verfun = 0
      If verfun > 1 Then verfun = 1
  Else
      verfun = 1
  End If

ElseIf vzres = 0 Then
  verfun = 1
End If

'---------------------------- Temperature function
If TMP <= TBD Or TMP >= TCD Then
  tempfun = 0
ElseIf TMP > TBD And TMP < TP1D Then
  tempfun = (TMP – TBD) / (TP1D – TBD)
ElseIf TMP > TP2D And TMP < TCD Then
  tempfun = ((TCD – TMP) / (TCD – TP2))
ElseIf TMP >= TP1D And TMP <= TP2D Then
  tempfun = 1
End If

'---------------------------- Photoperiod function
Pi = 3.141592654:       RDN = Pi / 180
ALPHA = 90 + SABH
SMA3 = 0.9856 * DOY – 3.251
LANDA = SMA3 + 1.916 * Sin(SMA3 * RDN) + 0.02 * Sin(2 * SMA3 * RDN) + 282.565
DEC = 0.39779 * Sin(LANDA * RDN)
DEC = Atn(DEC / Sqr(1 – DEC ^ 2))
DEC = DEC / RDN
TALSOC = 1 / Cos(lat * RDN)
CEDSOC = 1 / Cos(DEC * RDN)
```

Continued

Box 8.1. Continued.

```
  SOCRA = (Cos(ALPHA * RDN) * TALSOC * CEDSOC) - (Tan(lat * RDN) *
    Tan(DEC * RDN))
  pp = Pi / 2 - (Atn(SOCRA / Sqr(1 - SOCRA ^ 2)))
  pp = pp / RDN
  pp = 2 / 15 * pp

  If CBD >= bdBRP And CBD <= bdTRP Then
    If ppres = 1 Then
      If pp < cpp Then '<---- Long-day plant
        ppfun = 1 - ppsen * (cpp - pp)
      Else
        ppfun = 1
      End If
    ElseIf ppres = 2 Then '<---- Short-day plant
      If pp > cpp Then
        ppfun = 1 - ppsen * (pp - cpp)
      Else
        ppfun = 1
      End If
    End If
  Else
    ppfun = 1
  End If
'---------------------------- Biological day
  DTU = (TP1D - TBD) * tempfun 'daily temperature unit
  bd = verfun * tempfun * ppfun
  CBD = CBD + bd
  DAP = DAP + 1

'---------------------------- Days to stages
  If CBD < bdEMR Then dtEMR = DAP + 1
  If CBD < bdBRP Then dtBRP = DAP + 1
  If CBD < bdTRP Then dtTRP = DAP + 1
  If CBD < bdTLM Then dtTLM = DAP + 1
  If CBD < bdBSG Then dtBSG = DAP + 1
  If CBD < bdTSG Then dtTSG = DAP + 1
  If CBD < bdMAT Then dtMAT = DAP + 1

  If CBD > bdMAT Then MAT = 1
Return

DailyPrintOut:
  Sheet4.Cells (DAP + 2, 1) = Yr
  Sheet4.Cells (DAP + 2, 2) = DOY
  Sheet4.Cells (DAP + 2, 3) = DAP
  Sheet4.Cells (DAP + 2, 4) = TMP
  Sheet4.Cells (DAP + 2, 5) = tempfun
  Sheet4.Cells (DAP + 2, 6) = CUMVER
  Sheet4.Cells (DAP + 2, 7) = verfun
```

Continued

Box 8.1. Continued.

```
  Sheet4.Cells (DAP + 2, 8) = pp
  Sheet4.Cells(DAP + 2, 9) = ppfun
  Sheet4.Cells(DAP + 2, 10) = bd
  Sheet4.Cells(DAP + 2, 11) = CBD
  Sheet4.Cells(DAP + 2, 12) = SNOW
Return

SummaryPrintOut:
  Sheet1.[g7] = dtEMR
  Sheet1.[g8] = dtBRP
  Sheet1.[g9] = dtTRP
  Sheet1.[g10] = dtTLM
  Sheet1.[g11] = dtBSG
  Sheet1.[g12] = dtTSG
  Sheet1.[g13] = dtMAT
Return

End Sub
```

Table 8.2. Parameter estimates of phenology (sub-) model for wheat (cv. Tajan) (A. Soltani, unpublished data). BD is biological day requirement, TLM is termination leaf growth on main stem, BSG is beginning seed growth, TSG is termination seed growth, and MAT is harvest maturity.

Parameter	Wheat
Base temperature for development (°C)	0
Lower optimum temperature for development (°C)	24
Upper optimum temperature for development (°C)	28
Ceiling temperature for development (°C)	37
Beginning response to vernalization (BD)	5
Termination response to vernalization (BD)	22.5
Base temperature for vernalization (°C)	−1
Lower optimum temperature for vernalization (°C)	0
Upper optimum temperature for vernalization (°C)	8
Ceiling temperature for vernalization (°C)	12
Vernalization saturation (day)	30
Vernalization sensitivity coefficient	0.002
Beginning response to photoperiod (BD)	5
Termination response to photoperiod (BD)	22.5
Sun angle below horizon (°)	6
Critical photoperiod (h)	14
Photoperiod sensitivity coefficient	0.17
BD from sowing to emergence	5
BD from emergence to TLM	29.2
BD from TLM to BSG	16.7
BD from BSG to TSG	22.7
BD from TSG to MAT	11.9

Exercises

1. In Chapters 6 and 7, relational diagrams were presented for the models used. Try to draw a relational diagram similar to those diagrams for the effect of vernalization on development rate.
2. Calculate VERDAY for a wheat crop during 5 consecutive days with crown temperature of −10, 0, 1, 10, and 13°C. Equation 8.3 should be used for the calculation.
3. Calculate *verfun* for a wheat cultivar with VDSAT of 50 days and *vsen* of 0.03 day^{-1} for CUMVER values of 0, 10, 20, 40, and 60 days. Note that *verfun* should have a value between 0 and 1.

9 Crop Leaf Area

As illustrated in Chapter 2, leaf area development is critical for crop light interception and dry matter production, and hence has a substantial influence on crop yield (Sinclair, 1984). It is also an important determinant of crop water loss (Chapter 14). Therefore, the ability to predict leaf area development is important for crop simulation models. In this chapter, a simple method is presented for predicting leaf area development under non-limiting water and nutrients, and free of insects, diseases, and weeds. Also, a simple method is presented to simulate leaf area senescence, although a more complete treatment of leaf senescence involving the plant nitrogen balance is presented in Chapter 17.

Background

Leaf area development involves the appearance of new leaves and expansion of newly emerged leaves. A range of approaches with different levels of complexity have been used to predict leaf area development, from those dealing with appearance, expansion, and senescence of individual leaves (e.g. Hofstra *et al.*, 1977) to those predicting leaf area at the whole plant or crop level (e.g. Sinclair, 1984). Some others take a middle approach (e.g. Ranganathan *et al.*, 2001; Brown *et al.*, 2005). These methods can be divided into three categories: (i) carbon-based methods; (ii) temperature-based methods; and (iii) hybrid (combined) methods.

Carbon-based methods

These methods assume assimilate available for leaf growth is the most important limiting factor affecting the development of leaf area. That is, in a given day the rate of increase in plant leaf area depends on the amount of dry matter available for leaf growth that day. In these methods, first, daily dry matter production is calculated, then leaf area development is computed as the amount of dry matter partitioned to leaf growth (increase in leaf weight) times a specific leaf area (e.g. Lee and Heuvelink, 2003). Specific leaf area is the area of leaves per unit of leaf dry matter (m^2 g^{-1}). These methods are sensitive to the partitioning of dry matter to leaves and specific leaf area itself (van Delden *et al.*, 2001). However, as discussed later, specific leaf area is not constant and it is difficult to define this variable.

Temperature-based methods

These methods assume that leaf area development is not usually restricted by availability of assimilates, but it depends on temperature as the most limiting factor of leaf expansion (Amir and Sinclair, 1991; Chapman *et al*, 1993). This approach was used in the example in Chapter 2. These methods relate the amount of leaf area development to temperature directly or indirectly. Linking leaf area development to temperature unit using appropriate relationships is common (e.g. Williams *et al.*, 1989; Chapman *et al.*, 1993; Soltani *et al.*, 1999).

Hybrid methods

These methods are based on the assumption that solar radiation determines daily amount of photosynthate available for leaf expansion while temperature affects the rate of cell division and cell extension (Kropff and van Laar, 1993; van Delden *et al.*, 2001; Yin and van Laar, 2005). Therefore, leaf area development is computed as a function of both carbon available for leaf growth and temperature which determines leaf expansion. Boote *et al.* (1998) used the hybrid approach by estimating leaf area development as the smaller of the estimates obtained by the temperature and assimilate availability approaches.

The hybrid approach of Kropff and van Laar (1993) assumed that in the early stages of wheat development when the number of growing points is low and leaf area index is also low there is no mutual shading, therefore temperature determines leaf area development. The logic is that all leaves receive solar radiation for high photosynthesis rates and produce excessive assimilate for leaf expansion growth. However, they assumed that this stage ends when mutual shading increases as a result of increases in leaf area on the plant. Also, there is extension of stems and there are usually a number of growing points so assimilate supply to the leaves could be limiting. Thus, daily increase in leaf area is dependent on dry matter allocated to leaf growth and specific leaf area (Kropff and van Laar, 1993). The difficulty is that there is little experimental evidence to support such a complex view of leaf area development.

Leaf Area Submodel

A simple method is presented to predict development in leaf area through major developmental stages in grain crops. Phenological development predicts the time interval when leaf development occurs. In this chapter, it is assumed that the crop is well supplied with water and nutrients and does not suffer from insects, diseases, and weeds. This method takes into consideration the major physiological determinants of leaf area development under potential production conditions.

In grain crops, the main phase of leaf production and development begins after emergence of the plant from the soil and continues until cessation of *effective* leaf production on the main stem. The end of effective leaf production is labeled termination leaf growth on the main stem (TLM). From TLM to beginning growth of seeds (BSG), increase in leaf area may occur due to leaf development on branches or tillers. Following BSG, no further development of leaf area is assumed for major grain crops, but this assumption can be modified if required.

During the main phase of leaf area development from emergence to TLM, the leaf area prediction method is based on the method presented by Sinclair (1984). Main-stem leaf number is first predicted as a function of temperature based on the phyllochron concept, i.e. the temperature unit between appearance of two consecutive leaves (PHYL, °C per node). Individual plant leaf area is calculated from main-stem leaf number using an allometric relationship between mainstem leaf/node number and plant leaf area. Plant density (plants m^{-2}) can affect the allometric relationship. Daily increase in the crop leaf area index (LAI, leaf area per unit of soil surface area) is finally computed as the daily increase in individual plant leaf area multiplied by plant density.

From TLM until BSG, daily increase in leaf area depends on available dry matter for leaf growth estimated from daily increase in leaf dry matter multiplied by specific leaf area. After BSG, leaf senescence begins due to seed growth and leads to a linear decrease in LAI, until it reaches zero at maturity. Box 9.1 summarizes the procedure that is used here.

Leaf area development prior to TLM

In many crops, leaves on the main stem emerge in a predictable pattern. In these crops, if the number of leaves (or nodes) on the main stem is plotted versus temperature unit a trend like that indicated in Fig. 9.1 is obtained. The increase in main-stem leaf (node) number versus cumulative temperature unit during main phase of leaf development can be simplified and described using a linear relationship (Fig. 9.1).

The slope of the line will indicate leaf appearance rate (leaf per °C). The inverse of leaf appearance rate will give the duration of a phyllochron, i.e. temperature unit (°C) elapsed between the emergence of two consecutive leaves. Phyllochron values have been widely studied and reported in many crops (e.g. Wilhelm and McMaster, 1995; Pengelly *et al.*, 1999).

Box 9.1. Different steps in calculation of leaf area development and senescence.

Increase in Leaf Area

From sowing to emergence:

- No leaf development.

From emergence to termination leaf growth on mainstem (TLM):

- Fraction phyllochron experienced by the crop is calculated.
- Increase in leaf/node number on main stem from phyllochron.
- Total number of nodes/leaves on main stem.
- Increase in plant leaf area as a function of main-stem node number.
- Increase in LAI as a function of increase in plant leaf area and plant density.

From TLM to beginning seed growth (BSG):

- Increase in leaf dry matter (Chapter 11).
- Leaf area development from leaf growth and specific leaf area.

From BSG to crop harvest maturity:

- No leaf development.

Decrease in Leaf Area

From emergence to beginning seed growth (BSG):

- No leaf senescence.

From BSG to crop harvest maturity:

- LAI decreases linearly and reaches 0 at harvest maturity.

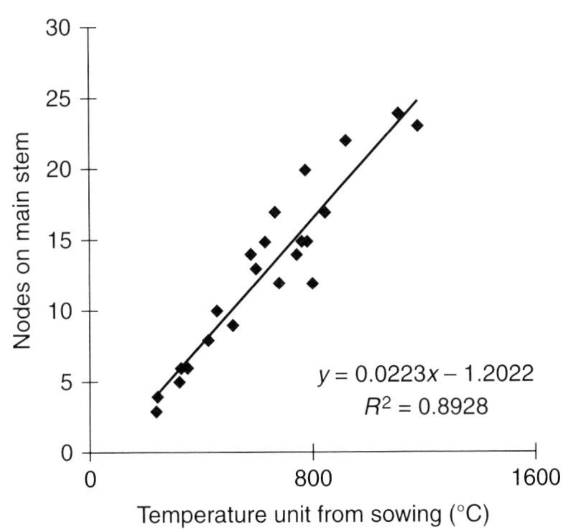

Fig. 9.1. Main-stem node appearance in pigeonpea as a function of cumulative temperature unit (Ranganathan *et al.*, 2001). Leaf appearance rate is 0.0223 leaf per °C and phyllochron is 44.8°C.

Based on the phyllochron concept, from emergence to TLM daily increase in main-stem leaf number (INODE, node #) can be calculated from daily temperature unit (DTU, °C; Chapter 6).

$$\text{INODE} = \text{DTU} / \text{PHYL} \quad (9.1)$$

The daily increase in node number, INODE, is used to track increasing total main-stem node number on the plant (MSNN, node #). This is done daily by adding the daily value of INODE to the previous day total main-stem node number (MSNN_{i-1}). Therefore, the total number of leaves on the main stem at any point up to TLM is MSNN_i.

$$\text{MSNN}_i = \text{MSNN}_{i-1} + \text{INODE} \quad (9.2)$$

The next step is to calculate plant leaf area (PLA, cm^2 per plant) from MSNN_i. Allometric relationships have been reported for most crop species between main-stem leaf number and plant leaf area during the major phase of leaf area development. For example, such allometric relationships have been reported in sorghum (Hammer *et al.*, 1993), wheat and barley (Wahbi and Sinclair, 2005), soybean (Sinclair, 1984), and chickpea (Soltani *et al.*, 2006c). The form of this relationship that has been most commonly reported is a power function (Fig. 9.2a):

$$\text{PLA} = \text{PLACON} \times \text{MSNN}^{\text{PLAPOW}} \quad (9.3)$$

where PLACON and PLAPOW are constants for each particular crop species, and, if available, each genotype within a species. Equation 9.3 is further simplified if we assume PLACON = 1 cm^2, which means a small value of 1 cm^2 for PLA when there is one node or leaf on the main stem. This assumption allows Eqn 9.3 to be simplified to:

$$\text{PLA} = \text{MSNN}^{\text{PLAPOW}} \quad (9.4)$$

At high plant densities in particular, smaller and/or fewer leaves and branches or tillers might be produced on the plant. In this case, the effect of plant density can be simulated by decreasing PLAPOW. That is, PLAPOW is defined as a function of plant density (Fig. 9.2b).

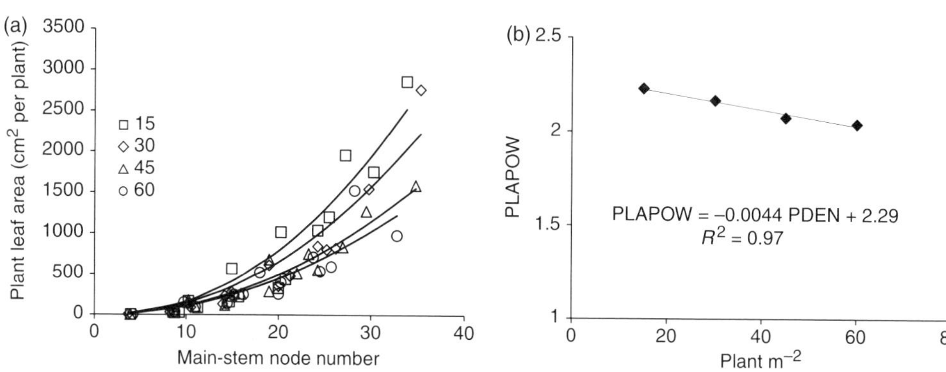

Fig. 9.2. (a) Plant leaf area as a function of main-stem node number described by a power function as PLA = $\text{MSNN}^{\text{PLAPOW}}$. Numbers indicate plant densities. (b) Dependency of the PLAPOW coefficient of the power function (Eqn 9.4) on plant density (PDEN) (Soltani *et al.*, 2006c).

Figure 9.2a indicates that the deviation between observed values for PLA and those represented by the power function is greater at larger MSNN than at low MSNN. This situation is generally satisfactory because the calculation of light interception is done using an exponential equation so errors in PLA, and thereby LAI, have little consequence at high PLA. When the crop LAI is greater than 3.0, virtually all the light is intercepted and any variation in LAI has little influence on this calculation (Sinclair, 1984).

Daily increase in crop LAI (GLAI, $m^2\ m^{-2}\ day^{-1}$) can be computed from plant leaf area increase and plant density (PDEN, plant m^{-2}) as:

$$GLAI = (PLA_i - PLA_{i-1}) \times PDEN / 10{,}000 \qquad (9.5)$$

where PLA_i is plant leaf area today, PLA_{i-1} the plant leaf area yesterday, and 10,000 is to convert PLA from cm^2 per plant to m^2 per plant. The flow diagram for the calculation of GLAI is shown in Fig. 9.3.

Leaf area between TLM and BSG

After termination of leaf development on the main stem (TLM) through to the beginning of seed growth (BSG), crop leaf area often continues to develop on branches of dicot plants and tillers of cereal plants. In this period of leaf development, daily increase in LAI is calculated based on available dry matter for leaf growth. It is obtained by multiplying daily increase in leaf dry matter (GLF, g $m^{-2}\ day^{-1}$) and specific leaf area (SLA, $m^2\ g^{-1}$):

$$GLAI = GLF \times SLA \qquad (9.6)$$

The value of SLA for the leaves produced during this period is an input to this submodel. The value for GLF is computed as daily dry matter produced by the crop times the fraction of the dry matter that is partitioned to leaves. This will be discussed in detail in Chapter 11. The flow diagram for the calculation between TLM and BSG is shown in Fig. 9.4.

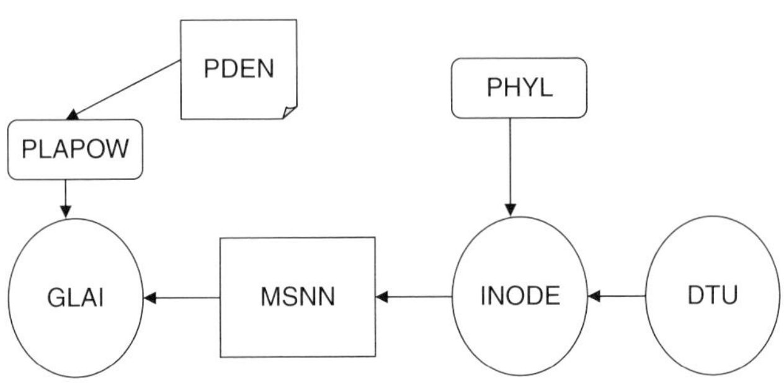

Fig. 9.3. Flow diagram depicting the calculation of the daily increase in LAI (GLAI, $m^2\ m^{-2}\ day^{-1}$) prior to TLM. DTU the daily temperature unit (°C), INODE the daily increase in leaf (node) number on main stem (# day^{-1}), MSNN the total cumulative number of leaf/node on main stem (#), PHYL the phyllochron (°C per leaf/node), PLAPOW the parameter in relationship between plant leaf area and main-stem leaf/node number, and PDEN the plant density (# m^{-2}).

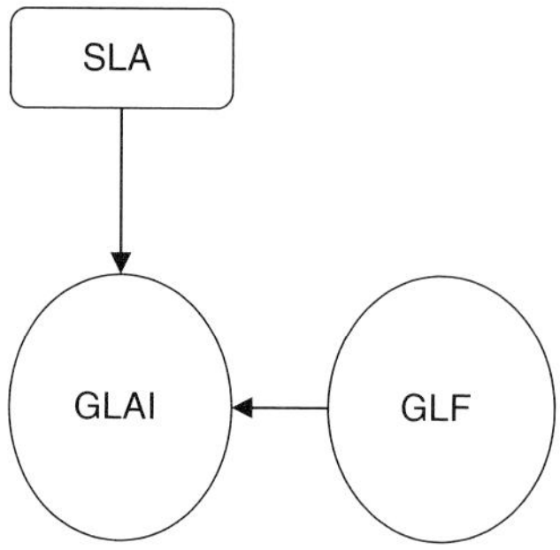

Fig. 9.4. Flow diagram depicting the calculation of the daily increase in LAI (GLAI, m^2 m^{-2} day^{-1}) between TLM and BSG depending on GLF the daily increase in leaf dry matter (g m^{-2} day^{-1}) and SLA the specific leaf area (m^2 g^{-1}).

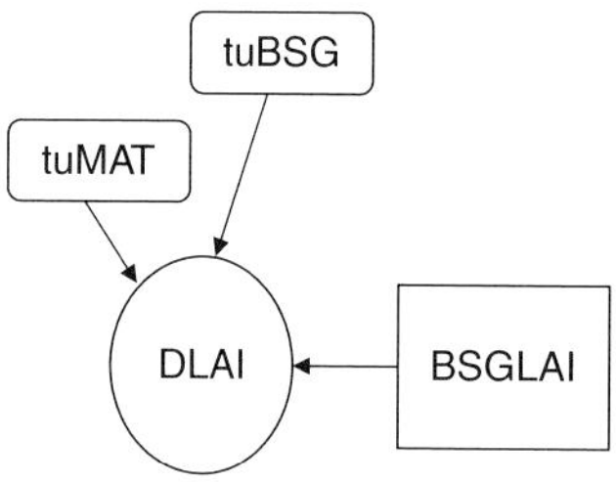

Fig. 9.5. Flow diagram illustrating the calculation of DLAI (m^2 m^{-2} day^{-1}) based on BSGLAI the crop LAI at beginning seed growth, tuBSG the temperature unit to beginning seed growth (°C), and tuMAT the temperature unit to maturity (°C).

Leaf area after BSG

In major grain crops, generally few if any new additional leaves are produced once seed growth begins (BSG). Therefore, changes in leaf area after BSG are characterized by leaf senescence resulting in a loss of leaf area. In grain crops, leaf senescence is mainly affected by the growth of seeds and resultant nitrogen mobilization from leaves to seeds (Sinclair and de Wit, 1976; Sinclair *et al.*, 2003; Yin *et al.*, 2003a). To account for this process directly it is necessary to track the nitrogen economy in the plant. Leaf senescence as a result of loss of leaf nitrogen is considered in Chapter 17. In models that do not include nitrogen or models that are developed for non-limiting nitrogen conditions (as is assumed in this chapter), leaf senescence is defined as a function of plant age, which indirectly includes grain growth and nitrogen mobilization.

The simplest approach to account for leaf senescence is a linear decrease in LAI during seed growth, reaching zero LAI at maturity. Therefore, daily decrease in LAI (DLAI, m^2 m^{-2} day^{-1}) is calculated as:

$$\text{DLAI} = \text{DTU} / (\text{tuMAT} - \text{tuBSG}) \times \text{BSGLAI} \qquad (9.7)$$

where DTU is daily temperature unit and (tuMAT – tuBSG) is the temperature unit from BSG to maturity. The value of LAI at BSG is BSGLAI (m^2 m^{-2}). The flow diagram for calculation of DLAI is presented in Fig. 9.5. In Chapters 15 and 17, the submodel is expanded to account for water deficit and nitrogen limitation, respectively, resulting in premature termination of seed growth.

Crop LAI

The estimates obtained in the above calculations can now be used to calculate current crop LAI (LAI_t) from the previous day LAI (LAI_{t-1}) and the daily increase in LAI (GLAI) and LAI senescence (DLAI).

$$LAI_t = LAI_{t-1} + GLAI - DLAI \tag{9.8}$$

Parameter Estimation

The required parameters for the leaf area submodel are: PHYL, PLACON, PLAPOW, and SLA. Estimates of these parameters are given in Table 9.1 for a number of field crops. All these variables can be obtained directly from experimental observations of the crops. For wide use of these parameters, the experiments should be done under potential growth conditions with optimal or near optimal water and nutrient conditions, free of insects, diseases, and weeds.

Temperature unit requirements from sowing to different stages are also required to simulate changes in LAI, but they are considered as parameters related to phenological development (discussed in detail in Chapter 6).

Phyllochron can be obtained from fitting a linear regression line to data of main-stem leaf/node number versus temperature unit (refer to Fig. 9.1). There

Table 9.1. Estimates of parameters relating to leaf area development and senescence in some crops (Penning de Vries *et al.*, 1989; Chapman *et al.*, 1993; Hammer *et al.*, 1993; Keating *et al.*, 2003; Wahbi and Sinclair, 2005; Yin and van Laar, 2005; A. Soltani, unpublished data).

Crop	PHYL	PLACON	PLAPOW	SLA
Wheat	120	1	2.464	0.021
Barley	75	1	2.341	0.031
Rice	83			0.023
Maize	50	1	3.050	0.022
Sorghum	50	1	2.890	0.025
Soybean	45	1	3.110	0.025
Peanut	56	1	2.750	0.020
Canola	75	1	2.921	0.030
Sunflower	40	1	3.119	0.025
Dry bean	60	1	3.325	0.030
Chickpea	46	1	2.158	0.021

PHYL: phyllochron (°C per leaf/node)
PLACON: constant in relationship between plant leaf area and main-stem leaf/node number, Eqn 9.3
PLAPOW: exponent in relationship between plant leaf area and main-stem leaf/node number, Eqn 9.3
SLA: specific leaf area ($m^2\ g^{-1}$)

are a number of criteria to measure leaf/node number on the main stem, e.g. Haun stage in wheat (Haun, 1973), counting leaf tips, or counting the number of fully expanded leaves. The number of fully expanded leaves/nodes is often more suitable for evaluation under field conditions.

The coefficients of the allometric relationship between plant leaf area and main-stem leaf number (PLACON and PLAPOW) can be found by fitting a power equation to the data of plant leaf area versus main-stem leaf number (Fig. 9.2a). The data must also come from non-stressed conditions from the time of emergence to TLM. If the data are available for different plant densities, their relation to plant density can be found as well (Fig. 9.2b). To do this, first a power equation is fitted to data of each plant density, then the relationship of the obtained coefficients and plant density is found. Figure 9.2 indicates a sample fit of the power equation for different plant densities in chickpea (Soltani *et al.*, 2006c).

Calculation of leaf area after TLM requires an estimate of SLA. The value of SLA can be determined by harvesting leaf disks of known area and measuring their dry weight. These data allow direct calculation of SLA as leaf dry weight per unit of leaf area (g m^{-2}). Another experimental approach is based on rearrangement of Eqn 9.6 such that SLA = GLAI / GLF. Therefore, the slope of a plot of LAI versus leaf dry weight provides an estimate of SLA (Fig. 9.6). Since early leaves are thin, data obtained only at higher LAIs should be used to avoid an overestimation of SLA.

Programming

The sub-program for simulation of changes in LAI through major developmental stages is presented in Box 9.2. For description of the variable names refer

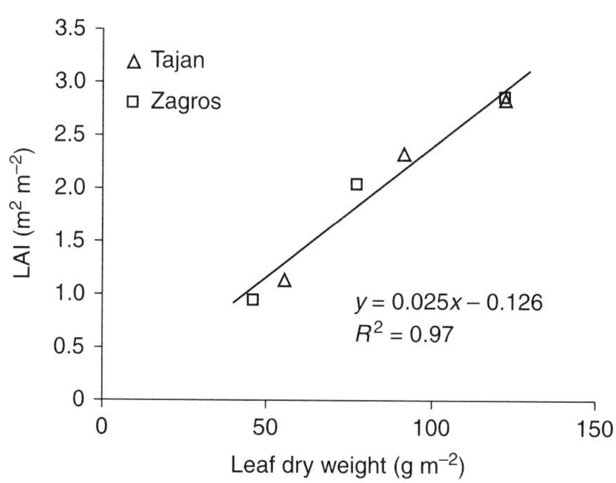

Fig. 9.6. LAI as a function of leaf dry weight in two wheat cultivars (Tajan and Zagros). The slope of the regression line is specific leaf area (0.025 m^2 g^{-1}) (V. Madah and A. Soltani, unpublished data).

Box 9.2. Program of submodel for simulating crop LAI. For a list of variables refer to Appendix III or the text.

```
CropLAI:
'------------------------------ LAI initials and pars
  If iniLAI = 0 Then
    PHYL = Sheet5.[b17]
    PLACON = Sheet5.[b18]
    PLAPOW = Sheet5.[b19]
    SLA = Sheet5.[b20]

    MSNN = 1:   PLA2 = 0:   PLA1 = 0:   LAI = 0:
    MXLAI = 0:  iniLAI = 1
  End If
'------------------------------ Yesterday LAI to intercept PAR today
  LAI = LAI + GLAI – DLAI
  If LAI < 0 Then LAI = 0
  If LAI > MXLAI Then MXLAI = LAI            'Saving maximum LAI
'------------------------------ Daily increase and decrease in LAI today
  If CTU <= tuEMR Then
    GLAI = 0:     DLAI = 0
  ElseIf CTU > tuEMR And CTU <= tuTLM Then
    INODE = DTU / PHYL
    MSNN = MSNN + INODE
    PLA2 = PLACON * MSNN ^ PLAPOW
    GLAI = ((PLA2 – PLA1) * PDEN / 10000)
    PLA1 = PLA2
    DLAI = 0
  ElseIf CTU > tuTLM And CTU <= tuBSG Then
    GLAI = GLF * SLA
    BSGLAI = LAI                       'Saving LAI at BSG
    DLAI = 0
  ElseIf CTU > tuBSG Then
    GLAI = 0
    DLAI = DTU / (tuMAT – tuBSG) * BSGLAI
  End If
Return
```

to the text in this chapter or Appendix III. Later in Chapter 12, this submodel is incorporated in a simulation model of potential production. The parameter estimates in this submodel are related to a wheat cultivar (cv. Tajan).

Additional Notes

Non-constant phyllochron

In the method presented here it was assumed that the value of the phyllochron is constant throughout the leaf production phase. However, there is some

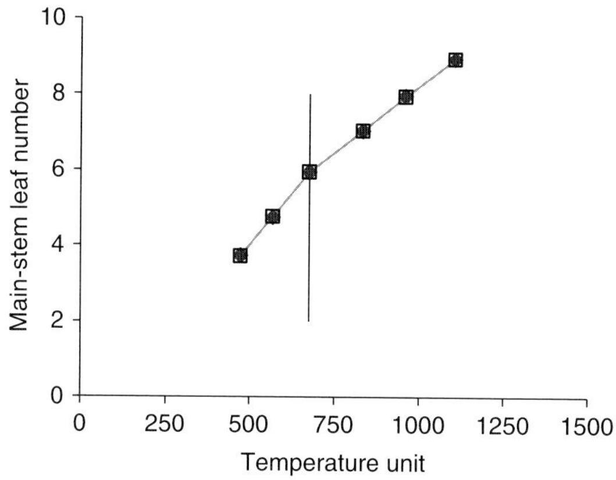

Fig. 9.7. A sample of change in phyllochron in wheat. The phyllochron is 90°C for the first line and 143°C for the second line. The vertical line is the temperature unit at inflection time (675°C). Phyllochron for the whole period is 120°C (V. Madah and A. Soltani, unpublished data).

experimental evidence that phyllochron may not be constant during the whole period (e.g. Jamieson *et al.*, 1995). Figure 9.7 shows as an example data of main-stem leaf number versus temperature unit in wheat (V. Madah and A. Soltani, unpublished data). It seems that before temperature unit of 675°C, phyllochron is smaller and equal to 90°C but afterwards phyllochron increases to 143°C. In cases that phyllochron is not constant, using two or more values of phyllochron might be necessary. The simulation program can be simply changed to include multi-phyllochron.

Other allometric equations

A power equation was used to describe the relationship of plant leaf area and main-stem leaf number. Some researchers have used other functions for this relationship. The exact type of the equation depends on the data and any allometric relationship that fits the data can be substituted in the simulation model. Parameters of the equations can be ascribed to plant density.

For example, Sinclair (1984) has used the following relation in soybean:

$$\mathrm{PLA} = a \times \mathrm{PHYL} + b \times (\exp(c \times \mathrm{PHYL}^{3/2}) - 1) \quad (9.9)$$

where *a*, *b*, and *c* are the coefficients of the equation. Wahbi and Sinclair (2005) used an exponential equation for this purpose in wheat and barley as (Fig. 9.8):

$$\mathrm{PLA} = a \times \exp(b \times \mathrm{MSLN}) \quad (9.10)$$

where MSLN is the number of leaves on main stem and *a* and *b* are the coefficients of the equation. Another function is an expo-linear function introduced by Goudriaan and Monteith (1990).

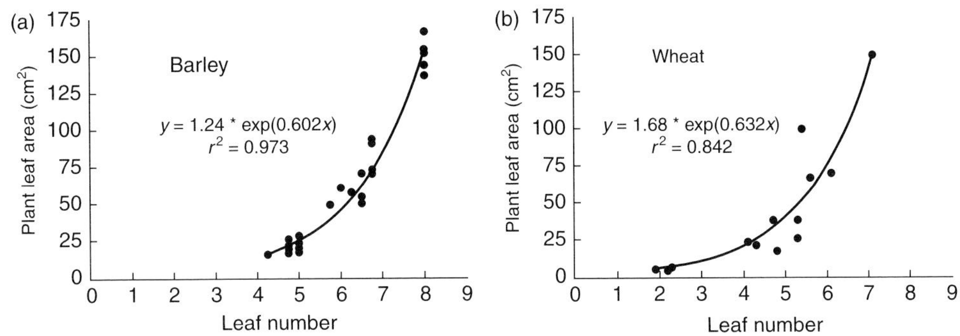

Fig. 9.8. Allometric relationship between plant leaf area and main-stem leaf number in barley (a) and wheat (b), as described by an exponential function (Wahbi and Sinclair, 2005).

A simple approach to parameter estimation

In the parameterization of Eqn 9.9, Sinclair (1984) found that the coefficient *a* in the equation can be simply estimated as the area of fully expanded soybean leaves at trifoliolate nodes 8 to 10. Further, he found that other coefficients in that equation, i.e. *b* and *c*, can be estimated from the coefficient *a* using regression equations. It may be possible to find similar estimation approximation for other parameters to be used in a simulation model. Such approximations greatly simplify model parameterization and application.

Non-constant SLA

A constant SLA was used here to simulate development in leaf area after TLM. This assumption of constant SLA is adequate for educational purposes and also for many simulations. However, sometimes constant SLA may not be adequate. In some crop models, the value of SLA is adjusted for current conditions of solar radiation and temperature or crop growth. For example, Fig. 9.9 indicates how SLA is adjusted for solar radiation and temperature in the CROPGRO-Soybean model (Boote *et al.*, 1998). Higher levels of solar radiation result in thicker leaves and thereby lower SLA (Fig. 9.9a). Inversely, higher temperatures have the inverse effect and result in thinner leaves with a higher SLA (Fig. 9.9b).

Shading and leaf senescence

Leaf shading may also cause leaf senescence. Some crop models account for the effects of shading when leaf area index increases beyond a certain level. Goudriaan and van Laar (1994) presented a simple method to account for the impact of shading on leaf senescence. This method has been used in some crop simulation models from Wageningen University. In this method, when LAI is less than a certain limit (mainly 4 or 5), shading is assumed to have no effect

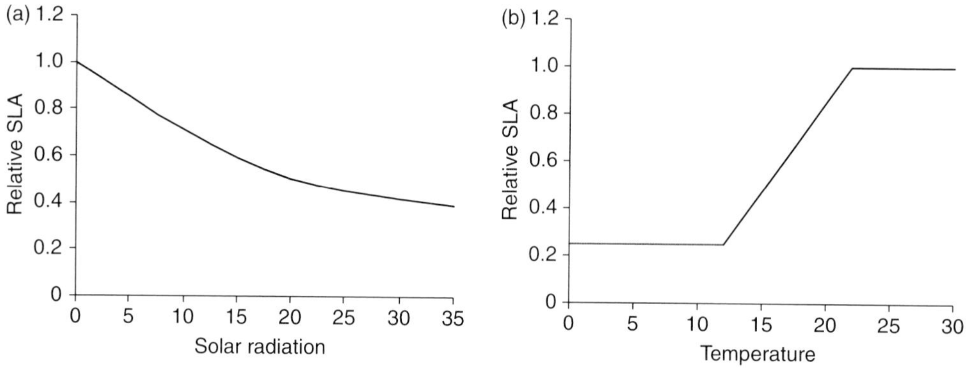

Fig. 9.9. (a) The effect of solar radiation (MJ m^{-2} day^{-1}) and (b) mean temperature (°C) on specific leaf area (SLA) in soybean as used in CROPGRO model (Boote *et al.*, 1998; Jones *et al.*, 2003).

on leaf senescence; however, with increase in LAI above this limit, shading results in leaf senescence. The fraction of LAI senesced due to shading each day (LDRSH) is calculated using a critical LAI (LAICR) by the following equation:

$$\text{LDRSH} = 0 \qquad \text{if LAI} \le \text{LAICR}$$
$$\text{LDRSH} = 0.03 \times (\text{LAI} - \text{LAICR}) / \text{LAICR} \qquad \text{if LAI} > \text{LAICR} \tag{9.11}$$

The 0.03 term in the above equation represents the maximum fraction of daily leaf senescence resulting from shading. According to Eqn 9.11, at LAI twice the value of LAICR, the value of LDRSH is equal to 0.03. By multiplying LDRSH with the current crop LAI, daily rate of decrease in LAI as a result of shading (DLAISH) is calculated:

$$\text{DLAISH} = \text{LAI} \times \text{LDRSH} \tag{9.12}$$

Unfortunately, we are not aware of any experimental approach to estimate the critical parameter of maximum leaf senescence (i.e. 0.03) and LAICR.

Freezing and leaf senescence

Chilling and freezing temperatures can result in LAI destruction. When the probability of such low temperatures is high, their effects must be taken into account. The relationship between minimum temperatures (TMIN, °C) and the fraction of LAI that is destroyed (LDRFR) can be described by the following equation:

$$\text{LDRFR} = -\text{TMIN} / (-\text{TMIN} + \exp(a - b \times -\text{TMIN})) \tag{9.13}$$

where *a* and *b* are the constants. Having two points from data of freezing temperature (TMIN) and fraction destroyed LAI (LDRFR), the constants can be found analytically (Kiniry *et al.*, 1992):

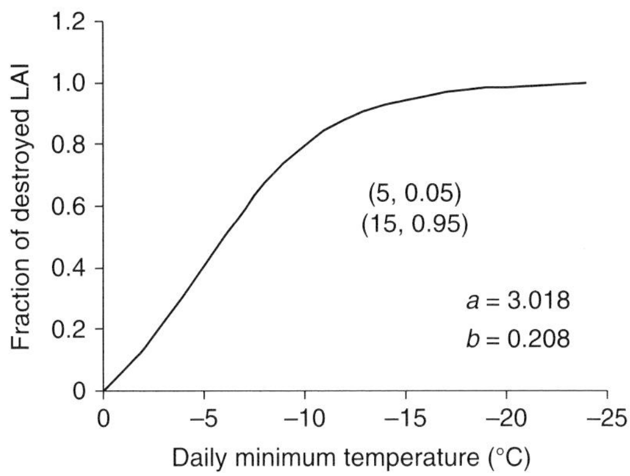

Fig. 9.10. Fraction of LAI destroyed by chilling or freezing temperatures as a function of daily minimum temperature below zero using Eqn 9.13.

$$b = \frac{\log\left(\frac{1/y_2 - 1}{1/x_2}\right) - \log\left(\frac{1/y_1 - 1}{1/x_1}\right)}{x_1 - x_2} \tag{9.14}$$

$$a = \log\left(\frac{1/y_1 - 1}{1/x_1}\right) + bx_1 \tag{9.15}$$

For example, Fig. 9.10 shows relationships between fraction destruction in LAI by freezing temperature in a crop that loses 5 and 95% of its LAI by –5 and –15°C, respectively. So, two points for the crop will be (–5, 0.05) as (x_1, y_1) and (–15, 0.95) as (x_2, y_2) and estimates of *a* and *b* are 3.018 and 0.208, respectively.

Exercises

1. The table below includes cumulative temperature unit, accumulated node number on main stem, and plant leaf area in a field crop during main phase of leaf area development. Using the data try to obtain estimates of PHYLL, PLACON, and PLAPOW for the crop.

CTU	MSNN	PLA
494	3.5	8.3
586	4.7	13.6
712	6.4	26.1
853	7.1	56.6
980	8.3	79.5
1119	9.3	109.8

2. Try to gather similar data to Exercise 1 for crops/cultivars grown in your area and then try to obtain appropriate estimates of leaf development parameters from that data.
3. Use parameter estimates of Exercise 1 or 2 to estimate plant leaf area and LAI as a function of thermal time. To do this, you will need temperature data. You can obtain the temperature data for growing season of your crop from a local weather station or from internet resources. One sample year will be fine for the practice.

10 Dry Matter Production

Simulation of dry matter production is a central part of crop growth models. The very simple sugarcane growth model presented in Chapter 1 illustrates an application of a dry matter production model. The objective of this chapter is to discuss and quantify in detail some of the key features in modeling dry matter production.

Background

Different methods have been developed and used for predicting dry matter production by crops. Some of these methods are based on detailed modeling of photosynthesis and growth and maintenance respiration. In these methods, radiation intercepted by leaves is calculated first. Then, gross photosynthesis is computed and finally daily dry matter is obtained after subtraction of maintenance and growth respiration from gross photosynthesis. Most of the Wageningen crop models and CROPGRO models consider the details of simulation of photosynthesis and respiration (Boote and Pickering, 1994; Goudriaan and van Laar, 1994; Boote *et al.*, 1998; Lizaso *et al.*, 2005).

Other models simulate dry matter production based on the concept of radiation interception and radiation use efficiency (RUE, g MJ^{-1}) (Sinclair, 1986, 1994). Radiation interception is expressed on the basis of energy unit, commonly photosynthetically active radiation (PAR, MJ m^{-2}). As will be discussed later, leaf area index (LAI) and extinction coefficient (KPAR) are determinants of PAR interception.

Plant and canopy characteristics are combined into a single composite property, the extinction coefficient (Hay and Porter, 2006). The most important variables influencing the extinction coefficient are the lack of randomness in the horizontal distribution of leaves (i.e. non-overlapping leaves in adjacent rows), sun angle, and leaf angles.

It is also influenced by crop management practices such as row spacing and plant density. Flenet *et al.* (1996) reported that KPAR decreases with increase in inter-row spacing. Maddonni *et al.* (2001) found that KPAR decreases with increase in plant population density. Similarly, Campbell and Norman (1998) and Sinclair (2006) also showed that KPAR decreases with increasing LAI (Fig. 10.1). While all these confounding factors can be taken into account to obtain KPAR, at this point it will be assumed that KPAR is held constant throughout the crop life cycle.

Since the extinction coefficient is influenced by solar elevation, it is not constant throughout the day and normally it has greater values in the morning and in the afternoon and a relatively constant value in the middle of the day (Fig. 10.2; Sinclair, 2006). It has been shown that extinction coefficient under

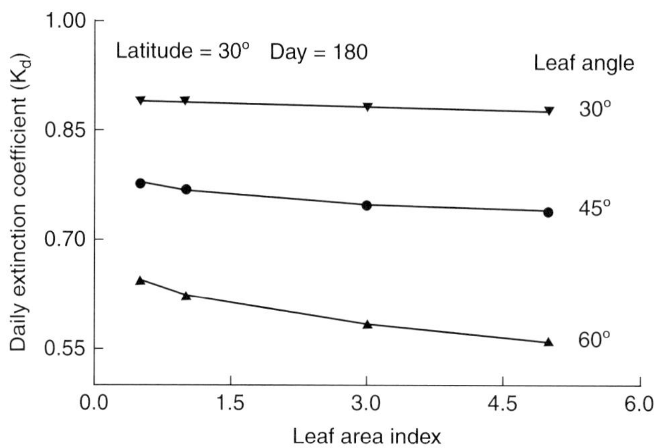

Fig. 10.1. Daily extinction coefficient (K_d) calculated at various leaf area indices and for three leaf angles (Sinclair, 2006).

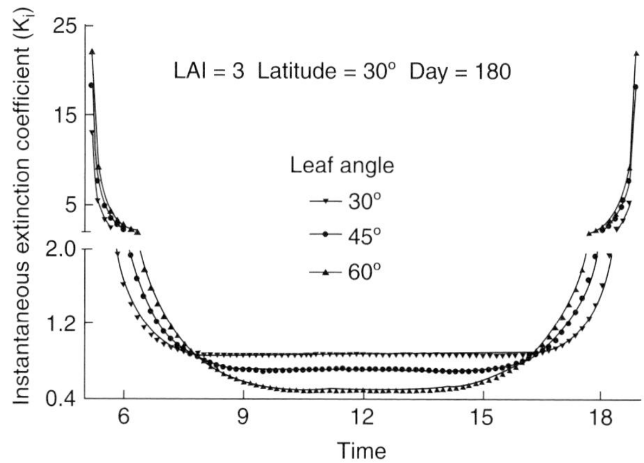

Fig. 10.2. Instantaneous extinction coefficients (K_i) throughout the day for canopies with a LAI of 3.0 and various leaf angles (Sinclair, 2006).

diffuse radiation is a good estimate of extinction coefficient for the whole day to be used in estimating daily PAR interception (Campbell and Norman, 1998).

At the crop level, the amount of dry matter produced for each unit of PAR intercepted by a crop can be calculated using RUE (Sinclair and Muchow, 1999). RUE is dry matter produced per unit of intercepted or absorbed solar radiation or photosynthetic active radiation (PAR) and its unit is grams dry matter produced per megajoules intercepted or absorbed solar radiation or PAR (g MJ^{-1}). RUE is a summary variable that represents the processes of photosynthesis, and maintenance and growth respiration in a single term (Sinclair and Horie, 1989; Sinclair and Muchow, 1999). Consequently, RUE depends on plant characteristics such as leaf maximum photosynthetic capacity and biochemical composition of produced dry matter (Sinclair and Horie, 1989; Sinclair, 1991). Each crop species has a unique potential RUE based on these characteristics (Table 1.1 in Chapter 1). If the plant has the C_4 photosynthetic pathway and a high photosynthetic rate, the value of RUE is higher than that of species with only the C_3 pathway. Figure 10.3 shows the dependency of RUE on maximum photosynthetic capacity of leaves.

Biochemical composition of the plant material being produced has a direct influence on RUE (Sinclair and Horie, 1989). Species producing mainly carbohydrates – sugar, starch – have larger values of RUE than those species producing plant tissue high in protein and lipids. The difference between maize and rice, and soybean in Fig. 10.3 is that soybean produces high amounts of protein and lipid in contrast to the two cereals. Sinclair (1991) presented a simple model to calculate potential RUE.

Experimentally, RUE is determined as the slope of the line that relates cumulative dry matter production (usually aboveground dry matter) to the cumulative intercepted or absorbed solar radiation or PAR (Fig. 10.4). As both total solar radiation and PAR have been used in calculation of RUE, it

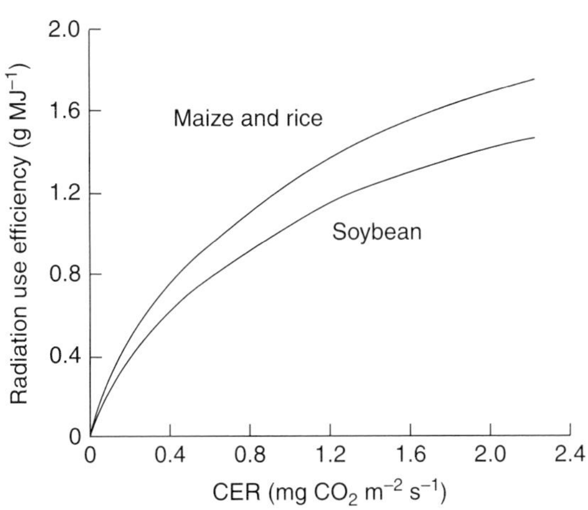

Fig. 10.3. Calculated crop radiation use efficiency of maize and rice, and soybean plotted as a function of leaf carbon dioxide exchange rate (CER) under light-saturated conditions (Sinclair and Horie, 1989).

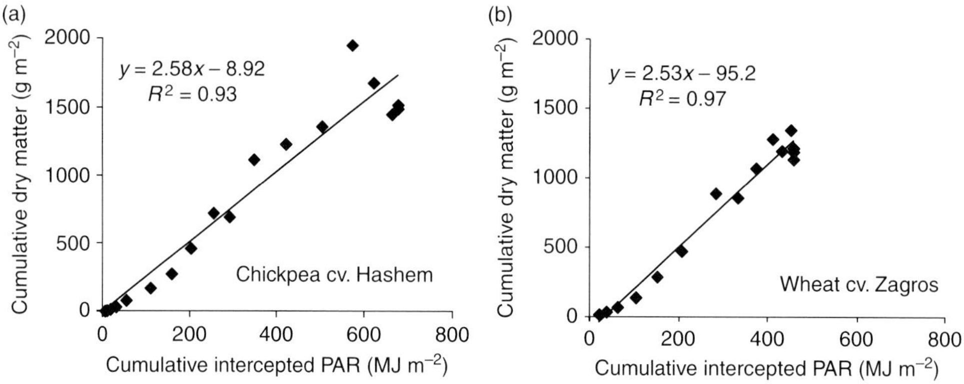

Fig. 10.4. Relationship between cumulative produced dry matter and cumulative intercepted PAR in (a) chickpea and (b) wheat (V. Madah and A.Soltani, unpublished data).

is necessary to specify explicitly the basis upon which RUE is calculated. In this book, RUE will be defined as grams of aboveground dry matter per MJ of intercepted PAR. Using absorbed PAR instead of intercepted PAR will also result in different RUE values. A detailed review on RUE is presented by Sinclair and Muchow (1999).

In this book, dry matter production is modeled in the simplest way using the RUE concept. The objective in this chapter is to estimate dry matter production by a crop canopy that is not subjected to water or nutrient stresses and its growth is not influenced by competition with insects, diseases, and weeds. The amount of intercepted PAR each day is obtained from current crop LAI (Chapter 9) and KPAR. A constant KPAR is used here throughout the crop life cycle. The amount of dry matter produced by the crop each day is obtained by multiplying intercepted PAR and RUE. RUE can be adjusted for mean daily temperature and CO_2 concentration. Box 10.1 presents a summary of the method used here to compute daily dry matter production.

Box 10.1. A summary of calculation method of daily dry matter production by crop canopies.

From sowing to emergence:

- No dry matter production!

From emergence to termination seed growth (TSG):

- Fraction intercepted PAR is calculated from LAI and extinction coefficient.
- Daily incident PAR is assumed to be half of daily total solar radiation.
- Actual RUE is obtained by adjusting potential RUE for daily mean temperature.
- Daily mass production is computed from intercepted PAR and RUE.

From TSG to crop harvest maturity:

- No dry matter production!

Functions

The daily amount of dry matter produced (DDMP, g m^{-2} day^{-1}) can be readily calculated (Sinclair, 1986) based on the amount of photosynthetically active radiation incident to the crop (PAR, MJ m^{-2} day^{-1}), the fraction of the incident radiation intercepted by the leaves (FINT), and the radiation use efficiency of the crop (RUE, g MJ^{-1}).

$$DDMP = PAR \times FINT \times RUE \tag{10.1}$$

The determination of each of these terms is discussed below.

Radiation interception

Weather records often report incident light as total solar radiation rather than PAR. PAR makes up about 0.48 of the energy of the total solar radiation, so the solar radiation data can be converted to PAR by multiplying by 0.48 (Monteith and Unsworth, 2007).

To calculate dry matter production each day, it is necessary to calculate the proportion of incident PAR that is intercepted by the crop canopy (FINT). This fraction is simply calculated using an exponential radiation-interception equation analogous to the Beer–Bouguer–Lambert Law (Sinclair, 2006) based on KPAR and LAI.

$$FINT = 1 - \exp(-KPAR \times LAI) \tag{10.2}$$

Figure 10.5 presents a schematic representation of FINT over a range of LAI values and two values of KPAR, i.e. 0.65 and 0.85.

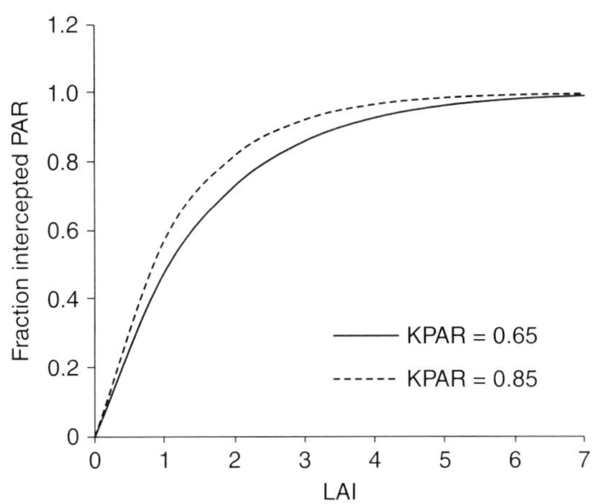

Fig. 10.5. Fraction of intercepted PAR by two canopies with KPAR of 0.65 and 0.85 over different LAIs.

Radiation use efficiency

Radiation use efficiency (RUE, g MJ^{-1}) is the critical parameter to calculate crop growth from the amount of radiation intercepted. A crop potential RUE under optimal temperature, water, and nitrogen conditions is a crop parameter that needs to be inputted. Potential RUE is unique for each crop species as shown in Table 1.1 in Chapter 1. Actual RUE can be decreased below potential RUE by non-optimal temperatures. As illustrated in Fig. 10.3, any factor that alters leaf photosynthetic rate results in a decrease in RUE, including temperature, atmospheric CO_2 concentration, water-deficit stress (Chapter 15), and leaf nitrogen concentration (Chapter 17).

The temperature effect on RUE can be described by a 3-segment function (Fig. 10.6). The cardinal temperatures for RUE response need to be provided: TBRUE, TP1RUE, TP2RUE, and TCRUE are base, lower optimum, upper optimum, and ceiling temperature in response of RUE to temperature, respectively. Each day an adjusting factor (TCFRUE) is calculated depending on mean temperature that day (TMP). TCFRUE is a scalar factor that varies between 0 and 1.

$$\begin{aligned}
&\text{TCFRUE} = 0 && \textit{if}\ \text{TMP} \leq \text{TBRUE}\\
&\text{TCFRUE} = (\text{TMP}-\text{TBRUE}) / (\text{TP1RUE}-\text{TBRUE}) && \textit{if}\ \text{TBRUE} < \text{TMP} < \text{TP1RUE}\\
&\text{TCFRUE} = 1 && \textit{if}\ \text{TP1RUE} \leq \text{TMP} \leq \text{TP2RUE}\\
&\text{TCFRUE} = (\text{TCRUE}-\text{TMP}) / (\text{TCRUE}-\text{TP2RUE}) && \textit{if}\ \text{TP2RUE} < \text{TMP} < \text{TCRUE}\\
&\text{TCFRUE} = 0 && \textit{if}\ \text{TMP} \geq \text{TCRUE} \qquad (10.3)
\end{aligned}$$

The actual RUE is calculated by multiplying the potential RUE by TCFRUE calculated above. According to Eqn 10.3, at temperatures lower than TBRUE or higher than TCRUE, RUE is zero. With increase in temperature from TBRUE, RUE increases linearly and reaches its maximum value at TP1RUE. It remains

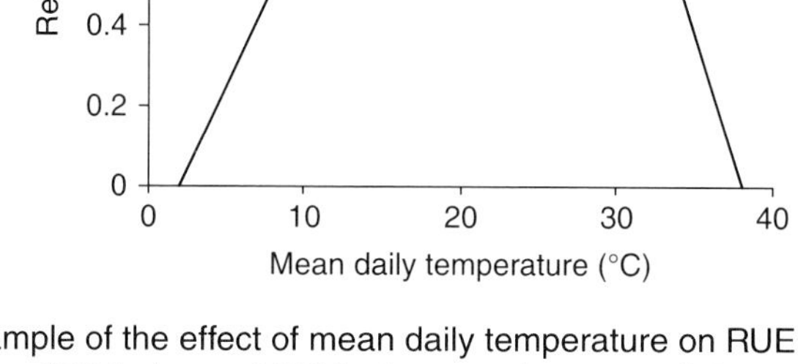

Fig. 10.6. A sample of the effect of mean daily temperature on RUE. Cardinal temperatures are 2°C for base, 14°C for lower optimum, 30°C for upper optimum and 38°C for ceiling temperatures.

at its maximum at temperatures between the lower and upper optimums. With further increase in temperature after TP2RUE, RUE is again declined and reaches zero at TCRUE.

Therefore, each day RUE is adjusted based on the physiological potential RUE (IRUE, g MJ^{-1}) and the average daily temperature of that day.

$$RUE = IRUE \times TCFRUE \quad (10.4)$$

In the simplest model, the value of IRUE can be assumed to be constant during the whole crop cycle from emergence to termination seed growth. However, changes in RUE during grain filling may occur for the following reasons:

1. In some crops, such as oil crops, biochemical composition of seed tissue is different from that of vegetative tissue. Therefore, photosynthates during the grain filling period are used to build energy-rich compounds such as oils and proteins compared to carbohydrates. This results in a lower RUE (e.g. Hammer *et al.*, 1995). This phenomenon is accounted for in the models presented here by introducing a grain conversion coefficient, which is the ratio of biochemical value of vegetative tissues to the seeds (Sinclair, 1986; Hammer *et al.*, 1995; see Chapter 11).
2. In many crop plants, nitrogen mobilization occurs from leaves with commencement of seed growth. Nitrogen mobilization can result in leaf senescence and/or decline in leaf nitrogen concentration per unit leaf area (Sinclair, 1986; Sinclair *et al.*, 2003). If nitrogen mobilization results in leaf loss rather than decline in leaf nitrogen concentration, RUE remains constant (e.g. Soltani *et al.*, 2006d). On the other hand, if this mobilization results in decreased leaf nitrogen concentration, it will decrease RUE (e.g. Birch *et al.*, 1999). Here, it is assumed that nitrogen mobilization is fully accounted for by leaf loss, so there is not an influence of nitrogen mobilization on RUE, but it can be considered if required (refer to Chapter 17).

Since increases in atmospheric CO_2 concentration increase leaf photosynthetic rates, simulations of future climates should also account for increases in RUE. There are several approaches to adjust RUE for higher (and lower) atmospheric CO_2 concentration. A simple method has been presented by Penning de Vries *et al.* (1989) for the effect of CO_2 on daily net assimilation. This method can be used to adjust RUE value for CO_2 concentration whenever it is required.

$$RUE_x = RUE_o\,(1 + b \times \ln (C_x / C_o)) \quad (10.5)$$

where RUE_x is the value of RUE at CO_2 concentration lower or higher than the reference level, RUE_o is the value of RUE at the reference CO_2 concentration, C_o is the reference CO_2 concentration (350 μmol mol^{-1}) and C_x is the target CO_2 concentration (μmol mol^{-1}). The coefficient *b* is a constant that has a value of 0.4 in C_4 plants and 0.8 in C_3 plants.

Crop growth

Having defined each of the variables in Eqn 10.1 to calculate daily dry matter production (DDMP), it is now possible to determine the accumulation of crop mass. Since RUE was defined on the basis of aboveground mass, the daily and total aboveground mass accumulated by the crop (WTOP, g m^{-2}) is calculated. Therefore, WTOP does not include root mass. The current value of $WTOP_i$ is obtained by adding DDMP to the total crop mass on the previous day ($WTOP_{i-1}$):

$$WTOP_i = WTOP_{i-1} + DDMP \quad (10.6)$$

Figure 10.7 shows a flowchart of calculations.

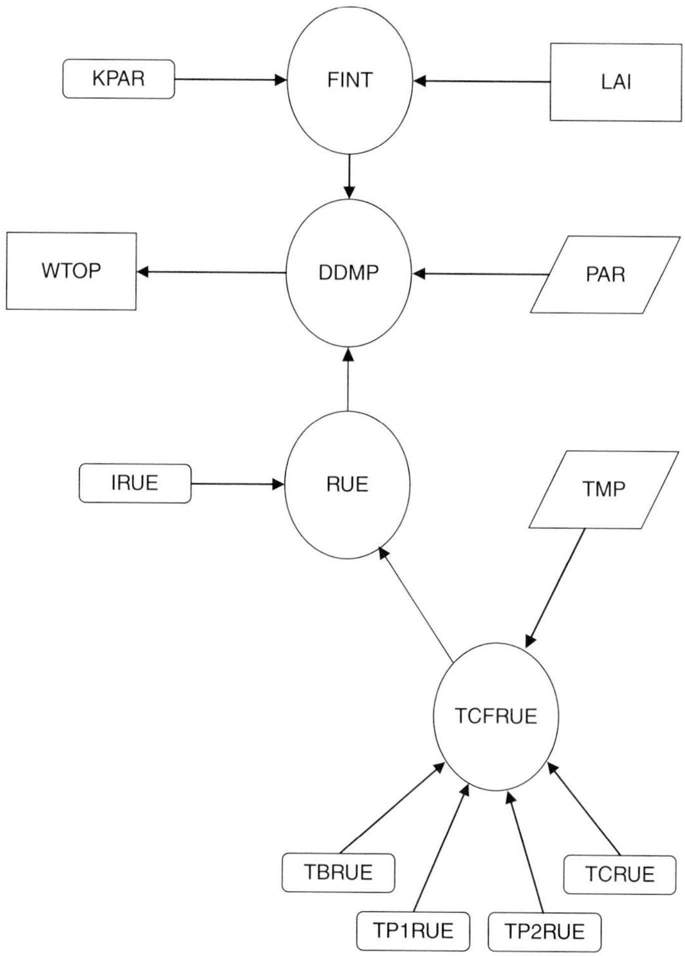

Fig. 10.7. Relational diagram of dry matter production submodel. LAI the leaf area index (m^2 m^{-2}), PAR the photosynthetic active radiation (MJ m^{-2} day^{-1}), FINT the fraction intercepted PAR, KPAR the extinction coefficient for PAR, DDMP the daily dry matter production (g m^{-2} day^{-1}), WTOP the total aboveground crop mass (g m^{-2}), TMP the mean daily temperature (°C), RUE the radiation use efficiency (g MJ^{-1}), IRUE the potential RUE under optimal conditions (g MJ^{-1}), TCFRUE the temperature correction factor for RUE, and TBRUE, TP1RUE, TP2RUE, and TCRUE are the base, lower optimum, upper optimum, and ceiling temperatures for RUE (°C), respectively.

Parameter Estimation

To simulate dry matter production by crop canopies, estimates of the following parameters are required:

- RUE under optimal conditions;
- cardinal temperatures (base, lower and upper optimum and ceiling) for the response of RUE to mean daily temperature; and
- extinction coefficient (KPAR).

Table 10.1 includes estimates of these parameters in important crops. For values of RUE in some crops refer to Table 1.1 (Chapter 1).

To estimate RUE experimentally, dry matter production by a crop canopy needs to be measured in conjunction with simultaneous measurements of PAR interception by the canopy. PAR interception data can be obtained by measuring PAR above and below the crop canopy. Then, RUE can be found as the slope of linear regression line between cumulative dry matter and cumulative intercepted PAR (Fig. 10.4). For more details and examples of RUE determinations refer to Turpin *et al.* (2002), Lindquist *et al.* (2005), Albrizio and Steduto (2005) and Brown *et al.* (2006).

Since temperature may not always be optimum for photosynthesis, cardinal temperatures for RUE must be determined. Andrade *et al.* (1993) studied the effect of temperature on RUE in maize using crop growth data of several years and sowing dates. They indicated that average daily temperatures lower than 21°C result in lower RUE and presented an equation that relates RUE to

Table 10.1. Parameter estimates relating to dry matter production (Penning de Vries *et al.*, 1989; Williams *et al.*, 1989; Chapman *et al.*, 1993; Villalobos *et al.*, 1996; Keating *et al.*, 2003; Soltani *et al.*, 2006c). Values of RUE for different crops were given in Table 1.1.

Crop	TBRUE	TP1RUE	TP2RUE	TCRUE	KPAR
Wheat	2	15	25	35	0.65
Barley	2	10	30	37	0.60
Rice	10	25	35	42	0.60
Maize	10	25	35	45	0.60
Sorghum	8	20	35	50	0.60
Cotton	10	20	40	50	0.60
Soybean	10	20	30	40	0.70
Peanut	10	21	30	40	0.60
Canola	2	10	25	35	0.75
Sunflower	8	17	27	45	0.90
Chickpea	2	14	30	38	0.65
Pea	2	15	30	40	0.65

TBRUE: base temperature for RUE (°C)
TP1RUE: lower optimum temperature for RUE (°C)
TP2RUE: upper optimum temperature for RUE (°C)
TCRUE: ceiling temperature for RUE (°C)
KPAR: extinction coefficient

temperature. Such data are rarely available so another, easier approach is to apply the temperature response of leaf photosynthesis to RUE. However, caution is required because leaf and canopy responses to environmental factors are not always the same (Hay and Porter, 2006). Penning de Vries *et al.* (1989) presented data on response of leaf photosynthesis to temperature in different crops.

Another alternative to estimate the RUE response to temperature is to estimate the response using detailed models that simulate photosynthesis and respiration in response to temperatures. These models include response functions of leaf gross photosynthesis and respiration to temperature. Soltani *et al.* (2007) described a simple model, based on the RUE model of Sinclair (1991), which is able to calculate response of RUE to temperature and CO_2 concentration. For other examples of such models that can be used to obtain response of RUE to temperature refer to Boote and Pickering (1994), Goudriaan and van Laar (1994), and Hammer and Wright (1994).

The extinction coefficient, KPAR, can be obtained experimentally from measurements of PAR interception and crop LAI. PAR interception above the layer of dead leaves should be used because PAR intercepted by dead (senesced) leaves does not contribute to dry matter production. Simultaneous measurements of PAR interception and LAI every 7 to 10 days before canopy closure will provide a data set that can be used to estimate KPAR. Fitting Eqn 10.2 to the data of fraction PAR interception versus LAI results in an estimate of KPAR (Fig. 10.8).

An integrated approach to measuring PAR interception by the leaves can be obtained from photographs pointing upward in the canopy. The fraction of the photograph obscured by leaves is used to obtain an approximation of PAR interception. For more information refer to Andrieu *et al.* (1997) and Purcell (2000).

It has been shown that KPAR under indirect, diffuse radiation will give an estimate of daily KPAR that is required to calculate PAR interception (Campbell and Norman, 1998). To reach a direct estimate of KPAR, PAR measurements can be performed under overcast conditions or under an umbrella. Alternatively,

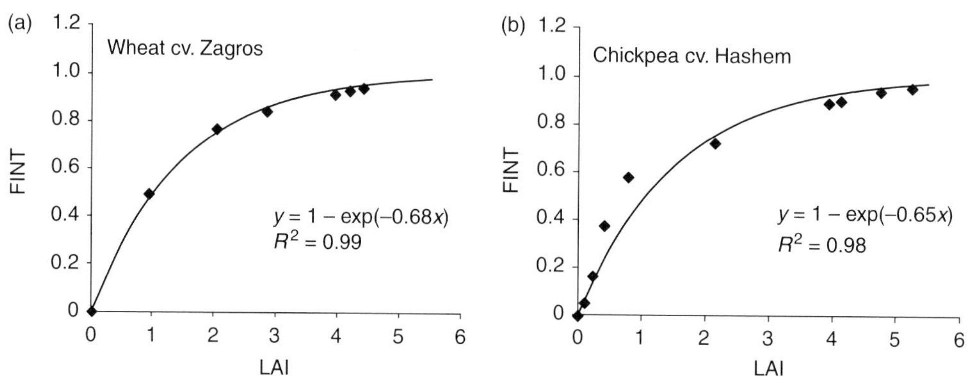

Fig. 10.8. Sample fit of Eqn 10.2 to fraction intercepted PAR versus LAI in wheat (a) and chickpea (b). PAR interception has been measured under an umbrella (V. Madah and A. Soltani, unpublished data).

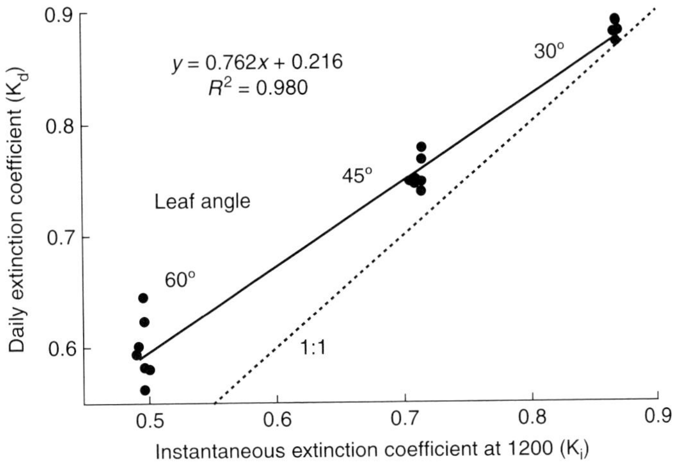

Fig. 10.9. Daily extinction coefficient (K_d) plotted against midday extinction coefficient (K_i) calculated for the same conditions. The dashed line represents the line of equality for the two variables and the solid line is the regression between the two variables (Sinclair, 2006).

instantaneous values of PAR for around noon can be used to estimate daily KPAR. Sinclair (2006) provided a relationship to estimate daily KPAR from its instantaneous measurement at around noon (Fig. 10.9).

Programming

A submodel prepared for dry matter production according to principles described in this chapter is given in Box 10.2. For a list of variables in the submodel refer to Appendix III. This submodel will be incorporated in a simulation model of potential production in Chapter 12. The parameter estimates in this submodel relate to a wheat cultivar (Tajan).

Box 10.2. Program of submodel to simulate dry matter production. For a list of variables refer to the text or Appendix III.

```
DMProduction:
'---------------------------- Parameters and Initials
  If iniDMP = 0 Then
    TBRUE = Sheet5.[b22]
    TP1RUE = Sheet5.[b23]
    TP2RUE = Sheet5.[b24]
    TCRUE = Sheet5.[b25]
    KPAR = Sheet5.[b26]
    IRUE = Sheet5.[b27]

    iniDMP = 1:
  End If
```

Continued

Box 10.2. Continued.

```
'------------------------------ Adjustment of RUE
  If TMP <= TBRUE Or TMP >= TCRUE Then
    TCFRUE = 0
  ElseIf TMP > TBRUE And TMP < TP1RUE Then
    TCFRUE = (TMP – TBRUE) / (TP1RUE – TBRUE)
  ElseIf TMP > TP2RUE And TMP < TCRUE Then
    TCFRUE = (TCRUE – TMP) / (TCRUE – TP2RUE)
  ElseIf TMP >= TP1RUE And TMP <= TP2RUE Then
    TCFRUE = 1
  End If

  RUE = IRUE * TCFRUE

'------------------------------ Daily dry matter production
  FINT = 1 – Exp(–KPAR * LAI)
  DDMP = SRAD * 0.48 * FINT * RUE
Return
```

Exercises

1. Cardinal temperatures for dry matter production in different field crops are presented in Table 10.1. Try to classify the crops into two groups based on their cardinal temperatures. Is this classification similar to any other classification you know? Explain it.

2. Potential RUEs of important field crops are included in Table 1.1. Classify the crops in this table into three groups based on their potential RUE. Explain how the groups are different.

3. Is there any relationship between KPAR presented in Table 10.1 and leaf characteristics (e.g. angle, shape, etc)? Explain it.

4. The table below includes LAI, fraction intercepted PAR (FINT), cumulative intercepted PAR (CumIPAR, MJ m^{-2}), and total crop mass (WTOP, g m^{-2}) measured in a crop from sowing to canopy closure. Using these data obtain KPAR and RUE for the crop. Is this a C_3 or C_4 crop? Why?

LAI	FINT	CumIPAR	WTOP
0.00	0.00	0.0	0.0
0.94	0.49	51.6	66.3
2.03	0.77	76.9	136.4
2.85	0.84	108.0	285.6
3.95	0.92	139.0	469.3
4.19	0.93	185.1	886.8
4.42	0.94	209.6	858.4

5. In continuation of Exercise 3, Chapter 9, calculate dry matter production for the crop of Exercise 4 in this chapter for your crop of interest.

11 Dry Matter Distribution and Yield Formation

In Chapter 10, the accumulation of total crop mass was simulated. A key task in simulating grain crop yield is to simulate dry matter distribution among plant tissues throughout the growing season, including mass accumulation by the grain.

Background

Distribution of dry matter has been simulated using various methods. In the simplest case only the distribution of dry matter between grain and non-grain organs is considered based on the harvest index (HI, ratio of grain mass to total plant mass, g g^{-1}) concept. In this case, total dry matter production is calculated and then grain yield is obtained as product of total dry matter and harvest index (e.g. Williams *et al.*, 1989; Kemanian *et al.*, 2007).

Another simple approach, also related to harvest index, is to predict yield formation based on a linear increase in HI during seed growth. This method is based on the observations in soybean by Speath and Sinclair (1985) that HI increased in a highly linear manner over much of the seed growth period. That is, HI increased at a constant rate as a function of time after beginning seed growth (Sinclair, 1986). This response has also been observed in other crops (e.g. Moot *et al.*, 1996; Lecoeur and Sinclair, 2001; Turpin *et al.*, 2002). The strength of this approach lies in its simplicity by intrinsically combining the contribution of current and stored assimilate to seed yield. In this approach, complex assumptions about the development of seed number and size in the prediction of seed yield are avoided (Chapman *et al.*, 1993).

Those methods not based on harvest index rely on defining partitioning coefficients to simulate dry matter partitioning to different organs (e.g. Kropff and van Laar, 1993; Goudriaan and van Laar, 1994; Jones *et al.*, 2003). In these methods, different partitioning coefficients are assumed for each organ at different developmental (phenological) stages. Dry matter partitioned to each organ each day is obtained by multiplying the relevant coefficient and the dry matter produced on that day. Therefore, grain growth is directly linked to defining accurately the partitioning coefficient for grain.

An additional consideration in calculating yield formation is the relation of mass supply to the grains and the demand of the grains. The approaches considering the mass balance can be divided into three categories, i.e. source-limited, sink-limited, and combined methods (Ritchie, 1991). Source-limited approaches assume that grain growth is not limited with respect to absorption of available assimilates and that seed growth and grain yield is determined by the ability of leaves and other organs in providing assimilates to fulfill grain demand (e.g. Goudriaan and van Laar, 1994; Jamieson *et al.*, 1998). Therefore, yield accumulation each day is determined by available assimilates.

Sink-limited approaches assume that there is no limitation in assimilate supply for grain growth and that the capacity of the grains to absorb the available assimilates determines the grain yield. In these methods, grain number and potential grain size (weight) are first calculated and then accumulated grain yield each day during the grain filling period is obtained by multiplying the number of grains and potential seed growth. The wheat model of Stapper (1984) belongs to this category. Under normal field conditions, there appears to be little evidence for sink limitations in soybean (Sinclair, 2004) and in wheat (Sinclair and Jamieson, 2006).

In combined approaches, assimilate supply for grain growth and the capacity of the grain to absorb the available assimilates are calculated separately. Then, actual grain growth is considered to be equal to the minimum value of these two amounts (e.g. Villalobos *et al.*, 1996; Jones *et al.*, 2003).

Model

In this book, a simple method is used to simulate dry matter partitioning and yield formation. In each phenological stage, specific sinks for dry matter need to be specified. Three possible sinks are considered:

1. Leaves.
2. "Stems".
3. Grains or other storage organs.

The leaf component is the leaf blades. The stem component includes the actual stems and any other organs other than leaf blades and grains. The type and the number of sinks that need to be considered depends on objectives of the model, requirements for component mass for calculating other processes in the model, and accounting for differences in biochemical composition among the organs. For example, vegetative organs are separated between leaves and stems because they are required for simulation of nitrogen uptake (Chapter 17). Distinction is not usually required among stems, petioles, and pod-walls in legumes or stems, leaf sheaths, and rachises in cereals because

Table 11.1. Crop phenological intervals and active sinks for dry matter during crop life cycle.

Phenological interval	Active sinks
Sowing to emergence	None
Emergence to termination leaf growth on main stem (TLM)	Leaf, "stem"
TLM to beginning seed growth (BSG)	Leaf, "stem"
BSG to termination seed growth (TSG)	Grain
TSG to maturity	None

these tissues within each species have similar biochemical composition (including nitrogen content).

Table 11.1 indicates the sinks that are active during each development interval used in this book for grain crops. From emergence to beginning seed growth (BSG), leaves and stems are active organs. However, after termination leaf growth on main stem (TLM) the proportion of dry matter partitioned to the leaves decreases dramatically and stems (including seed-bearing organs such as pods or rachises) become the major destination for new dry matter. From emergence to TLM, fraction of dry matter partitioned to the leaves is predicted based on allometric relationships between leaf and total aboveground dry matter (Soltani *et al.*, 2006d).

From BSG to termination of seed growth (TSG), grains become the dominant sink for the photosynthates and it is assumed that all the photosynthates produced during the seed-filling period goes to the grains. It is also possible that vegetative organs (leaves and stems) become significant sources of assimilate for grain growth via mobilization. Therefore, it is assumed that crop yield is source-limited. Box 11.1 presents an overview of the method that is used here.

Vegetative organ growth

From emergence to BSG, Stage 1 of dry matter distribution, daily increase in leaf mass (GLF, g m^{-2} day^{-1}) is simulated as a function of daily total crop growth (DDMP, g m^{-2} day^{-1}, Chapter 10) multiplied by the partitioning fraction of daily mass production for leaf growth (FLF, g g^{-1}). Daily mass not partitioned to the leaves is deposited in the stems (GST, g m^{-2} day^{-1}).

$$GLF = DDMP \times FLF \tag{11.1}$$

$$GST = DDMP - GLF \tag{11.2}$$

To use Eqns 11.1 and 11.2, FLF must be known. However, FLF can vary with plant development. Within the period from emergence to TLM two phases for FLF are often observed. This can be visualized by plotting cumulative leaf dry matter against cumulative total dry matter from emergence to TLM. Commonly,

Box 11.1. A summary of calculation method used for dry matter distribution and yield formation.

From sowing to emergence:

- No dry matter production and distribution!

From emergence to beginning seed growth (BSG):

- Partitioning coefficient to leaves is obtained.
- Daily dry matter distributed to leaves is calculated as the product of leaf partitioning coefficient and daily dry matter production.
- Remaining dry matter is partitioned to "stems".

From BSG to termination seed growth (TSG):

- All current dry matter produced is partitioned to the seeds.
- Dry matter available for seed growth from mobilization is considered.
- Seed growth rate is adjusted for biochemical difference between seed and vegetative organs.

From TSG to harvest maturity:

- No dry matter production and distribution.

a biphasic pattern is apparent as illustrated in Fig. 11.1 for chickpea (Soltani *et al.*, 2006d). A similar pattern has also been identified in wheat, soybean, and maize (A. Soltani, unpublished data).

The biphasic pattern of dry matter partitioning of dry matter to the leaves during the vegetative growth results in a high proportion of dry matter allocated to the leaves at low total accumulated dry matter (WTOP, g m^{-2}, Chapter 10) and a low proportion of dry matter allocated to leaves at high WTOP. These patterns are identified as phase A (FLF1A) and phase B (FLF1B) within Stage 1 of dry matter distribution. The inflection point between the stages (WTOPL) usually occurs when total crop mass is around 150 to 200 g m^{-2} as illustrated in Fig. 11.1.

From TLM to BSG, Stage 2 of dry matter distribution, the fraction of daily dry matter production that is partitioned to the leaves (FLF2) decreases dramatically relative to that partitioned to the stems. Therefore, leaf growth between TLM to BSG is predicted based on FLF2 and daily dry matter (Eqn 11.1).

From BSG to TSG, it is assumed there is no leaf growth, but leaf growth during this period can be easily included if this is necessary.

Grain growth and yield formation

Grain growth occurs in the period from BSG to TSG and is calculated using a method similar to that used by the Sirius wheat model (Jamieson *et al.*, 1998). Grain growth is simulated by assuming that from BSG all new dry matter produced goes to the grains. The second source for grain growth is translocation from the vegetative organs.

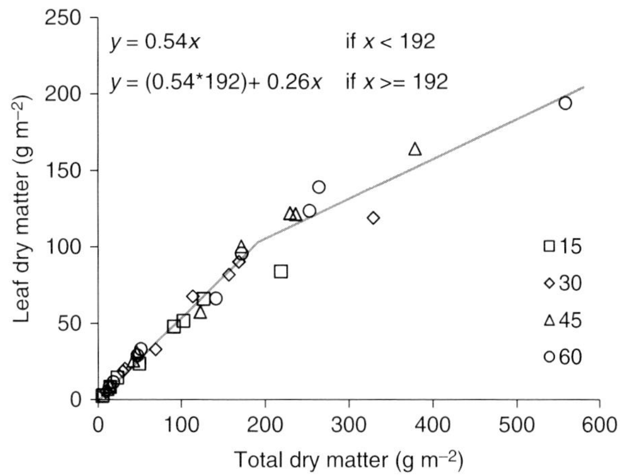

Fig. 11.1. Relationship between cumulative leaf dry matter versus total dry matter from sowing to first-pod (TLM) in chickpea (Soltani *et al*., 2006d). Numbers indicate plant densities (plants m^{-2}). The slopes of the first and second lines are 0.54 and 0.26 g g^{-1}, respectively. The inflection point is 192 g m^{-2}.

The amount of dry matter for seed growth via translocation from the vegetative organs is calculated using a simple approach (Goudriaan and van Laar, 1994; Jamieson *et al*., 1998). A fraction (FRTRL, g g^{-1}) of the total crop mass present at BSG is available for transfer to the grains. FRTRL is a crop parameter and needs to be provided to the model. Total amount of dry matter available for seed growth via mobilization (TRLDM, g m^{-2}) is then calculated by multiplying the crop mass at BSG (BSGDM, g m^{-2}) and FRTRL:

$$TRLDM = BSGDM \times FRTRL \tag{11.3}$$

During the grain filling period, each day a portion of TRLDM is transferred to the grains with the rate (TRANSL, g m^{-2} day^{-1}) proportional to daily temperature unit (DTU) divided by the potential duration from BSG to TSG (tuTSG – tuBSG; Chapter 6). Therefore, by the potential end of grain filling all of TRLDM will be transferred to the grain (Jamieson *et al*., 1998). The following equation is used to calculate TRANSL:

$$TRANSL = DTU / (tuTSG - tuBSG) \times TRLDM \tag{11.4}$$

To calculate daily seed growth (SGR, g m^{-2} day^{-1}), it is also necessary to account for the possibility that the energy content per unit seed mass is greater than the mass being transferred to the seed. For example, a seed synthesizing high amounts of protein and lipid will produce less seed mass than the mass received by the seed from the vegetative tissues. In this case, a grain conversion coefficient (GCC, g g^{-1}) must be included, which is the ratio of energy content of vegetative tissues to that of grain (Sinclair, 1986; Hammer *et al*., 1995). GCC will be lower than 1.0 when seeds of high protein and lipid content are being produced. Sinclair and de Wit (1975) calculated energy content for grain in different crops (see Table 11.2). In cases where grain and vegetative tissues are equal with respect to energy content, GCC would be equal to 1. The equation to calculate SGR is:

$$SGR = (DDMP + TRANSL) \times GCC \quad (11.5)$$

The flow diagram of computation related to dry matter distribution and yield formation is shown in Fig. 11.2.

Parameter Estimation

Parameters needed to simulate dry matter distribution and yield formation are:

- FLF1A;
- FLF1B;
- WTOPL;
- FLF2;
- FRTRL; and
- GCC.

Table 11.2. Parameter estimates relating to dry matter distribution and yield formation in some field crops (Sinclair and de Wit, 1976; Sinclair, 1986; Penning de Vries *et al.*, 1989; Muchow and Sinclair, 1991; Chapman *et al.*, 1993; Hammer and Muchow, 1994; Hammer *et al.*, 1995; Jamieson *et al.*, 1998; Soltani *et al.*, 2006d; A. Soltani, unpublished data).

Crop	FLF1A	FLF1B	WTOPL	FLF2	FRTRL	GCC
Wheat	0.60	0.30	160	0.10	0.22	1.00
Barley	0.90	0.40	160	0.05	0.22	1.00
Rice				0.05	0.22	1.00
Maize	0.70	0.15	210	0.05	0.22	1.00
Sorghum				0.05	0.22	1.00
Soybean	0.70	0.40	150	0.10	0.22	0.77
Peanut	0.70	0.50	150	0.15	0.22	0.67
Canola	0.70	0.40	160	0.05	0.22	0.77
Sunflower				0.05	0.22	0.77
Dry bean	0.65	0.65	160	0.34	0.22	1.00
Chickpea	0.53	0.28	180	0.13	0.22	1.00

FLF1A: Leaf partitioning coefficient from emergence to TLM at lower level of crop mass (g g^{-1})
FLF1B: Leaf partitioning coefficient from emergence to TLM at higher level of crop mass (g g^{-1})
WTOPL: Crop mass when leaf partitioning coefficient shifts from FLF1A to FLF1B (g m^{-2})
FLF2: Leaf partitioning coefficient from TLM to BSG (g g^{-1})
FRTRL: Fraction of crop mass present at BSG that is transferred to the grains (g g^{-1})
GCC: Grain conversion coefficient (g g^{-1})

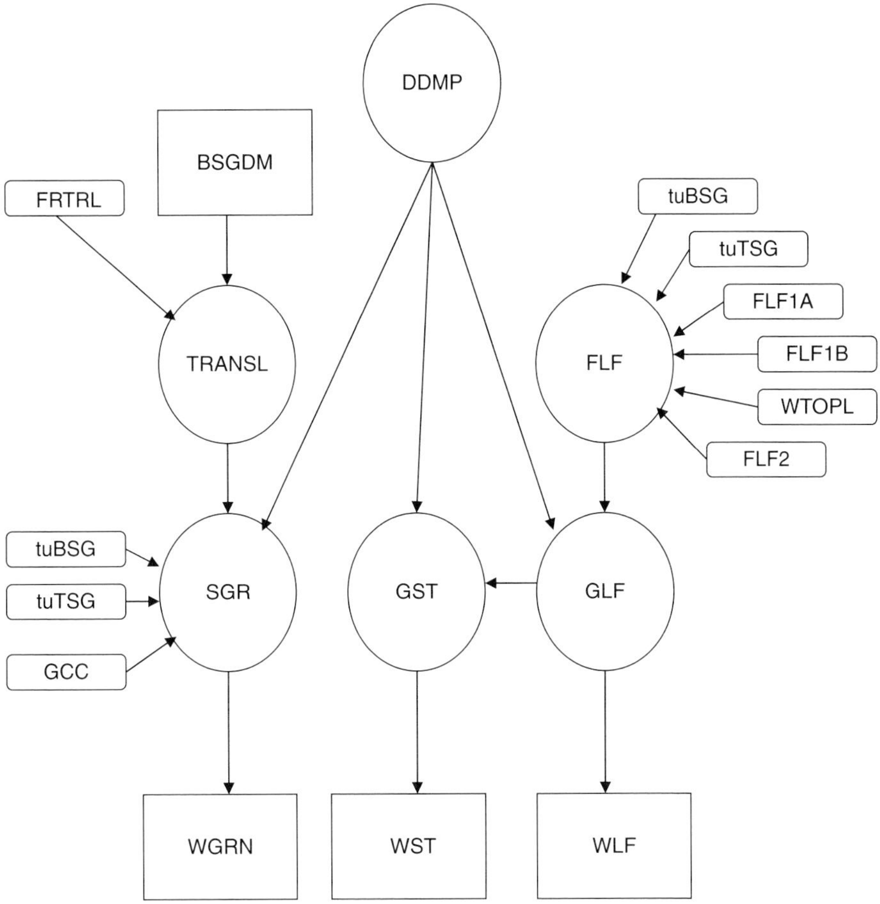

Fig. 11.2. Relational diagram of dry matter distribution submodel. DDMP the daily dry matter production (g m^{-2} day^{-1}), FLF the leaf partitioning coefficient (g g^{-1}), tuBSG the temperature unit to beginning seed growth (BSG) (°C), tuTSG the temperature unit to termination seed growth (°C), FLF1A the leaf partitioning coefficient from emergence to termination leaf growth on main stem (TLM) at lower level of crop mass (g g^{-1}), FLF1B the leaf partitioning coefficient from emergence to TLM at higher level of crop mass (°C), WTOPL the crop mass when leaf partitioning coefficient shifts from FLF1A to FLF1B (g m^{-2}), FLF2 the leaf partitioning coefficient from TLM to BSG (g g^{-1}), GLF the daily growth in leaf dry matter (g m^{-2} day^{-1}), GST the daily growth in stem dry matter (g m^{-2} day^{-1}), SGR the seed growth rate (g m^{-2} day^{-1}), TRANSL the daily translocation from vegetative organs to the grains (g m^{-2} day^{-1}), BSGDM the crop mass at beginning seed growth (g m^{-2}), FRTRL the fraction of crop mass present at BSG that is transferred to the grains, GCC the grain conversion coefficient (g g^{-1}), WLF the cumulative leaf dry matter (g m^{-2}), WST the cumulative stem dry matter (g m^{-2}), and WGRN the cumulative grain dry matter (g m^{-2}).

All the parameters must be estimated experimentally based on data sets obtained under optimal growth conditions. It is also important that attached and detached senesced leaves be included in leaf and total dry matter measurements. Estimates of the parameters are included in Table 11.2 for some crops.

FLF1A, FLF1B and WTOFL can be obtained using a data set of cumulative leaf dry matter versus cumulative total dry matter during the period from emergence to TLM. The three parameters are estimated by fitting the equation below to the data (Fig. 11.1).

$$\begin{aligned} WLF &= FLF1A \times WTOP \quad \textit{if} \quad WTOP \leq WTOPL \\ WLF &= FLF1A \times WTOPL + FLF1B \times (WTOP - WTOPL) \\ & \qquad \textit{if} \quad WTOP > WTOPL \end{aligned} \quad (11.6)$$

where WTOP is the total aboveground dry matter, WLF the leaf dry matter, WTOPL the inflection point between the two phases of dry matter partitioning (within Stage 1 of dry matter distribution), FLF1A the partitioning coefficient to leaves during phase A, and FLF1B the same as FLF1A for phase B.

Having data of leaf and total dry matter at TLM and BSG, FLF2 can be computed as:

$$FLF2 = (BSGLDM - TLMLDM) / (BSGDM - TLMDM) \quad (11.7)$$

where BSGLDM and TLMLDM are leaf dry weight at BSG and TLM and BSGDM and TLMDM are crop dry weight at BSG and TLM, respectively. The mass terms in Eqns 11.6 and 11.7 are in g m^{-2}.

Similarly, FRTRL is computed using observations on total crop dry matter at BSG (BSGDM) and grain yield (GYLD), non-grain (NGYLD), and total crop (BYLD) dry weights at physiological maturity as:

$$FRTRL = (BSGDM - NGYLD) / BSGDM \quad (11.8)$$

$$NGYLD = BYLD - GYLD \quad (11.9)$$

Although FRTRL may vary with cultivar and growth conditions, errors in the above assumptions are unlikely to be a large contributor to errors in predicted yield. Jamieson *et al.* (1998) indicated that in wheat an error of the order of 20% will contribute a maximum error of 10% in yield.

FRTRL can also be estimated as minimum harvest index, which usually has a value about 0.2 in several crops such as sunflower (Chapman *et al.*, 1993), peanut (Hammer *et al.*, 1995), sorghum (Hammer and Muchow, 1994), maize (Muchow and Sinclair, 1991), and chickpea (Soltani *et al.*, 1999).

Temperature unit requirements from sowing to TLM, BSG, and TSG are also required, but they are considered as parameters related to phenological development and were explained in Chapter 6.

Briefly, TLM is the temperature unit when leaf (node) production on main stem is effectively terminated (refer to Fig. 6.6). BSG and TSG are thermal units from sowing to when effective seed growth starts and ends, respectively. BSG occurs slightly after anthesis and TSG occurs slightly before maturity. Regarding problems with measuring seed growth, a better and straightforward method to obtain BSG and TSG is to estimate them from the end points of the plot of the increase in harvest index (Bindi *et al.*, 1999; Soltani *et al.*, 2004a). BSG is considered as the time or temperature unit when the linear increase in harvest index extrapolates to zero harvest index. TSG is the extrapolation of the linear increase in harvest index until the harvest index is equal to the final harvest index (refer to Fig. 6.7).

Programming

A submodel prepared for dry matter distribution and yield formation according to principles described in this chapter is presented in Box 11.2. For a list of variables in the submodel refer to Appendix III. In Chapter 12, when a simulation model of potential production is developed, this submodel will be used to simulate dry matter distribution. The parameter estimates in this submodel are from a wheat cultivar (cv. Tajan).

Box 11.2. Program of the submodel to simulate dry matter distribution and yield formation. For the list of variables refer to the text or Appendix III.

```
DMDistribution:
'------------------------------ Parameters and Initials
  If iniDMD = 0 Then
    FLF1A = Sheet5.[b29]
    FLF1B = Sheet5.[b30]
    WTOPL = Sheet5.[b31]
    FLF2 = Sheet5.[b32]
    FRTRL = Sheet5.[b33]
    GCC = Sheet5.[b34]

    WLF = 0.5:    WST = 0.5:    WVEG = WLF + WST:
    WGRN = 0:   iniDMD = 1:
  End If

'------------------------------ Biomass partitioning and yield formation
  If CTU <= tuEMR Or CTU > tuTSG Then
    DDMP = 0:    GLF = 0:    GST = 0:    TRANSL = 0:    SGR = 0
  ElseIf CTU > tuEMR And CTU <= tuTLM Then
    If WTOP < WTOPL Then FLF1 = FLF1A Else FLF1 = FLF1B
    GLF = FLF1 * DDMP
    GST = DDMP – GLF
    SGR = 0
  ElseIf CTU > tuTLM And CTU <= tuBSG Then
    GLF = FLF2 * DDMP
    GST = DDMP – GLF
    SGR = 0
  BSGDM = WTOP                    'Saving WTOP at BSG
```

Continued

Box 11.2. Continued.

```
  ElseIf CTU > tuBSG And CTU <= tuTSG Then
    GLF = 0:
    GST = 0:
    TRLDM = BSGDM * FRTRL
    TRANSL = DTU / (tuTSG – tuBSG) * TRLDM
    SGR = (DDMP + TRANSL) * GCC
  End If
  WLF = WLF + GLF
  WST = WST + GST
  WGRN = WGRN + SGR
  WVEG = WVEG + DDMP – (SGR / GCC)
  WTOP = WVEG + WGRN
Return
```

Additional Notes

Constant FRTRL

We used a constant for the fraction of crop total dry matter at the time of BSG that is available for re-mobilization (FRTRL) for simulation of yield formation. It has been shown that, at least in wheat, model predictions are not very sensitive to changes in FRTRL (Fletcher and Jamieson, 2006). In chickpea, Soltani and Sinclair (2011) indicated that FRTRL can be related to crop total dry matter at BSG (Fig. 11.3). If required, therefore, FRTRL can be adjusted for crop growth condition before seed growth.

Yield formation based on linear increase in harvest index

As mentioned earlier in this chapter, in some crop models, seed yield accumulation is calculated as the product of dry matter accumulation and harvest index, and harvest index is assumed to increase linearly as a function of time after beginning seed growth with a constant rate (dHI/dt) (Sinclair, 1986). This method is based on the concept that HI linearly increases with time over much of the seed growth period. This response has been observed in a number of crops (e.g. Muchow, 1988; Moot *et al.*, 1996; Lecoeur and Sinclair, 2001; Turpin *et al.*, 2002). Furthermore, it has been shown that dHI/dt remains stable over a range of growth conditions such as variations in sowing date, irrigation treatments, and N level (Moot *et al.*, 1996; Bindi *et al.*, 1999; Lecoeur and Sinclair, 2001).

Based on the method, each day during the grain filling period, first daily dry matter produced is added to crop vegetative mass, then seed growth rate is calculated by multiplying daily rate of increase in harvest index (dHI/dt, g g^{-1} day^{-1}) and crop vegetative mass. Daily seed growth is then subtracted from crop vegetative mass.

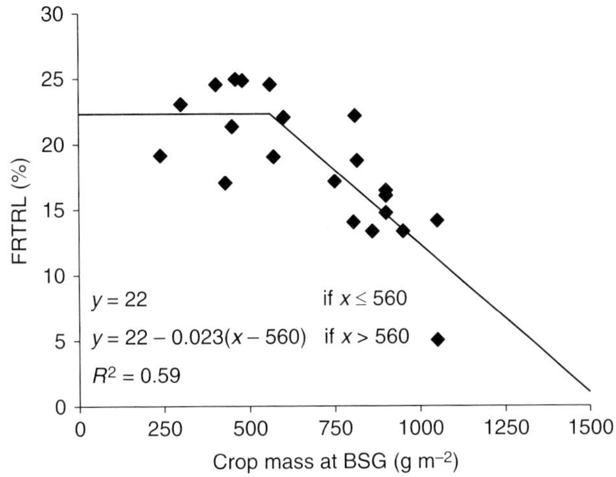

Fig. 11.3. Fraction of crop mass at beginning seed growth (BSG) that is transferred to the grains as a function of crop dry matter at BSG itself (Soltani and Sinclair, 2011).

A problem sometimes reported with this method is that dHI/dt is not constant over a wide range of conditions (e.g. Hammer *et al.*, 1995; Bange *et al.*, 1998; Hammer and Broad, 2003; Soltani *et al.*, 2005). One solution for the problem is that dHI/dt has to be adjusted at BSG for the conditions that crop has experienced during pre-seed growth period. For instance, in the chickpea model of Soltani and Sinclair (2011), dHI/dt is calculated as a function of crop mass at BSG, which reflects conditions experienced by the crop up to BSG.

Exercises

1. The table below includes accumulated total (WTOP, g m^{-2}) and leaf (WLF, g m^{-2}) mass of a crop from sowing to TLM. Obtain leaf partitioning coefficients, i.e. FLF1A, FLF1B, and WTOPL. You will need appropriate statistical software to fit Eqn 11.6 to the data.

WTOP	WLF
14.3	12.4
32.6	26.5
66.3	46.7
136.4	92.9
285.6	154.2
469.3	185.7

2. Try to obtain estimates of dry matter distribution and yield formation using Eqns 11.6 to 11.9 for crops/cultivars grown in your area.

12 A Model for Potential Production

In previous chapters submodels (subprograms) were developed to simulate different crop processes including phenology, crop leaf area, dry matter production, and dry matter distribution and yield formation. In this chapter, those submodels are integrated into a crop model to predict potential production, i.e. crop production under optimal nutrient and water conditions free from insects, diseases, and weeds. As quantitative information about phenological development as affected by both temperature and photoperiod are scarce, the phenology submodel based only on temperature is used here.

Model Structure

The integrated model is programmed as a subroutine (macro) written in Visual Basic for Application (VBA) in Excel. It includes a main part and the submodels. The main part calls different submodels as needed (Box 12.1). Crop submodels and other submodels that will be described below are subroutines within this main subroutine. Figure 12.1 shows Excel's macro window as it appears once opened.

The flow diagram of the overall model is presented in Fig. 12.2. The model is divided into the main process submodels and the "administrative" submodels. The following are the process submodels within the macro:

- *"Phenology"* as developed in Chapter 6.
- *"CropLAI"* as developed in Chapter 9.
- *"DMProduction"* as developed in Chapter 10.
- *"DMDistribution"* as developed in Chapter 11.

Below are additional "administrative" submodels within the macro that facilitate the running of the model.

Box 12.1. Main part of the crop model. Submodels are called by the main part when necessary.

```
'---------------- Main program
  GoSub ManagInputs
  GoSub InitialsHeaders
  GoSub FindSowingDate
  Do Until MAT = 1
    GoSub Weather
    GoSub Phenology
    GoSub CropLAI
    GoSub DMProduction
    GoSub DMDistribution
    GoSub DailyPrintOut
  Loop
  GoSub SummaryPrintOut
  Exit Sub
'---------------- End of main program
```

- *"ManagInputs"*: this submodel reads management inputs from the "Run" sheet from the Excel file containing the model. The structure of the Excel file and its sheets will be explained in the next section.
- *"InitialsHeaders"*: this submodel initializes crop submodels and prints headers to the "Output" sheet. Initialization is a process by which a value is devoted to a state variable or other variables at the beginning of simulation. Five variables, i.e. MAT, iniPheno, iniLAI, iniDMP, and iniDMD are set to 0. MAT = 0 means that crop has not reached maturity. Once cumulative temperature unit (CTU) passes the temperature unit requirement for maturity (tuMAT), this variable is set to 1, which means crop has reached maturity. With other variables equal to 0, related crop submodels start initialization at the first time they are called. After initialization in each submodel, its related *ini*-variable is set to 1. Thus, initialization is done only once and at the first time that a submodel is called.
- *"FindingSowingDate"*: this submodel searches in the weather data sheet to find sowing date. Sowing date must be given by users. However, it is possible to add codes to this submodel to calculate or predict sowing date. For example, sowing date for an early-spring crop can be defined as the 5th day in a 5-day period after 1 April with mean temperature greater than 7.5°C.
- *"Weather"*: this submodel reads daily weather data. If required, codes can be added to this submodel to modify weather data, for instance for climate change studies.
- *"DailyPrintOut"*: this submodel prints daily outputs in the "Output" sheet at the end of each day of simulation.
- *"SummaryPrintOut"*: this submodel transfers summary outputs to the "Run" sheet at the end of a crop simulation. Summary outputs are the most important crop characteristics simulated by the model.

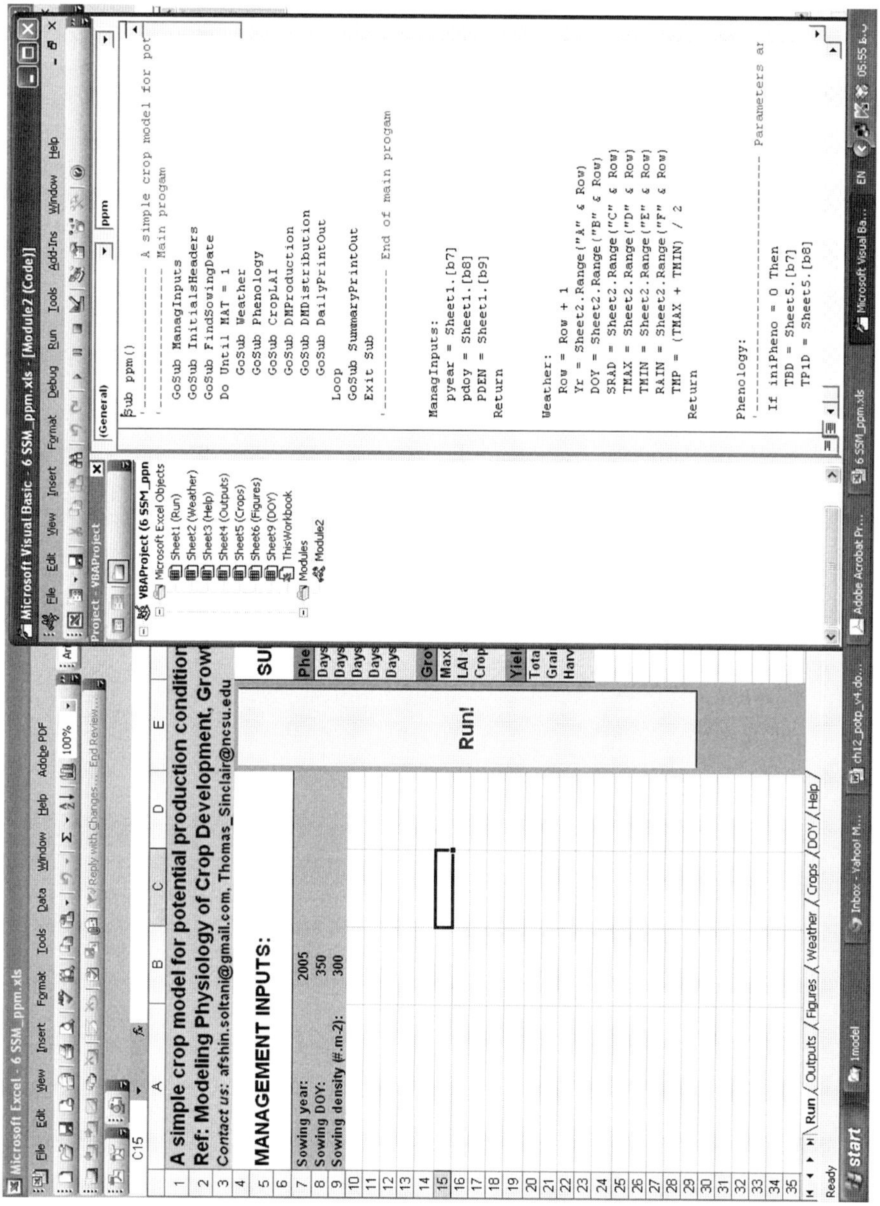

Fig. 12.1. Appearance of Excel file containing the model. Excel's macro window that includes the model is on the right.

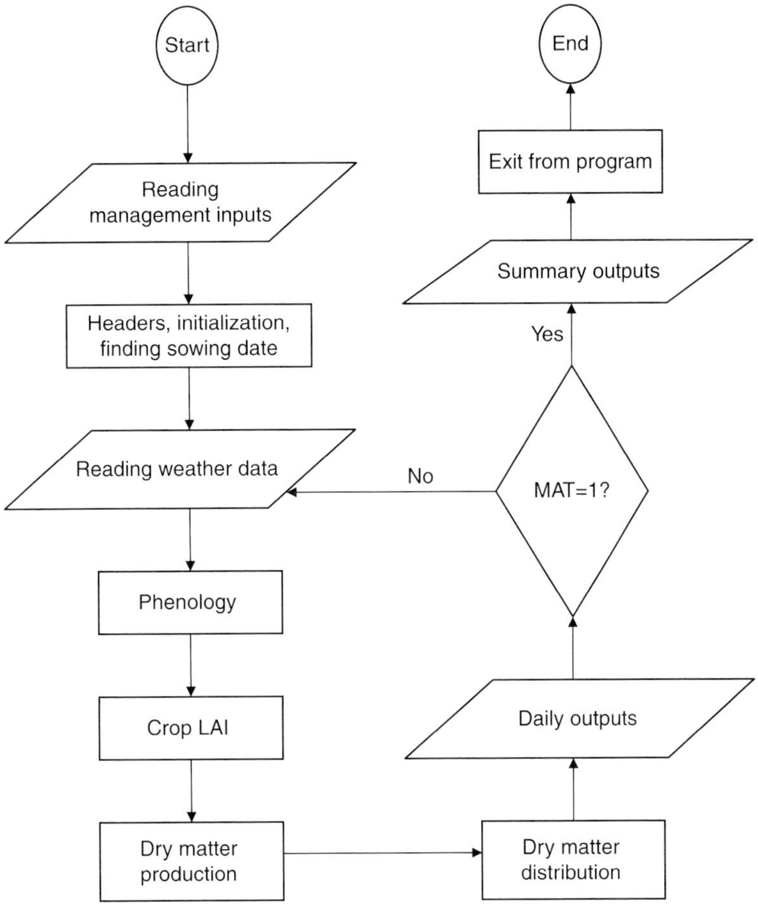

Fig. 12.2. Flow chart of the model for potential production.

The sequences of events executed in the model are listed below. Also, these are shown in the flow diagram in Fig. 12.2.

1. Management inputs are read.
2. Headers are printed and crop submodels are initialized.
3. Weather sheet is searched and sowing date is found.
4. Daily loop of calculations is started including daily calculation of phenological development, crop leaf area, daily dry matter production, and daily distribution of dry matter. By the end of the daily calculations, daily outputs are printed in the "Output" sheet. The daily loop of calculation is terminated once MAT = 1.
5. Summary outputs are printed in the "Run" sheet at crop maturity when daily calculation is finished.

Box 12.2 includes the model program to simulate potential production. Parameter estimates in this program belong to a wheat cultivar (cv. Tajan). The complete list of variables is presented in Appendix III.

Box 12.2. Program to simulate crop development, growth, and yield under potential production. The model can be requested from the authors or can be downloaded from the book's website (https://sites.google.com/site/CropModeling).

```
Sub ppm()
'---------------- A simple crop model for potential production conditions
'---------------- Main program
  GoSub ManagInputs
  GoSub InitialsHeaders
  GoSub FindSowingDate
  Do Until MAT = 1
     GoSub Weather
     GoSub Phenology
     GoSub CropLAI
     GoSub DMProduction
     GoSub DMDistribution
     GoSub DailyPrintOut
  Loop
  GoSub SummaryPrintOut
  Exit Sub
'---------------- End of main program

ManagInputs:
  pyear = Sheet1.[b7]
  pdoy = Sheet1.[b8]
  PDEN = Sheet1.[b9]
Return

Weather:
    Row = Row + 1
    Yr = Sheet2.Range("A" & Row)
    DOY = Sheet2.Range("B" & Row)
    SRAD = Sheet2.Range("C" & Row)
    TMAX = Sheet2.Range("D" & Row)
    TMIN = Sheet2.Range("E" & Row)
    RAIN = Sheet2.Range("F" & Row)
    TMP = (TMAX + TMIN) / 2
Return

Phenology:
'------------------------------- Parameters and Initials
  If iniPheno = 0 Then
    TBD = Sheet5.[b7]
    TP1D = Sheet5.[b8]
    TP2D = Sheet5.[b9]
    TCD = Sheet5.[b10]
    tuSOWEMR = Sheet5.[b11]
    tuEMRTLM = Sheet5.[b12]
    tuTLMBSG = Sheet5.[b13]
    tuBSGTSG = Sheet5.[b14]
    tuTSGMAT = Sheet5.[b15]
```

Continued

Box 12.2. Continued.

```
    tuEMR = tuSOWEMR
    tuTLM = tuEMR + tuEMRTLM
    tuBSG = tuTLM + tuTLMBSG
    tuTSG = tuBSG + tuBSGTSG
    tuMAT = tuTSG + tuTSGMAT

    DAP = 0:   CTU = 0:   iniPheno = 1
  End If

'------------------------------ Temperature unit calculation
  If TMP <= TBD Or TMP >= TCD Then
    tempfun = 0
  ElseIf TMP > TBD And TMP < TP1D Then
    tempfun = (TMP – TBD) / (TP1D – TBD)
  ElseIf TMP > TP2D And TMP < TCD Then
    tempfun = (TCD – TMP) / (TCD – TP2D)
  ElseIf TMP >= TP1D And TMP <= TP2D Then
    tempfun = 1
  End If

  DTU = (TP1D – TBD) * tempfun
  CTU = CTU + DTU
  DAP = DAP + 1

  If CTU < tuEMR Then DTEMR = DAP + 1 'Saving days to EMR
  If CTU < tuTLM Then DTTLM = DAP + 1 'Saving days to TLM
  If CTU < tuBSG Then DTBSG = DAP + 1 'Saving days to BSG
  If CTU < tuTSG Then DTTSG = DAP + 1 'Saving days to TSG
  If CTU < tuMAT Then DTMAT = DAP + 1 'Saving days to MAT

  If CTU > tuMAT Then MAT = 1
Return

CropLAI:
'------------------------------ LAI initials and pars
  If iniLAI = 0 Then
    PHYL = Sheet5.[b17]
    PLACON = Sheet5.[b18]
    PLAPOW = Sheet5.[b19]
    SLA = Sheet5.[b20]

    MSNN = 1:   PLA2 = 0:   PLA1 = 0:   LAI = 0:
    MXLAI = 0:   iniLAI = 1
  End If

'------------------------------ Yesterday LAI to intercept PAR today
  LAI = LAI + GLAI – DLAI
  If LAI < 0 Then LAI = 0
  If LAI > MXLAI Then MXLAI = LAI          'Saving maximum LAI

'------------------------------ Daily increase and decrease in LAI today
  If CTU <= tuEMR Then
```

Continued

Box 12.2. Continued.

```
    GLAI = 0: DLAI = 0
  ElseIf CTU > tuEMR And CTU <= tuTLM Then
    INODE = DTU / PHYL
    MSNN = MSNN + INODE
    PLA2 = PLACON * MSNN ^ PLAPOW
    GLAI = ((PLA2 – PLA1) * PDEN / 10000)
    PLA1 = PLA2
    DLAI = 0
  ElseIf CTU > tuTLM And CTU <= tuBSG Then
    GLAI = GLF * SLA
    BSGLAI = LAI                    'Saving LAI at BSG
    DLAI = 0
  ElseIf CTU > tuBSG Then
    GLAI = 0
    DLAI = DTU / (tuMAT – tuBSG) * BSGLAI
  End If
Return

DMProduction:
'------------------------------- Parameters and Initials
  If iniDMP = 0 Then
    TBRUE = Sheet5.[b22]
    TP1RUE = Sheet5.[b23]
    TP2RUE = Sheet5.[b24]
    TCRUE = Sheet5.[b25]
    KPAR = Sheet5.[b26]
    IRUE = Sheet5.[b27]

    iniDMP = 1:
  End If

'------------------------------- Adjustment of RUE
  If TMP <= TBRUE Or TMP >= TCRUE Then
    TCFRUE = 0
  ElseIf TMP > TBRUE And TMP < TP1RUE Then
    TCFRUE = (TMP – TBRUE) / (TP1RUE – TBRUE)
  ElseIf TMP > TP2RUE And TMP < TCRUE Then
    TCFRUE = (TCRUE – TMP) / (TCRUE – TP2RUE)
  ElseIf TMP >= TP1RUE And TMP <= TP2RUE Then
    TCFRUE = 1
  End If

  RUE = IRUE * TCFRUE

'------------------------------- Daily dry matter production
  FINT = 1 – Exp(–KPAR * LAI)
  DDMP = SRAD * 0.48 * FINT * RUE
Return

DMDistribution:
```

Continued

Box 12.2. Continued.

```
'------------------------------- Parameters and Initials
  If iniDMD = 0 Then
    FLF1A = Sheet5.[b29]
    FLF1B = Sheet5.[b30]
    WTOPL = Sheet5.[b31]
    FLF2 = Sheet5.[b32]
    FRTRL = Sheet5.[b33]
    GCC = Sheet5.[b34]

    WLF = 0.5:    WST = 0.5:    WVEG = WLF + WST:
    WGRN = 0:    iniDMD = 1:
  End If
'------------------------------- Biomass partitioning and yield formation
  If CTU <= tuEMR Or CTU > tuTSG Then
    DDMP = 0:    GLF = 0:    GST = 0:    TRANSL = 0:    SGR = 0
  ElseIf CTU > tuEMR And CTU <= tuTLM Then
    If WTOP < WTOPL Then FLF1 = FLF1A Else FLF1 = FLF1B
    GLF = FLF1 * DDMP
    GST = DDMP – GLF
    SGR = 0
  ElseIf CTU > tuTLM And CTU <= tuBSG Then
    GLF = FLF2 * DDMP
    GST = DDMP – GLF
    SGR = 0
    BSGDM = WTOP                    'Saving WTOP at BSG
  ElseIf CTU > tuBSG And CTU <= tuTSG Then
    GLF = 0:
    GST = 0:
    TRLDM = BSGDM * FRTRL
    TRANSL = DTU / (tuTSG – tuBSG) * TRLDM
    SGR = (DDMP + TRANSL) * GCC
  End If

  WLF = WLF + GLF
  WST = WST + GST
  WGRN = WGRN + SGR
  WVEG = WVEG + DDMP – (SGR / GCC)
  WTOP = WVEG + WGRN
Return

FindSowingDate:
  Row = 10
  Do
    Row = Row + 1
    Yr = Sheet2.Range("A" & Row)
    DOY = Sheet2.Range("B" & Row)
    SRAD = Sheet2.Range("C" & Row)
    TMAX = Sheet2.Range("D" & Row)
    TMIN = Sheet2.Range("E" & Row)
```

Continued

Box 12.2. Continued.

```
    RAIN = Sheet2.Range("F" & Row)
  Loop Until Yr = pyear And DOY = pdoy
Return

InitialsHeaders:
'-------------------------------- Initials
  MAT = 0
  iniPheno = 0
  iniLAI = 0
  iniDMP = 0
  iniDMD = 0
  iniSW = 0
  iniPNB = 0
  iniSNB = 0
'-------------------------------- Headers
  Sheet4.Cells(2, 1) = "Year"
  Sheet4.Cells(2, 2) = "DOY"
  Sheet4.Cells(2, 3) = "DAP"
  Sheet4.Cells(2, 4) = "TMP"
  Sheet4.Cells(2, 5) = "DTU"
  Sheet4.Cells(2, 6) = "CTU"
  Sheet4.Cells(2, 7) = "MSNN"
  Sheet4.Cells(2, 8) = "GLAI"
  Sheet4.Cells(2, 9) = "DLAI"
  Sheet4.Cells(2, 10) = "LAI"
  Sheet4.Cells(2, 11) = "TCFRUE"
  Sheet4.Cells(2, 12) = "FINT"
  Sheet4.Cells(2, 13) = "DDMP"
  Sheet4.Cells(2, 14) = "GLF"
  Sheet4.Cells(2, 15) = "GST"
  Sheet4.Cells(2, 16) = "SGR"
  Sheet4.Cells(2, 17) = "WLF"
  Sheet4.Cells(2, 18) = "WST"
  Sheet4.Cells(2, 19) = "WVEG"
  Sheet4.Cells(2, 20) = "WGRN"
  Sheet4.Cells(2, 21) = "WTOP"
Return

SummaryPrintOut:
  Sheet1.[g8] = DTEMR
  Sheet1.[g9] = DTTLM
  Sheet1.[g10] = DTBSG
  Sheet1.[g11] = DTTSG
  Sheet1.[g12] = DTMAT
  Sheet1.[g15] = MXLAI
  Sheet1.[g16] = BSGLAI
  Sheet1.[g17] = BSGDM
  Sheet1.[G20] = WTOP
  Sheet1.[G21] = WGRN
```

Continued

Box 12.2. Continued.

```
  Sheet1.[G22] = WGRN / WTOP * 100
Return

DailyPrintOut:
  Sheet4.Cells(DAP + 2, 1) = Yr
  Sheet4.Cells(DAP + 2, 2) = DOY
  Sheet4.Cells(DAP + 2, 3) = DAP
  Sheet4.Cells(DAP + 2, 4) = TMP
  Sheet4.Cells(DAP + 2, 5) = DTU
  Sheet4.Cells(DAP + 2, 6) = CTU
  Sheet4.Cells(DAP + 2, 7) = MSNN
  Sheet4.Cells(DAP + 2, 8) = GLAI
  Sheet4.Cells(DAP + 2, 9) = DLAI
  Sheet4.Cells(DAP + 2, 10) = LAI
  Sheet4.Cells(DAP + 2, 11) = TCFRUE
  Sheet4.Cells(DAP + 2, 12) = FINT
  Sheet4.Cells(DAP + 2, 13) = DDMP
  Sheet4.Cells(DAP + 2, 14) = GLF
  Sheet4.Cells(DAP + 2, 15) = GST
  Sheet4.Cells(DAP + 2, 16) = SGR
  Sheet4.Cells(DAP + 2, 17) = WLF
  Sheet4.Cells(DAP + 2, 18) = WST
  Sheet4.Cells(DAP + 2, 19) = WVEG
  Sheet4.Cells(DAP + 2, 20) = WGRN
  Sheet4.Cells(DAP + 2, 21) = WTOP
Return

End Sub '-------------------------------
```

Structure of Excel File Containing the Model

The Excel file containing the model has several sheets and a module, i.e. macro program or crop model. The program uses Excel's sheets for inputs and outputs. Figure 12.3 shows how the module and the sheets interact.

The individual sheets are as follows.

"Run"

Management inputs are specified in this sheet (Fig. 12.4). Management inputs to simulate potential production are:

- year of sowing;
- day of year (DOY) of sowing; and
- plant density (plants m^{-2}).

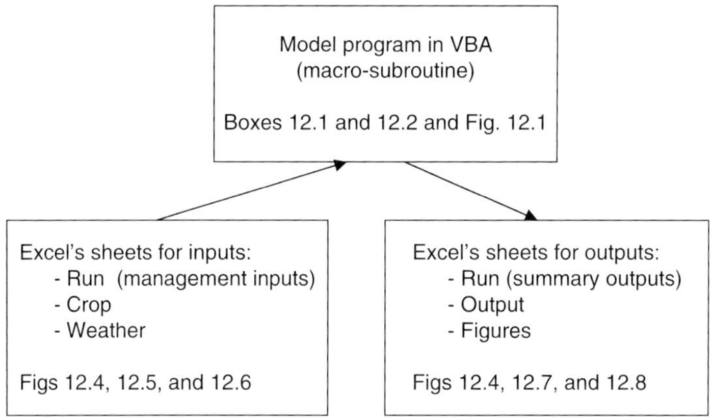

Fig. 12.3. Structure of the Excel file containing the model program and Excel sheets. The program uses Excel sheets for input and output.

Appendix II presents DOYs for all calendar days of a non-leap year and a leap year.

Summary outputs are also published in this sheet by the model. They are:

- days to emergence;
- days to termination leaf growth on main stem;
- days to beginning seed growth;
- days to termination seed growth;
- days to maturity;
- maximum LAI;
- LAI at beginning seed growth;
- crop mass at beginning seed growth;
- total crop mass at maturity;
- grain yield; and
- harvest index.

"Crops"

Estimates of crop parameters that are required for simulation of crop growth and yield are entered in this sheet in specific locations, i.e. value of the first parameters appears in cell B7. Parameters of different crops or cultivars may be stored in this sheet (Fig. 12.5).

"Weather"

Required weather data for crop simulation must be inserted in this sheet. The minimum weather data set required to run a crop model for potential production are: daily minimum and maximum temperatures and daily solar

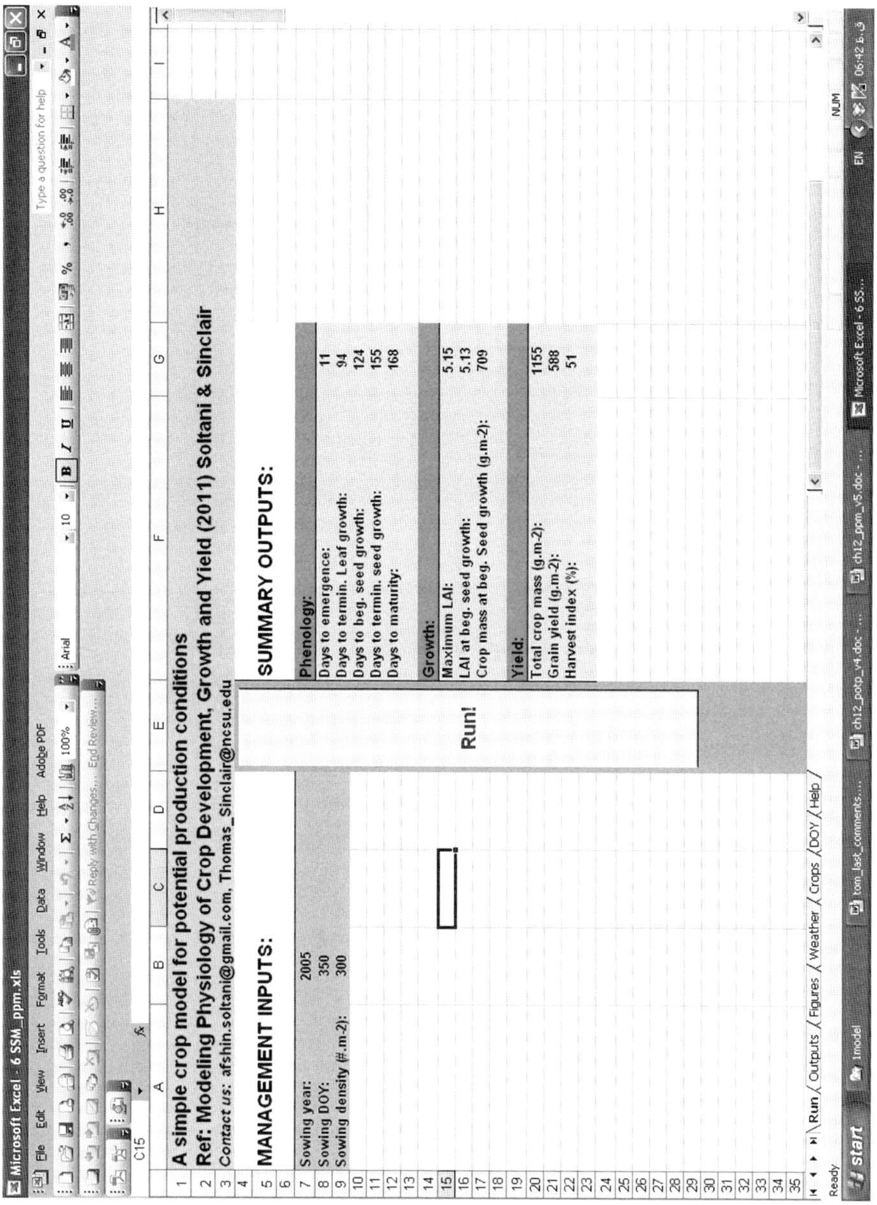

Fig. 12.4. Appearance of "Run" sheet in Excel file containing the model.

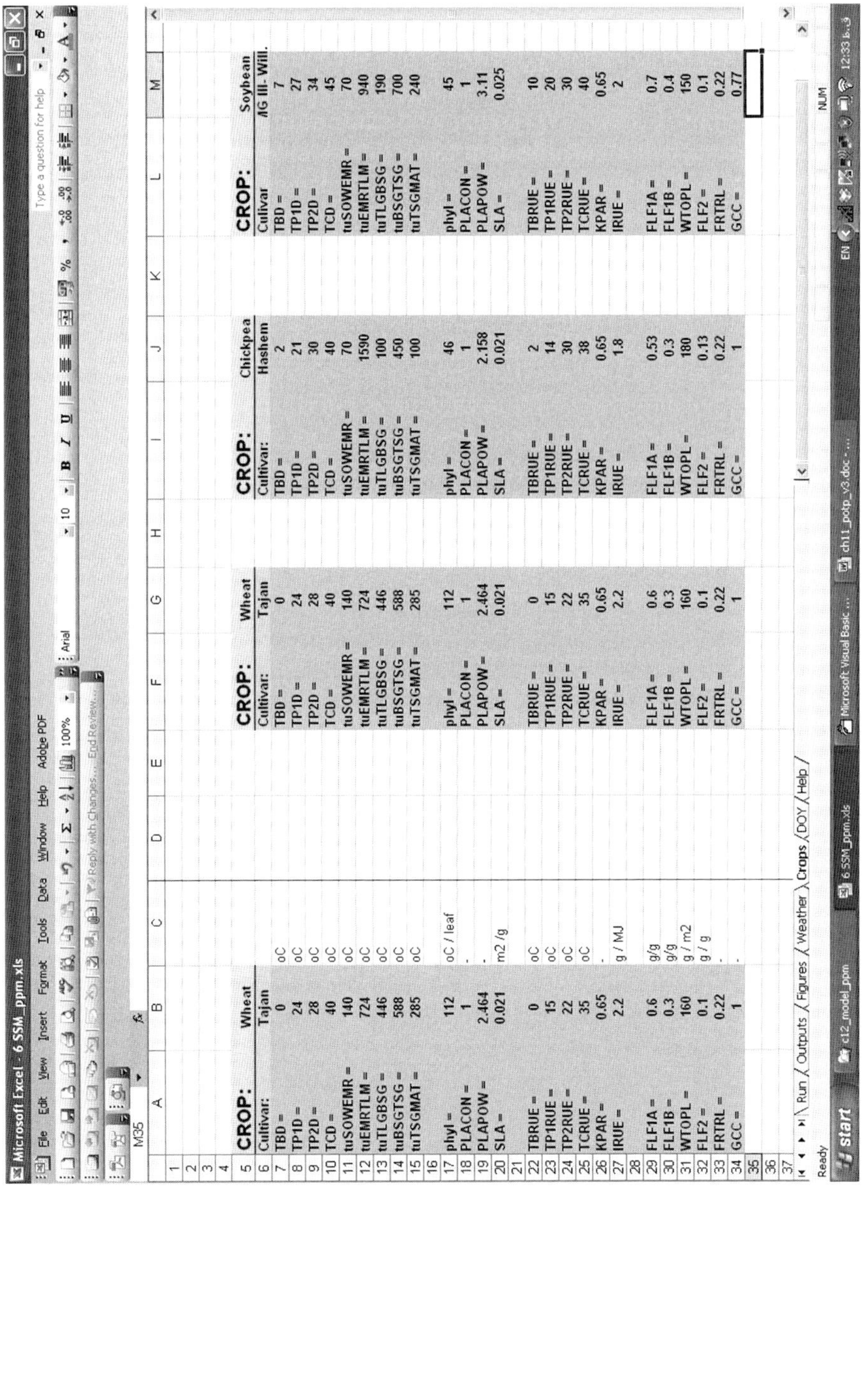

CROP:	Wheat		CROP:	Wheat	CROP:	Chickpea	CROP:	Soybean
Cultivar:	Tajan		Cultivar:	Tajan	Cultivar:	Hashem	Cultivar	AG III. Will.
TBD =	0	oC	TBD =	0	TBD =	2	TBD =	7
TP1D =	24	oC	TP1D =	24	TP1D =	21	TP1D =	27
TP2D =	28	oC	TP2D =	28	TP2D =	30	TP2D =	34
TCD =	40	oC	TCD =	40	TCD =	40	TCD =	45
tuSOWEMR =	140	oC	tuSOWEMR =	140	tuSOWEMR =	70	tuSOWEMR =	70
tuEMRTLM =	724	oC	tuEMRTLM =	724	tuEMRTLM =	1590	tuEMRTLM =	940
tuTLGBSG =	446	oC	tuTLGBSG =	446	tuTLGBSG =	100	tuTLGBSG =	190
tuBSGTSG =	588	oC	tuBSGTSG =	588	tuBSGTSG =	450	tuBSGTSG =	700
tuTSGMAT =	285	oC	tuTSGMAT =	285	tuTSGMAT =	100	tuTSGMAT =	240
phyl =	112	oC / leaf	phyl =	112	phyl =	46	phyl =	45
PLACON =	1	.	PLACON =	1	PLACON =	1	PLACON =	1
PLAPOW =	2.464	.	PLAPOW =	2.464	PLAPOW =	2.158	PLAPOW =	3.11
SLA =	0.021	m2 /g	SLA =	0.021	SLA =	0.021	SLA =	0.025
TBRUE =	0	oC	TBRUE =	0	TBRUE =	2	TBRUE =	10
TP1RUE =	15	oC	TP1RUE =	15	TP1RUE =	14	TP1RUE =	20
TP2RUE =	22	oC	TP2RUE =	22	TP2RUE =	30	TP2RUE =	30
TCRUE =	35	oC	TCRUE =	35	TCRUE =	38	TCRUE =	40
KPAR =	0.65	.	KPAR =	0.65	KPAR =	0.65	KPAR =	0.65
IRUE =	2.2	g / MJ	IRUE =	2.2	IRUE =	1.8	IRUE =	2
FLF1A =	0.6	g/g	FLF1A =	0.6	FLF1A =	0.53	FLF1A =	0.7
FLF1B =	0.3	g/g	FLF1B =	0.3	FLF1B =	0.3	FLF1B =	0.4
WTOPL =	160	g / m2	WTOPL =	160	WTOPL =	180	WTOPL =	150
FLF2 =	0.1	g / g	FLF2 =	0.1	FLF2 =	0.13	FLF2 =	0.1
FRTRL =	0.22	.	FRTRL =	0.22	FRTRL =	0.22	FRTRL =	0.22
GCC =	1	.	GCC =	1	GCC =	1	GCC =	0.77

Fig. 12.5. Appearance of “Crops” sheet in Excel file containing the model.

radiation. (A method for estimating daily solar radiation is presented below for instances where observational data are not available.) In following chapters, daily precipitation data will be required for simulation of crop production under water- and/or nitrogen-limited conditions. These data must be available for the entire simulation period from sowing time to maturity. Some simulations may be started before sowing to establish soil water conditions and in these cases weather data must be provided for this period. The data along with their corresponding year and date are entered in this sheet.

This sheet has these columns:

- year;
- date as day of year from 1 January (DOY);
- solar radiation (MJ m^{-2} day^{-1});
- minimum temperature (°C);
- maximum temperature (°C); and
- rainfall (mm), not necessary for potential production.

The first line of weather data should be started on line 11, column A. Figure 12.6 shows how an Excel sheet containing the weather appears. Lines 1 to 9 in this sheet can be used to insert user comments. Line 10 usually will be a label that specifies columns. Any numbers of years of weather data can be included in this sheet.

"Outputs"

Daily simulated crop characteristics describing crop development and growth are printed in this sheet. Data on this sheet can be used for further analysis if required. Figure 12.7 presents a sample of the "Outputs" sheet but not all variables are visible. Important daily outputs may include: daily and cumulative temperature unit, LAI, daily dry matter production and cumulative dry matter of leaves, stems, and grains.

"Figures"

After each run, some figures related to important crop variables can be found in this sheet (Fig. 12.8). More figures may be added in this sheet if necessary.

"Help"

Definitions of all variables in the model and "Outputs" sheet can be found in this sheet.

After preparing weather data and inputting crop parameters and management inputs, the model is simply run by clicking the run button in the "Run" sheet. Then, the "Run," "Output," and "Figures" sheets can be examined for simulation results. As instructed in the "Run" sheet, before each new simulation the "Outputs" sheet must be cleared.

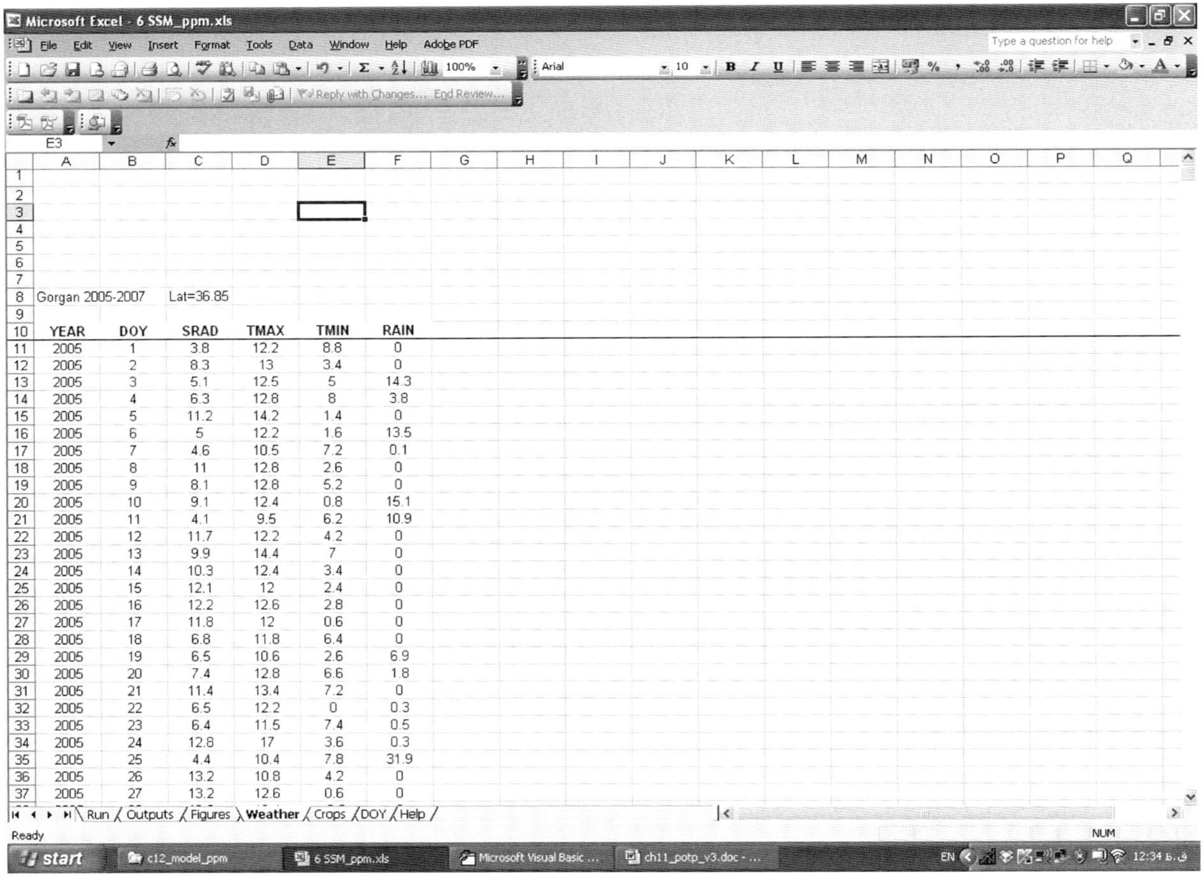

Fig. 12.6. Appearance of “Weather” sheet in Excel file containing the model.

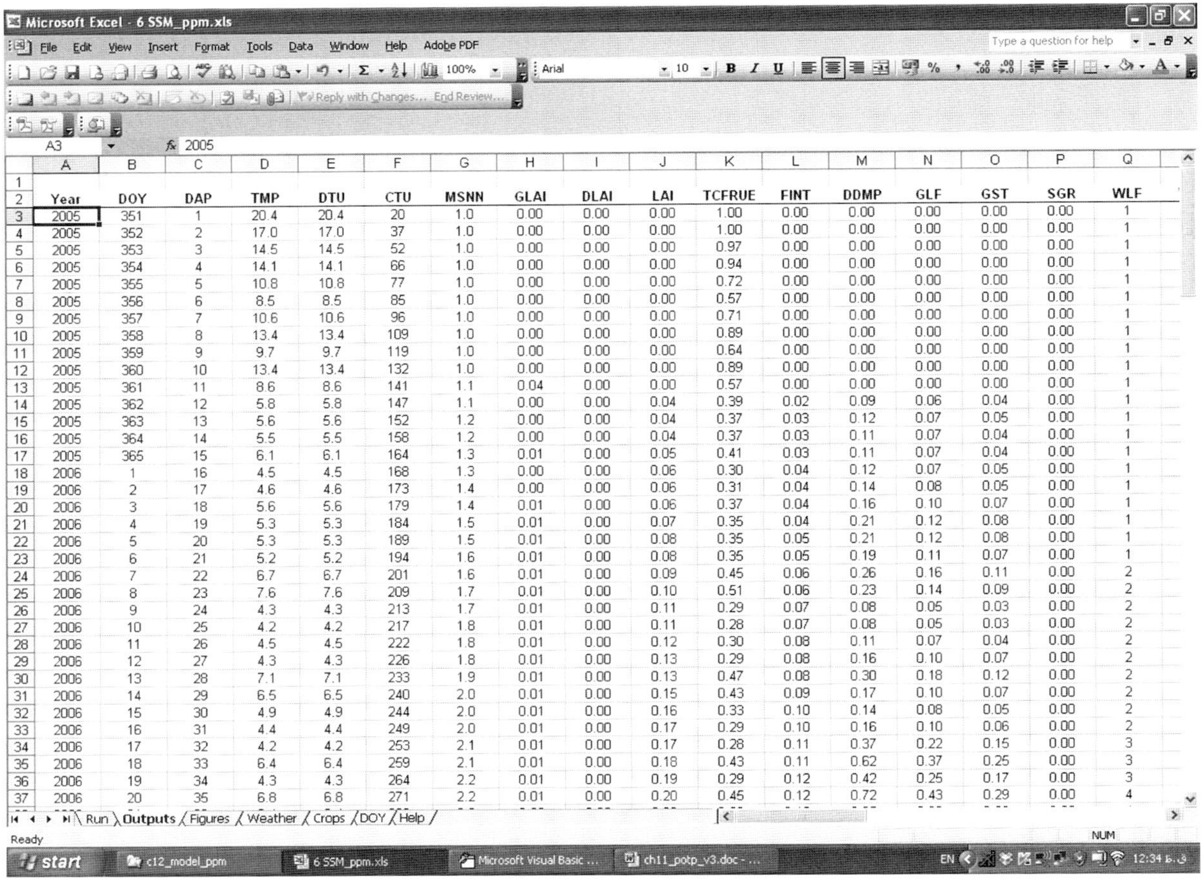

Year	DOY	DAP	TMP	DTU	CTU	MSNN	GLAI	DLAI	LAI	TCFRUE	FINT	DDMP	GLF	GST	SGR	WLF
2005	351	1	20.4	20.4	20	1.0	0.00	0.00	0.00	1.00	0.00	0.00	0.00	0.00	0.00	1
2005	352	2	17.0	17.0	37	1.0	0.00	0.00	0.00	1.00	0.00	0.00	0.00	0.00	0.00	1
2005	353	3	14.5	14.5	52	1.0	0.00	0.00	0.00	0.97	0.00	0.00	0.00	0.00	0.00	1
2005	354	4	14.1	14.1	66	1.0	0.00	0.00	0.00	0.94	0.00	0.00	0.00	0.00	0.00	1
2005	355	5	10.8	10.8	77	1.0	0.00	0.00	0.00	0.72	0.00	0.00	0.00	0.00	0.00	1
2005	356	6	8.5	8.5	85	1.0	0.00	0.00	0.00	0.57	0.00	0.00	0.00	0.00	0.00	1
2005	357	7	10.6	10.6	96	1.0	0.00	0.00	0.00	0.71	0.00	0.00	0.00	0.00	0.00	1
2005	358	8	13.4	13.4	109	1.0	0.00	0.00	0.00	0.89	0.00	0.00	0.00	0.00	0.00	1
2005	359	9	9.7	9.7	119	1.0	0.00	0.00	0.00	0.64	0.00	0.00	0.00	0.00	0.00	1
2005	360	10	13.4	13.4	132	1.0	0.00	0.00	0.00	0.89	0.00	0.00	0.00	0.00	0.00	1
2005	361	11	8.6	8.6	141	1.1	0.04	0.00	0.00	0.57	0.00	0.00	0.00	0.00	0.00	1
2005	362	12	5.8	5.8	147	1.1	0.00	0.00	0.04	0.39	0.02	0.09	0.06	0.04	0.00	1
2005	363	13	5.6	5.6	152	1.2	0.00	0.00	0.04	0.37	0.03	0.12	0.07	0.05	0.00	1
2005	364	14	5.5	5.5	158	1.2	0.00	0.00	0.04	0.37	0.03	0.11	0.07	0.04	0.00	1
2005	365	15	6.1	6.1	164	1.3	0.01	0.00	0.05	0.41	0.03	0.11	0.07	0.04	0.00	1
2006	1	16	4.5	4.5	168	1.3	0.00	0.00	0.06	0.30	0.04	0.12	0.07	0.05	0.00	1
2006	2	17	4.6	4.6	173	1.4	0.00	0.00	0.06	0.31	0.04	0.14	0.08	0.05	0.00	1
2006	3	18	5.6	5.6	179	1.4	0.01	0.00	0.06	0.37	0.04	0.16	0.10	0.07	0.00	1
2006	4	19	5.3	5.3	184	1.5	0.01	0.00	0.07	0.35	0.04	0.21	0.12	0.08	0.00	1
2006	5	20	5.3	5.3	189	1.5	0.01	0.00	0.08	0.35	0.05	0.21	0.12	0.08	0.00	1
2006	6	21	5.2	5.2	194	1.6	0.01	0.00	0.08	0.35	0.05	0.19	0.11	0.07	0.00	1
2006	7	22	6.7	6.7	201	1.6	0.01	0.00	0.09	0.45	0.06	0.26	0.16	0.11	0.00	2
2006	8	23	7.6	7.6	209	1.7	0.01	0.00	0.10	0.51	0.06	0.23	0.14	0.09	0.00	2
2006	9	24	4.3	4.3	213	1.7	0.01	0.00	0.11	0.29	0.07	0.08	0.05	0.03	0.00	2
2006	10	25	4.2	4.2	217	1.8	0.01	0.00	0.11	0.28	0.07	0.08	0.05	0.03	0.00	2
2006	11	26	4.5	4.5	222	1.8	0.01	0.00	0.12	0.30	0.08	0.11	0.07	0.04	0.00	2
2006	12	27	4.3	4.3	226	1.8	0.01	0.00	0.13	0.29	0.08	0.16	0.10	0.07	0.00	2
2006	13	28	7.1	7.1	233	1.9	0.01	0.00	0.13	0.47	0.08	0.30	0.18	0.12	0.00	2
2006	14	29	6.5	6.5	240	2.0	0.01	0.00	0.15	0.43	0.09	0.17	0.10	0.07	0.00	2
2006	15	30	4.9	4.9	244	2.0	0.01	0.00	0.16	0.33	0.10	0.14	0.08	0.05	0.00	2
2006	16	31	4.4	4.4	249	2.0	0.01	0.00	0.17	0.29	0.10	0.16	0.10	0.06	0.00	2
2006	17	32	4.2	4.2	253	2.1	0.01	0.00	0.17	0.28	0.11	0.37	0.22	0.15	0.00	3
2006	18	33	6.4	6.4	259	2.1	0.01	0.00	0.18	0.43	0.11	0.62	0.37	0.25	0.00	3
2006	19	34	4.3	4.3	264	2.2	0.01	0.00	0.19	0.29	0.12	0.42	0.25	0.17	0.00	3
2006	20	35	6.8	6.8	271	2.2	0.01	0.00	0.20	0.45	0.12	0.72	0.43	0.29	0.00	4

Fig. 12.7. Appearance of “Outputs” sheet in Excel file containing the model.

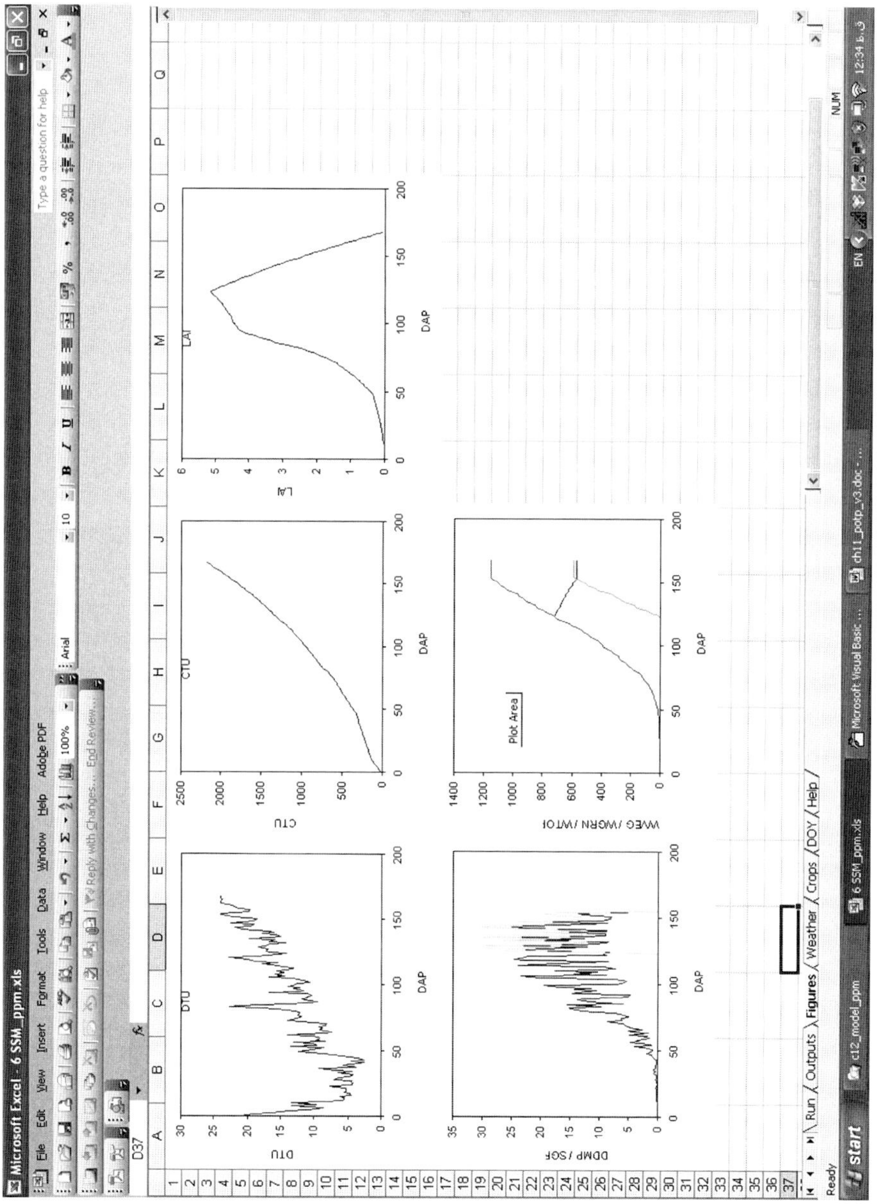

Fig. 12.8. Appearance of "Figures" sheet in Excel file containing the model.

If model predictions do not correspond to observed data, a practical guide presented in Appendix I can be used for troubleshooting.

Estimating Daily Solar Radiation

Solar radiation is not recorded in many weather stations around the world. If there is no solar radiation data, it can be estimated from sunshine-hour data or daily temperature data. This estimation is based on sun–earth geometry and an estimate of atmospheric transmission coefficient. First, solar declination, daylength, and solar radiation at the top of the atmosphere (extraterrestrial radiation; DSO, MJ m^{-2} day^{-1}) are estimated from latitude of the location and day of year (DOY). The calculation of DSO is based on Gouadriaan and van Laar (1994). Then, incident solar radiation (SRAD, MJ m^{-2} day^{-1}) is estimated from DSO and atmospheric transmission coefficient (ATMTR):

$$\text{SRAD} = \text{DSO} \times \text{ATMTR} \tag{12.1}$$

If daily sunshine-hours (SUNH) data are available, the value of ATMTR can be estimated using the Angstrom relationship based on daylength (DAYL):

$$\text{ATMTR} = a + b \times (\text{SUNH} / \text{DAYL}) \tag{12.2}$$

where *a* and *b* are empirical coefficients and roughly equal to 0.25 and 0.5, respectively, but may change depending on location and time during the year. Rietveld (1978) outlined a simple method to calculate Angstrom coefficients from daily sunshine hour data.

Air temperature data can be used to estimate ATMTR based on the difference between maximum and minimum daily temperature (Bristow and Campbell, 1984; Allen *et al.*, 1998). The difference between the maximum and minimum air temperature relates to the degree of cloud cover in a location (Allen *et al.*, 1998). Clear-sky conditions result in high temperatures during the day (i.e. higher maximum temperature) because the atmosphere is transparent to the incoming solar radiation, and in low temperatures during the night (i.e. minimum temperature) because less outgoing long-wave radiation is absorbed by the atmosphere. On the other hand, in overcast conditions, maximum temperature is relatively smaller because a significant part of the incoming solar radiation never reaches the earth's surface and is absorbed and reflected by the clouds. Similarly, minimum temperature will be relatively higher as the cloud cover acts as a blanket and decreases the net outgoing long-wave radiation. Therefore, the difference between the maximum and minimum air temperature (TMAX – TMIN) can be used as an indicator of the fraction of extra-terrestrial radiation that reaches the earth's surface (Hargreaves and Samani, 1982).

Bristow and Campbell (1984) suggested the following relationship for ATMTR, as a function of the difference between maximum (TMAX, °C) and minimum (TMIN, °C) temperatures:

$$\text{ATMTR} = \text{A}\ (1-\exp(-\text{B}(\text{TMAX}-\text{TMIN})^{\text{C}}) \tag{12.3}$$

where *A*, *B*, and *C* are empirical coefficients. Values most frequently reported for these coefficients are 0.7 for *A*, the range 0.004 to 0.010 for *B*, and 2.4 for *C*.

An alternative in estimating ATMTR was suggested by Allen *et al.* (1998):

$$ATMTR = K_r\,(TMAX - TMIN)^{0.5} \quad (12.4)$$

The empirical coefficient (K_r) is obtainable from the ratio of atmospheric pressure at the site (*P*, kPa) and at sea level (P_o, 101.3 kPa) as follows:

$$K_r = K_{ra}\,(P_o - P)^{0.5} \quad (12.5)$$

Atmospheric pressure at a site can be estimated from its elevation (ELV, m above sea level):

$$P = 101.3 - 0.01055 \times ELV \quad (12.6)$$

In their work, Allen *et al.* (1998) suggested values of 0.17 for interior regions and 0.20 for coastal regions for the empirical coefficient K_{ra}. Thus, using Eqn 12.5 accounts for proximity to a large body of water and elevation effects on the volumetric heat capacity of the atmosphere.

Two simple programs for calculating solar radiation from sunshine hours and temperature can be found on the book's website. Donatelli *et al.* (2003, 2006) developed free software to calculate solar radiation using different approaches (available at: www.sipeaa.it). There is other software available to fill gaps in weather data, generate daily weather data from monthly data, or generate long-term weather data (say >50 years) from shorter terms (say >10 years) (Soltani and Hoogenboom, 2003a, b; Liu *et al.*, 2009).

Sample Runs of the Model

A simple simulation run is presented to demonstrate how a model may be applied. The model was run for weather data of 2005 in Gorgan, in the north-east of Iran. Then the model was run for the same year but with 2 and 4°C increases in temperature to simulate possible climate change conditions. In these runs, daily maximum and minimum temperatures were increased by 2 and 4°C but management inputs, i.e. sowing date and density, and cultivar remained constant for the three temperature scenarios.

The summary of crop development, growth, and yield for the three model runs are shown in Table 12.1. Increases in temperature by 2 and 4°C resulted in accelerated phenological development as can be seen in shorter days to different crop stages. For example, days to beginning seed growth was 124 days that decreased to 111 days (a 10% decrease) and 99 days (a 20% decrease) in increased 2 and 4°C scenarios, respectively. The 168 days to maturity under existing temperature decreased to 155 and 143 days as a result of a 2 and 4°C increase, respectively.

Due to higher temperature, LAI showed a more rapid rise in 2 and 4°C scenarios compared to normal temperature (Fig. 12.9a). Higher temperature under optimal conditions can lead to faster leaf area development as discussed

Table 12.1. Summary of crop characteristics simulated for wheat in Gorgan during growing season of 2005/06 under different scenarios of increase in temperature.

	Increase in temperature (°C)		
	0	2	4
Phenology:			
Days to emergence	14	10	9
Days to termination leaf growth	95	82	73
Days to beginning seed growth	124	111	99
Days to termination seed growth	156	142	129
Days to maturity	168	155	143
Growth:			
Maximum LAI	6.06	5.80	5.46
Crop mass at beginning seed growth	762	698	636
Yield:			
Crop total mass (g m^{-2})	1227	1144	1039
Grain yield (g m^{-2})	628	589	522
Harvest index (%)	51	51	50

in Chapter 9. However, as higher temperatures shortened duration of leaf area development, maximum LAI was lower for increased 2 and 4°C scenarios although the amount of decrease was limited; 4% for increased 2°C and 10% for increased 4°C (Table 12.1). Total crop mass at the beginning of seed growth also showed a decline similar to that of maximum LAI. Higher temperatures under increased 2 and 4°C scenarios resulted in higher rates of daily dry matter production during the first 100 days of crop growth (Fig. 12.9b). These higher rates were due to higher LAI and/or higher radiation use efficiency during the period that coincides with winter.

Crop mass accumulation was faster but with a shorter duration in 2 and 4°C scenarios (Fig. 12.9c). Total crop mass at maturity was 1227 g m^{-2} for the existing temperature scenario. This figure was 1144 g m^{-2} (7% decrease) for the 2°C temperature-increase scenario and 1039 g m^{-2} (15% decrease) for the 4°C temperature-increase scenario. Finally, crop yield has decreased from 628 g m^{-2} to 589 and 522 g m^{-2} for the 2 and 4°C temperature-increase scenarios, respectively (Table 12.1).

Exercises

1. Try to parameterize the model for potential production for your crops/cultivars. To do this, parameter estimates presented in previous chapters can be used as default values, but precise estimates of required temperature units are necessary.
2. Use the model of Exercise 1 to estimate the crops/cultivars' potential yields at your location. Compare the yield with those that farmers harvest. Explain how these two are similar or different. Why?

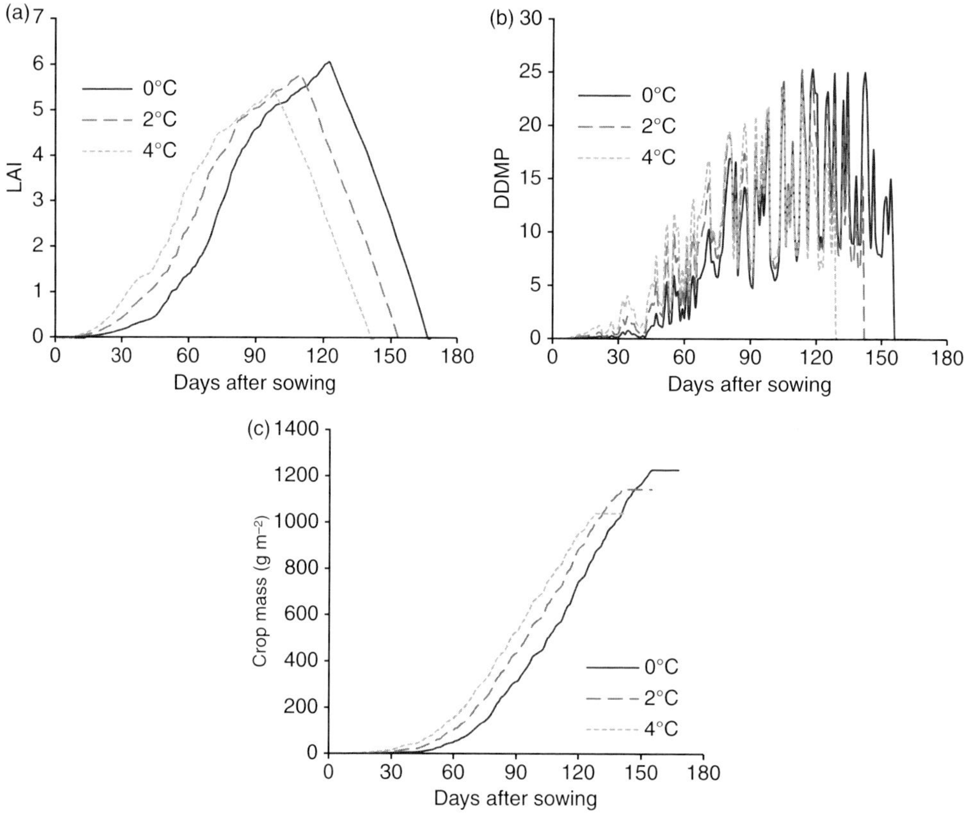

Fig. 12.9. Simulated changes in LAI (a), daily dry matter production (DDMP, g m^{-2} day^{-1}) (b) and crop mass (c) for wheat in Gorgan during growing season of 2005/06. Numbers indicate increase in temperature.

3. Prepare a list of applications for the model of Exercise 1. To complete the list, read Chapter 4 of the book to get better ideas.
4. Use the model of Exercise 1 to simulate the impact of higher temperatures on your crops/cultivars as done in this chapter.
5. Repeat simulations of Exercise 4 for higher concentrations of CO_2. Do not forget to correct radiation use efficiency for higher CO_2 concentrations.
6. Use the model of Exercise 1 for other applications you have listed in Exercise 3.

13 Soil Water

In many parts of the world where agriculture is important, crops usually encounter a large variation in water supply during their growing seasons. Therefore, water deficit is a major constraint to crop production, even in humid environments.

In Part II, a model of potential production was presented. The objective of Part III is to extend that model to water-limited production situations. For this objective, a basic knowledge of soil as a reservoir for water and plant responses to water deficit and flooding is necessary. The dynamic changes in soil water content through the season must be modeled since this is a vital input to understanding crop development and growth. Of course, the dependence of crop development and growth needs to be simulated as a function of daily soil water content.

This chapter deals with modeling soil as a reservoir for water. In the next chapter, Chapter 14, a simple soil water submodel will be presented, and Chapter 15 deals with plant responses to soil water deficit and flooding. In Chapter 16, the developed soil water submodel will be added to the potential production model and a model for water-limited conditions will be presented.

Soil as a Reservoir for Water

The concept of the soil as a water reservoir for plant growth is useful for calculating soil water balance and its impact on crop production (Ratliff *et al.*, 1983; Hochman *et al.*, 2001). Soil as a matrix can store water and then supply it to the plants. There is, of course, a limit to the water storage capacity of the soil. The maximum water that a soil can hold is called the drained upper limit (DUL). DUL is defined as the gravimetric ($g\ g^{-1}$) or volumetric ($m^3\ m^{-3}$) water content in the field after completely wetting the profile and then allowing water to drain to a steady state under gravity. In this book, volumetric water content will be used in defining soil water limits. Once DUL and soil depth are defined the total capacity of soil as a reservoir for water storage can be calculated. For example, if DUL is 0.264 mm^3 mm^{-3} and soil depth is 1200 mm, then total water storage capacity is obtained by multiplying the two variables resulting in 316.8 mm (= 0.264 × 1200).

Not all water stored in a soil is available to crops. Part of the stored water is unavailable for the plants because it is held tightly by the soil. This limit is characterized by a lower limit (LL), which is also defined in terms of gravimetric (g g^{-1}) or volumetric (m^3 m^{-3}) water content. There are several ways to define LL. These definitions include the soil water content when a healthy crop with uninterrupted root development has died, nearly died, or become dormant (Ratliff *et al.*, 1983), or when the crop has reached maturity under soil water-limited conditions (Hochman *et al.*, 2001), or when the transpiration of the drought-stressed plants has decreased to 10% or less of that of well-watered plants (Sinclair and Ludlow, 1985). Having defined LL, it is then possible to segregate the total soil water between that available to crop plants to support growth and that unavailable to support growth. The point of demarcation between the two is LL multiplied by soil depth, which gives directly the total amount of unavailable soil water. For example, if LL is 0.134 mm mm^{-1} and soil depth is 1200 mm, total amount of unavailable soil water is 160.8 mm (= 0.134 × 1200). Thus, the amount of soil water available for the crop or transpirable soil water is the difference between total soil water at DUL and LL and for the example it is obtained as 156 mm (=316.8 – 160.8).

The difference between DUL and LL is called extractable soil water (EXTR), which is another important soil water limit. Again, EXTR may be expressed in gravimetric or volumetric unit, but in this book the volumetric unit is used. It has been shown that EXTR is conservative for many agricultural soils except sandy soils, and has a value of approximately 0.13 mm mm^{-1} (Ratliff *et al.*, 1983; Ritchie *et al.*, 1999; and see below). Therefore, total plant available soil water or total transpirable soil water can be estimated directly from EXTR. For example, if EXTR is 0.13 mm and soil depth is 1200 mm, then total transpirable (available) soil water will be 156 mm (= 0.13 × 1200).

Typically, 2 days to as many as 12 days are required in the field for a saturated soil to drain to reach DUL. Fine-textured soils with restrictive sub-layers may even require up to 20 days of drainage (Ratliff *et al.*, 1983). Hence, an appreciable quantity of water between saturation and DUL might be available to the plants before drainage from the soil stops. Therefore, another soil water limit is required of fully saturated soil (SAT), which is the soil water content when all soil pores are filled with water after a heavy rainfall or irrigation. Again, SAT can be expressed either as a gravimetric (g g^{-1}) or volumetric (m^3 m^{-3}) water content. Soil saturation is also important in relation to flooding and the impact of resultant hypoxia on crops and water infiltration into the soil (runoff). The value of SAT can be used to obtain the total amount of water in the saturation zone of soils. For example, if SAT is 0.360 mm^3 mm^{-3} and soil depth is 1200 mm, then total soil water at saturation is 432 mm. Assuming that total soil water at DUL is 316.8 mm for this soil (see above), total soil water between DUL and SAT is 115.2 mm (= 432 − 316.8).

Therefore, the various water components of the example soil to calculate its stages of water content are:

LL = 0.134 mm mm^{-1}
DUL = 0.264 mm mm^{-1}
EXTR = 0.13 mm mm^{-1}
SAT = 0.360 mm mm^{-1}
Soil depth = 1200 mm

The various water contents are, therefore:

Total soil water at LL or total unavailable soil water = 160.8 mm
Total soil water at DUL = 316.8 mm
Total soil water at SAT = 432 mm
Total transpirable (available) soil water = 156 mm
Total soil water between DUL and SAT or gravitational soil water = 115.2 mm

Measuring Soil Water Limits

As described above, accurate estimation of soil water requires good estimates of the limits for the various stages of soil water content. The values used must be appropriate to the soil in the field, especially in situations where crop production is water limited (Ritchie *et al.*, 1999). In addition to using these limits to determine how much water in the soil is affecting plant growth, the amount of water in the soil also affects water runoff from the soil surface, soil evaporation, and deep drainage (refer to Chapter 14). This section indicates how these important limits can be obtained experimentally for use in a soil water submodel.

Laboratory measurements of permanent wilting point and field capacity have frequently been used to determine soil water limits. The most common procedure for estimating DUL is to extract water from a disturbed or undisturbed soil sample using a soil water extraction apparatus or "pressure chamber" (Ratliff *et al.*, 1983). A matric potential of −0.033 MPa is used for moderately coarse- and finer-textured soils whereas a −0.01 MPa potential is used for coarse-textured soils. LL is also estimated using the pressure chamber at a matric potential of −1.5 MPa.

The above measures of limits have become accepted as the "conventional" limits, although from a plant perspective these limits are not constant and vary among soils and crops. Laboratory-measured DUL and LL do not always coincide with field observations and have frequently proved inaccurate for establishing field limits of water availability (Ratliff *et al.*, 1983; Ritchie *et al.*, 1999). The basis for the criticism is that annual crops differ in their capacity to exploit water at depths. They also differ in their rooting pattern. Consequently, different LL values might be expected for different crop species grown on the same soil (Hochman *et al.*, 2001). Further, it has been argued that some plants remove water from the soil at matric potentials <-1.5 MPa. Other plants may not remove water to a matric potential of -1.5 MPa (Ratliff *et al.*, 1983).

Given the uncertainty in extrapolating laboratory measurements to field soils, limits derived from field measurements are best for calculating the soil water stages. In the field, DUL is derived from successive measurements of soil water content with depth after the soil has been thoroughly wetted and allowed to drain. LL is derived from successive measurements of soil water content with depth when a field crop is allowed to extract water to a very severe stress (Ratliff *et al.*, 1983; Hochman *et al.*, 2001).

Hochman *et al.* (2001) presented a cost-saving approach for measuring DUL and LL directly in the field. They measured DUL as the volumetric water content in the field after thoroughly wetting the profile and then allowing water to drain to constant soil water content. LL was measured as the volumetric soil water retained by the soil after a healthy crop, with uninterrupted root development, has stopped growing under soil water-limited conditions. Hochman *et al.* (2001) used this method to characterize soil water limits in 83 soil–crop combinations. They included different crops such as cotton, wheat, sorghum, fababean, chickpea, barley, and mungbean in their study. They further developed an equation to estimate LL from DUL in Black and Grey Vertisols in Australia's northeastern grain region. Their information can be found at the APSIM website (www.apsim.info).

Estimating Soil Water Limits

If soil water limits are not available, estimation methods will be needed. These alternative approaches estimate the limits from basic soil data that are more widely available through soil surveys. There are many estimation procedures. Gijsman *et al.* (2003) compared eight methods for estimating DUL, LL, and SAT. They concluded that discrepancy between estimation methods was so big that it was hard to make recommendations on which method to use for which soil. However, they concluded that the Saxton method (Saxton *et al.*, 1986) performed best for loamy sand, sandy loam, loam, and silt loam soils. They used this method to calculate soil water limits for default soils in the database of the DSSAT models. Recently, Saxton and Rawls (2006) presented an improved procedure to estimate soil water characteristics based on texture and organic matter.

Here, the estimation method presented by Ritchie *et al.* (1999) is used to estimate DUL and EXTR because it is simple, straightforward, and field-based. They developed simple, generic equations to estimate the field-measured limits of the soil water reservoir based on soil survey data such as texture and bulk density. Their method was based on the database of field-measured soil water limits described in Ratliff *et al.* (1983). The database contained 401 soil samples from 15 states in the USA from seven soil orders. DUL was derived from successive measurements of soil water content with depth after the soil had been thoroughly wetted and allowed to drain. Soils with a water table shallower than 2 m at the time DUL was measured were excluded. Some soil sites were covered with rainfall shelters or plastic sheeting, which prevented evaporation losses or precipitation gains of water. Other plots were uncovered and were subjected to rains and evaporation from the soil surface. Typically, 2 to 12 days were required for soils to reach DUL. The LL was derived from successive measurements of soil water content with depth when a field crop grown on the soil was subjected to severe drought stress. Water content measurements were continued until the plant died, nearly died, or became dormant. Data from adequately fertilized field plots in which plants had reached maximum vegetative growth before undergoing severe water stress were preferentially selected over data from plots inadequately fertilized or stressed early in the growing season.

In the method of Ritchie *et al.* (1999), a linear regression is used to estimate the gravimetric DUL (DULg, g g^{-1}) from the sand (%) to clay (%) ratio (Fig. 13.1):

$$\text{DULg} = 0.186\ (\text{sand/clay})^{-0.141} \qquad (13.1)$$

To convert the gravimetric DUL to the more useful volumetric DUL (m^3 m^{-3} or mm mm^{-1}) used in models it is necessary to multiply DULg by the soil bulk density (BD, g cm^{-3}):

$$\text{DUL} = \text{DULg} \times \text{BD} \qquad (13.2)$$

A non-linear regression is used to estimate EXTR based on sand (%) content (Fig. 13.2):

$$\text{EXTR} = 0.132 - 2.5 \times 10 - 6\ e^{0.105\ \text{sand}} \qquad (13.3)$$

As illustrated by Eqn 13.3 and in Fig. 13.2, the mean value for EXTR is approximately 0.13 until the sand content of the soil exceeds about 75%.

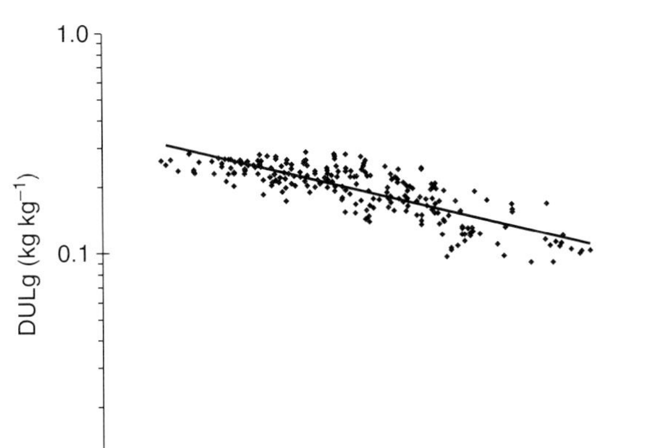

Fig. 13.1. Gravimetric soil water content at drained upper limit (DULg) as a function of sand to clay ratio (*x*). The equation of the line is $y = 0.186(x)^{-0.141}$ (Ritchie *et al.*, 1999).

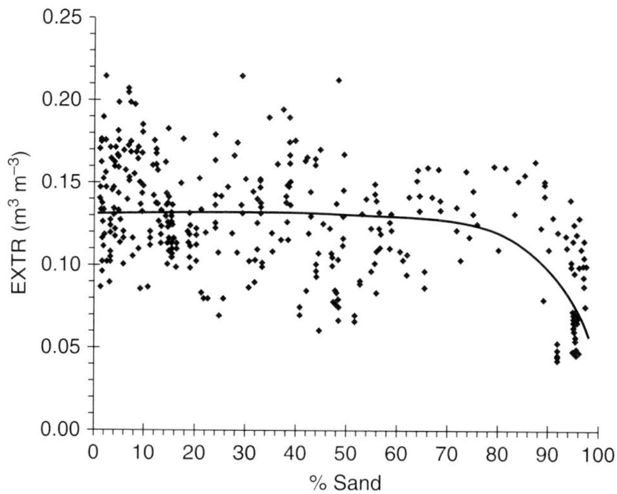

Fig. 13.2. Volumetric extractable soil water content as a function of sand percentage (*x*) in the soil. The equation of the line is $y = 0.132 - 2.5 \times 10 - 6\ e^{0.105x}$ (Ritchie *et al.*, 1999).

The lower limit can be estimated as the difference between DUL and EXTR:

$$LL = DUL - EXTR \quad (13.4)$$

To estimate SAT, a simple method described by Dalgliesh and Foal (2005) is used. SAT (m^3 m^{-3}) limit is obtained from total porosity of a soil (PO, fraction). Since some of the very tiny air spaces are not normally filled with water under saturation (e, fraction), this fraction needs to be subtracted from PO (Dalgleish and Foal, 2005):

$$SAT = PO - e \quad (13.5)$$

where *e* varies between 0.03 (heavy clay soils) and 0.07 (sandy soils).

PO can be estimated from soil bulk density:

$$PO = 1 - BD/2.65 \quad (13.6)$$

where BD/2.65 is the fraction of the soil volume occupied by solid (sand, silt, and clay) particles, based on an assumed density of 2.65 g cm^{-3} for the solid matter in the soil. A density value of 2.65 g cm^{-3} holds for a wide range of soils, except for organic soils (lower value required) and oxisols (higher value required).

If PO and BD are not available for soils, it is possible to estimate PO from the texture of a soil. Based on the fraction of sand (%) and clay (%) content of a soil, PO can be calculated using the following equation (Saxton *et al.*, 1986):

$$PO = 0.332 - 7.251 \times 10^{-4}\,(\text{sand}) + 0.1276\,\log_{10}(\text{clay}) \quad (13.7)$$

Table 13.1 includes estimates of soil water limits for some default soils. The default soils and their clay, silt, and sand contents and bulk densities are those

Table 13.1. Estimates of soil water limits from texture and bulk density.

	Silty clay	Silt loam	Sandy loam	Sand
Clay (%)	50	10	10	5
Sand (%)	5	30	60	90
Silt (%)	45	60	30	5
BD	1.32	1.37	1.61	1.66
PO	0.502	0.483	0.392	0.374
e	0.03	0.05	0.06	0.07
SAT	0.472	0.433	0.332	0.304
DUL	0.340	0.218	0.233	0.205
EXTR	0.132	0.132	0.131	0.100
LL	0.208	0.086	0.102	0.105

BD: Soil bulk density (g cm^{-3})
PO: Soil porosity (fraction or m^3 m^{-3})
e: Fraction of air spaces that are not filled with water under saturation (fraction or m^3 m^{-3})
SAT: Volumetric water content when the soil is fully saturated with water (m^3 m^{-3})
DUL: Volumetric water content at drained upper limit (m^3 m^{-3})
EXTR: Volumetric water content extractable by the crops (m^3 m^{-3})
LL: Volumetric water content at lower limit (m^3 m^{-3})

used in DSSAT (Gijsman *et al.*, 2003). However, soil water limits are calculated using the equations presented above. These limits may be used, especially for educational purposes, if measured limits are not available.

Soil Water Limits from Databases

There are existing databases that can be used as a source for estimating soil water limits. The APSIM website provides information for many Australian sites. The database described by Ratliff *et al.* (1983) can be used for sites within the USA. There are also some newer attempts to develop databases for the USA. For example, Ali *et al.* (2004) described the construction of a web-based soil physical properties database to meet data requirements of cotton and soybean models GOSSYM and GLYCIM plus providing a generic data file of 1074 soil horizons (about 300 sample sites) collected from farmers' fields in the USA. Across the USA, the STATSGO database of the Natural Resources Conservation Service of USDA provides soil information (Soil Survey Staff, 2011).

Recently, Gijsman *et al.* (2007) used the "World Inventory of Soil Emission Potentials" (WISE) database and converted 1125 soil profiles from around the world into a format that can be used as input data to models. If little or nothing is known about the soil profile for a particular location, a soil database can be used to estimate some of its parameters, based on a comparison with other soils from the same region. The WISE database is one of the most comprehensive soil databases, with samples well distributed in the world. Gijsman *et al.* (2007) prepared two soil files based on WISE that are available and can be downloaded from the ICASA web page (www.icasa.net). Figure 13.3 shows the coverage of the database.

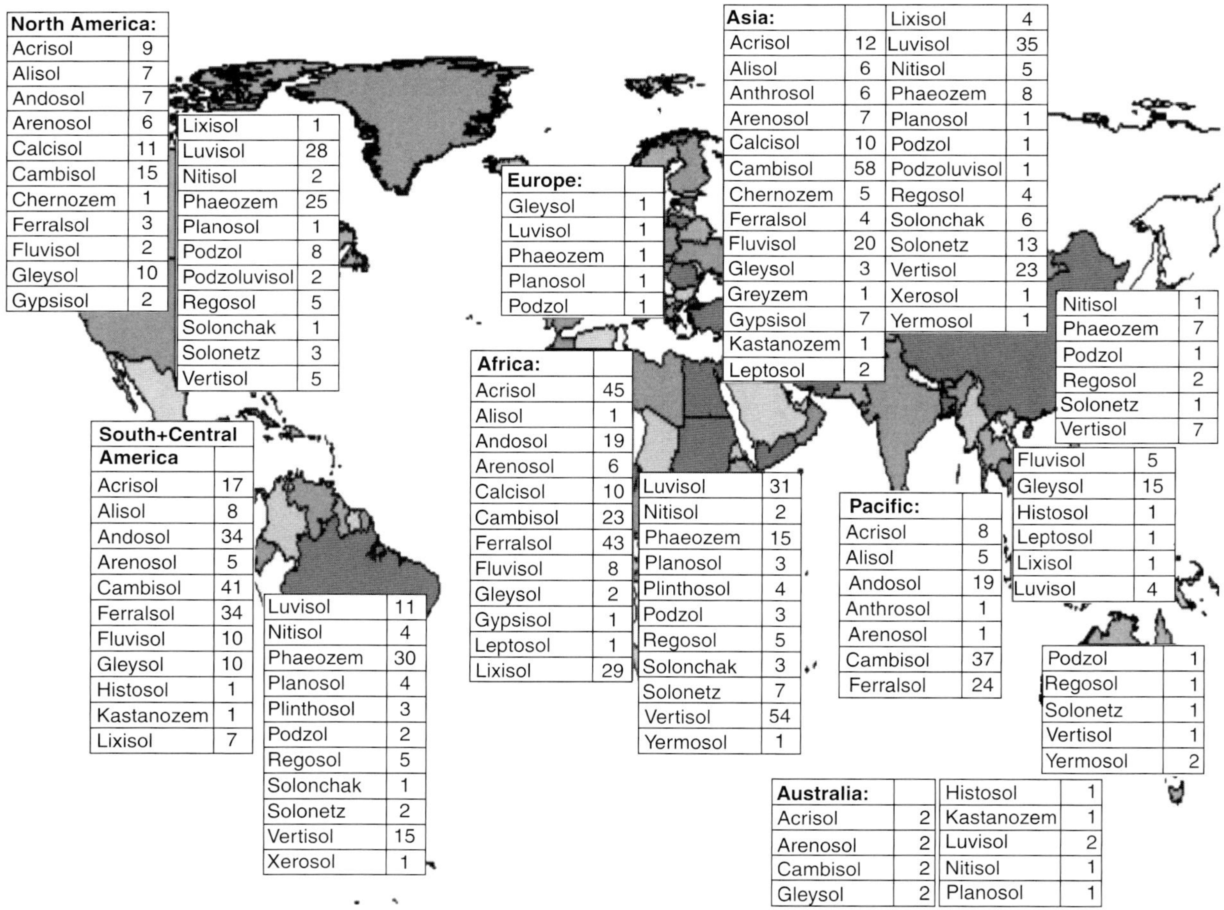

Fig. 13.3. Coverage of WISE database with the soil classes (equivalent to order) summarized by continent or region. The numbers indicate the number of profiles in the WISE database (Gijsman *et al.*, 2007).

Exercises

1. Calculate soil water limits for a soil with 59% sand, 27% clay, and bulk density of 1.4 g cm^{-3}. Use a value of 0.05 for *e*.

2. For the soil of Exercise 1, calculate soil water at LL, DUL, and SAT, assuming that soil depth is 600 mm. Also, calculate total available and unavailable soil water.

14 Soil Water Balance

In the last chapter, the concept of soil as a reservoir for water was presented. Crop growth and yield under water-limited conditions are highly dependent on plant-available soil water. Therefore, accurate prediction of dynamic changes in soil water is necessary for an acceptable prediction of crop growth and yield. Any error in prediction of available soil water certainly results in biased estimation of crop yield.

The objective of this chapter is to explain the concept of soil water balance and to present simple methods to calculate it. An understanding of dynamic changes in soil water balance is required for crop modeling under water-limited conditions. Soil water calculations allow an estimation of degree of water deficit/excess for plant processes. Plant responses to soil water deficit and flooding are dealt with in Chapter 15. Further, these calculations would help in evaluation and understanding of crop growth and yield under water-limited conditions and how they are affected by management practices and crop characteristics.

The simple soil water submodel of this chapter is based on Amir and Sinclair (1991), which has been used successfully in several other models and studies (e.g. Hammer *et al.*, 1995; Soltani *et al.*, 1999; Sinclair *et al.*, 2010).

Soil Water Balance Concept

Soil water balance can be defined as the amount of soil water or available soil water stored in a given layer of the soil at a given time. Similar to a bank account, such that balance is a result of deposits and withdrawals, soil water balance depends on quantities of water added to or removed from the soil layer, i.e.:

$$\text{Soil water balance} = \text{Inputs of water} - \text{Removals of water} \quad (14.1)$$

To obtain soil water balance, estimates of different processes of water losses and additions into the soil layer are needed, which finally influence the amount of soil water and available soil water for the crop (Fig. 14.1). Water is added to the soil through these processes:

- precipitation;
- irrigation;
- increasing soil layer thickness due to root growth; and
- capillary rise of water may occur if there is a high water table in the soil, but it is not considered here.

Water is removed from the soil via:

- deep drainage below the soil root zone;
- soil surface runoff;
- soil evaporation; and
- plant transpiration.

In modeling soil moisture, the soil is often divided into a number of layers and the water balance is calculated. A multi-layer soil approach may be needed for models that are focused on soil processes. However, for models attempting to simulate crop growth and yield as is the objective of this book, a two-layered soil or even a one-layer soil seems satisfactory (Robertson and Fukai, 1994). For more details about different approaches in simulating soil water balance refer to Robertson and Fukai (1994), Jara and Stockle (1999), and Brown *et al.* (2009).

For the objective of this book, the soil profile can be simply divided into two layers: a top layer of usually 150 to 600 mm depth from which water can

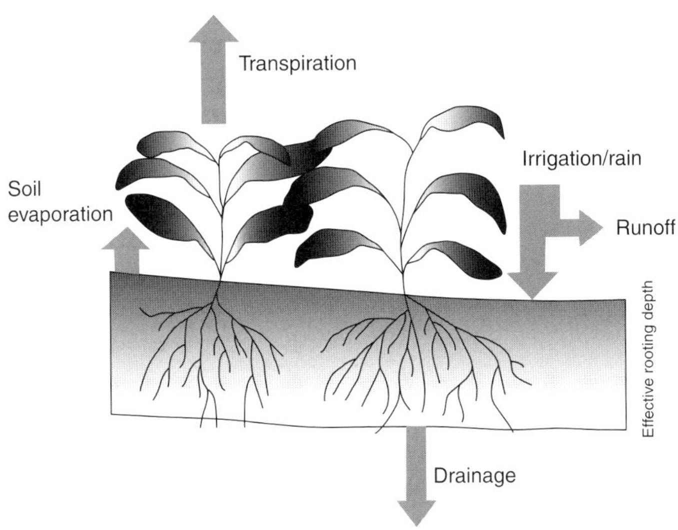

Fig. 14.1. Schematic representation of soil water balance and its components.

be readily removed by soil evaporation, and the total depth of a root layer that increases as a result of root growth and includes the top soil layer. Soil water balance in each layer is tracked separately. The same set of soil inputs, such as soil water limits (Chapter 13), is applied for both layers, but this can be changed if required.

Available soil water in the top layer each day ($ATSW1_i$, mm) is obtained by adding water to its value from the previous day ($ATSW1_{i-1}$) as a result of precipitation (RAIN, mm) and irrigation (IRGW, mm) and removal or withdrawal from the layer by drainage (DRAIN1, mm), runoff (RUNOF, mm), soil evaporation (SEVP, mm), and that part of plant transpiration water removed from the top layer (TR1, mm), i.e.:

$$ATSW1_i = ATSW1_{i-1} + RAIN + IRGW - DRAIN1 - RUNOF - SEVP - TR1 \quad (14.2)$$

where i and i−1 denote today and yesterday, respectively.

From the discussion in Chapter 13, total transpirable soil water of the top layer (TTSW1) is a product of the thickness of the top layer (DEP1, mm) and volumetric available soil water (EXTR, $mm^3\ mm^{-3}$):

$$TTSW1 = DEP1 \times EXTR \quad (14.3)$$

Fraction transpirable soil water in this top layer (FTSW1), which is required later in this chapter, is calculated by dividing ATSW1 by TTSW1:

$$FTSW1 = ATSW1 / TTSW1 \quad (14.4)$$

Available soil water balance for the total root zone ($ATSW_i$, mm) is computed using the same concept as used for $ATSW1_i$ in Eqn 14.2. For the total root zone, drainage is from the bottom of the root zone (DRAIN, mm) and transpiration is the total amount removed from the soil by the crop (TR, mm). Also, additional water may be added to the soil water balance as a result of root extension into deeper, wet soil (EWAT, mm). Therefore, the water balance for $ATSW_i$ is calculated as (Fig. 14.2):

$$ATSW_i = ATSW_{i-1} + RAIN + IRGW + EWAT - DRAIN - RUNOF - SEVP - TR \quad (14.5)$$

Total transpirable soil water in the root zone (TTSW, mm) each day is recomputed each day as the product of current crop root depth (DEPORT, mm) and EXTR. Calculation of current crop rooting depth will be discussed later in this chapter.

$$TTSW = DEPORT \times EXTR \quad (14.6)$$

Fraction of transpirable soil water in the root layer (FTSW), which will be very useful in Chapter 15 to model crop responses to soil water, is again computed from ATSW and TTSW in the root layer:

$$FTSW = ATSW / TTSW \quad (14.7)$$

Water Inputs

Precipitation (RAIN)

Daily rainfall is obtained from weather data. Thus, weather data must include daily rainfall. The same method presented in Chapter 8 is used to calculate snow cover and melt if necessary.

Irrigation (IRGW)

To simulate observed crops, actual dates and amounts of irrigation will be needed. Amounts of irrigation water need to be added to daily rainfall in the weather data sheet on the exact dates of irrigation.

The model can also be used to explore the consequences of various irrigation schemes. A trigger point of soil water content can be used to trigger the application of water in the model. Under this option, FTSW each day is compared with a user predefined FTSW as the level of irrigation. If FTSW is lower than the predefined FTSW and the crop has not reached termination of seed growth (physiological maturity), irrigation is performed.

The amount of irrigation can be decided by several criteria. One possibility is a fixed amount of water at each irrigation, which may be defined by the capacity of the irrigation system. Another possibility is to add sufficient water to return the root layer to a specific level, e.g. the drained upper limit. In the case of irrigating to the drained upper limit, irrigation water added to the soil is calculated as the difference between TTSW and ATSW in the root zone:

$$IRGW = TTSW - ATSW \tag{14.8}$$

The model can report the number and amount of the irrigations.

Root extension (EWAT)

As the depth of the soil explored by roots increases, there is an increase in the amount of water available to the crop if the deeper soil contains water. Water extraction depth in the model is increased with a constant rate (GRTD, mm day^{-1}), which is a crop characteristic and needs to be provided to the model. Values of GRTD for some crops are presented in Table 14.1. Therefore, extraction depth each day ($DEPORT_i$, mm) is calculated by adding GRTD to the depth on the previous day ($DEPORT_{i-1}$). That is:

$$DEPORT_i = DEPORT_{i-1} + GRTD \tag{14.9}$$

The value of DEPORT at crop emergence must be provided to the model. It is normally between 150 to 400 mm depending on crop species and soil conditions.

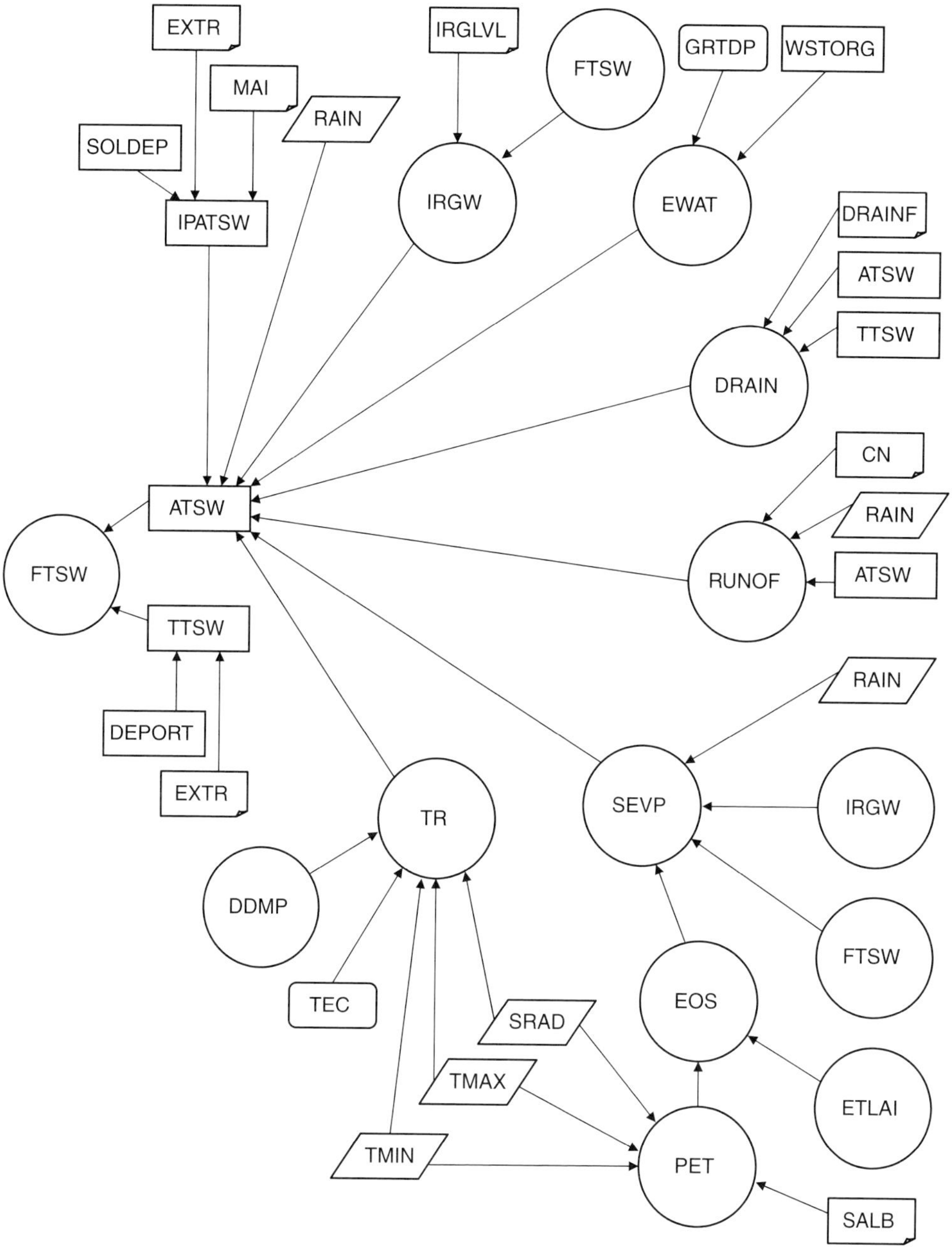

Fig. 14.2. Relational diagram of the soil water submodel. FTSW is the fraction transpirable soil water, ATSW the actual transpirable soil water (mm), TTSW the total transpirable soil water (mm), EXTR the volumetric extractable soil water (mm mm^{-1}), SOLDEP the soil depth (mm), MAI the soil moisture availability index, IPATSW the initial ATSW at sowing time (mm), RAIN the daily rainfall (mm), IRGLVL the irrigation level, IRGW the irrigation water (mm), GRTDP the daily rate of root depth growth (mm day^{-1}), WSTORG the soil water stored below rooting zone (mm), EWAT the amount of water becoming available to the crop due to root growth

Table 14.1. Estimates of potential daily rate of increase in rooting depth (GRTD, mm day^{-1}), maximum effective depth of water extraction from the soil (MEED, mm), and transpiration efficiency coefficient (TEC, Pa) in some field crops (Amir and Sinclair, 1991; Chapman *et al.*, 1993; Soltani *et al.*, 1999; Robertson *et al.*, 2002a; Keating *et al.*, 2003; Sinclair *et al.*, 2003; Stockle *et al.*, 2003; Wahbi and Sinclair, 2005).

Crop	GRTDP	MEED	TEC
Wheat	30	1200	5.8
Barley	30	1200	5.8
Rice	35		5.8
Maize	33	1300	9.0
Sorghum	27	1300	9.0
Soybean	30	1200	4.5
Peanut	35	1200	4.5
Canola	44	1100	4.5
Sunflower	38	1300	4.5
Dry bean	35	1000	5.0
Chickpea	17	1200	5.0

In the generic model, the use of Eqn 14.9 begins at emergence and ends at the termination of seed growth, but these stages can be changed if warranted. Thus, GRTD is set to 0 before emergence and after beginning seed growth. Similarly, GRTD is set to 0 once root depth reached either the maximum effective extraction depth defined for the crop (MEED, mm), or the maximum depth allowed by physical or chemical constraints in the soil (SOLDEP, mm). Another factor that inhibits root growth is a lack of water in the soil below the current root depth. That is, if there is no available soil water below the root depth, GRTD is set equal to 0.

MEED is a crop parameter and SOLDEP is a soil input, and both must be provided to the model. MEED as used here is different from maximum rooting depth that is used in some crop models. MEED is rooting depth with a root length density of greater than 0.1 cm cm^{-3} that provides at least 95% of crop water demand (Dardanelli *et al.*, 2004). Some crop roots might penetrate into a deeper layer but supply the crop with negligible water.

MEED can be estimated from measurements from a neutron meter during the crop growing season (e.g. Silim and Saxena, 1993). Baker and Ahern (1989)

Fig. 14.2 Continued.
(mm), DRAINF the drainage factor (mm mm^{-1}), DRAIN the daily rate of drainage (mm day^{-1}), CN the curve number, RUNOF the daily runoff (mm day^{-1}), EOS the daily potential soil evaporation (mm day^{-1}), SEVP the actual soil evaporation (mm day^{-1}), ETLAI the leaf area index effective in evapotranspiration (m^2 m^{-2}), SALB the soil albedo, PET the potential evapotranspiration (mm day^{-1}), SRAD the daily solar radiation (MJ m^{-2} day^{-1}), TMAX the daily maximum temperature (°C), TMIN the daily minimum temperature (°C), TEC the transpiration efficiency coefficient (Pa), DDMP the daily dry matter production (g m^{-2} day^{-1}), TR the daily transpiration (mm day^{-1}), and DEPORT the current depth of root (mm).

compared eight methods of estimating effective rooting (water extraction) depth. Using maximum rooting depth instead of MEED will result in overestimation of TTSW. Estimates of MEED are presented in Table 14.1 for some field crops.

The estimate of water input by root extension (EWAT) is calculated as the product of GRTD and volumetric available soil water content (EXTR). However, EWAT cannot exceed the amount of water stored in newly explored soil layer (WSTORG, mm). Since water may be added to WSTORG by drainage, it is necessary to also update WSTORG (see Eqn 14.14).

$$\mathrm{EWAT} = \min(\mathrm{GRTD} \times \mathrm{EXTR}, \mathrm{WSTORG}) \qquad (14.10)$$

Water Removals

Drainage (DRAIN)

Drainage will happen from both layers, i.e. the top layer (DRAIN1) and the root zone layer (DRAIN), when available transpirable soil water in these layers is higher than total transpirable soil water that can be stored in each of the respective layers (i.e. TTSW). The water in excess of the storable transpirable soil water drains to lower soil layers as a result of gravitational force. The rate of drainage depends on soil texture and nature of the soil into which the water is draining. Existence of a limiting subsurface layer restricts downward movement of water and hence drainage. In sandy soils, large pores allow fast drainage, but in clay soils small pores hinder drainage and drainage is slower.

Drainage rates are calculated using an empirical relation that evaluates field drainage reasonably well (Ritchie, 1998). For both layers, if the current amount of transpirable soil water (ATSW_i) in the layer is more than total transpirable soil water that can be stored in the layer (TTSW, mm), then a fraction (DRAINF) of the excess water will drain. This fraction is called the drainage factor. If ATSW_i is lower than TTSW, no drainage takes place.

$$\begin{aligned} \mathrm{DRAIN} &= 0 && \textit{if} \quad \mathrm{ATSW} \leq \mathrm{TTSW} \\ \mathrm{DRAIN} &= (\mathrm{ATSW}_i - \mathrm{TTSW}) \times \mathrm{DRAINF} && \textit{if} \quad \mathrm{ATSW} > \mathrm{TTSW} \qquad (14.11) \end{aligned}$$

DRAINF is dependent on soil texture and existence of a limiting subsurface layer. A DRAINF of 1 means that all the extra water will drain in 1 day (24 h). In the same way, a DRAINF of 0.3 means that 30% of extra water will drain each day. Table 14.2 presents DRAINF for several soils and three soil depths each. Figure 14.3 shows drainage from three soils with different values for DRAINF.

The same relationships used in Eqn 14.11 are used to estimate drainage from the top layer (DRAIN1) in which ATSW_i and TTSW are substituted with $\mathrm{ATSW1}_i$ and TTSW1, respectively.

For the root layer, not all the drained water (DRAIN) below the root layer may be considered a water loss. All or part of the drained water to deeper soil may be exploited later by the crop due to root growth. Therefore, the amount

Table 14.2. Soil inputs of depth (SOLDEP, mm), drainage factor (DRAINF), curve number (CN), and albedo (SALB) for 12 default soils, based on Gijsman *et al.* (2003), as are used in the DSSAT model (Jones *et al.*, 2003).

	SOLDEP	DRAINF	CN	SALB
Silty clay	210	0.3	85	0.11
	150	0.2	87	0.11
	60	0.1	89	0.11
Silty loam	210	0.4	77	0.12
	150	0.3	79	0.12
	60	0.2	81	0.12
Sandy loam	210	0.5	68	0.13
	150	0.5	70	0.13
	60	0.4	74	0.13
Sand	210	0.6	65	0.15
	150	0.5	70	0.15
	60	0.4	75	0.15

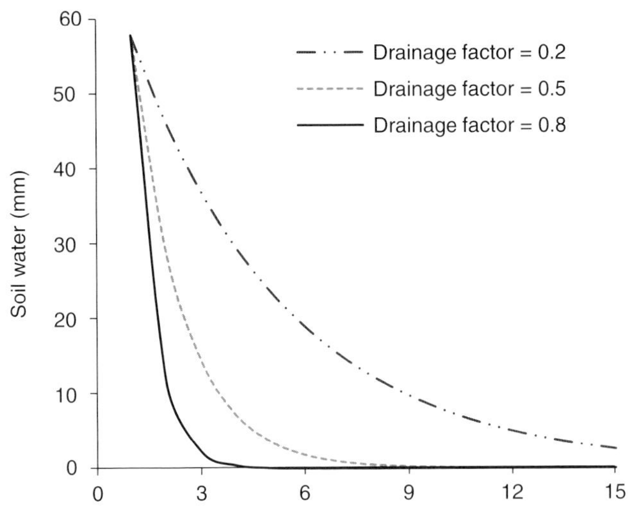

Fig. 14.3. Changes in soil water (mm) over drained upper limit in three soils with drainage factors of 0.2, 0.5, and 0.8. It was assumed that drainage is the only method of water loss.

of available water below the root layer (WSTORG, mm) is tracked and its changes due to drainage (DRAIN, mm) and root exploitation (EWAT, mm) are calculated.

$$WSTORG_i = WSTORG_{i-1} + DRAIN - EWAT \tag{14.12}$$

Final total drainage will be the amount of WSTORG at the end of simulation (growing season). That is, unused, drained water below the root zone.

Runoff (RUNOF)

Runoff is the amount of water that does not infiltrate into the soil after rainfall or irrigation. However, here it is assumed that usually the irrigation regime is well managed and no runoff occurs. Thus, runoff calculation is bypassed if irrigation is practiced. Rain runoff water is considered a water loss. The amount of runoff depends on rainfall intensity, soil texture, soil water, and vegetation cover.

Daily runoff (RUNOF, mm) is calculated using a simplified curve number procedure developed by scientists at USDA-Soil Conservation Service (SCS) (Williams, 1991). This procedure is widely used for calculation of runoff and a modified version of the method is used here.

In the curve number method, daily surface runoff is calculated as a function of daily rainfall (RAIN, mm) and a soil retention parameter (S).

$$\text{RUNOF} = (\text{RAIN} - 0.2 \times \text{S})^2 / (\text{RAIN} + 0.8 \times \text{S}) \quad \textit{if} \quad \text{RAIN} > 0.2 \times \text{S}$$
$$\text{RUNOF} = 0 \quad \textit{if} \quad \text{RAIN} \leq 0.2 \times \text{S} \qquad (14.13)$$

The retention parameter (S) is related to curve number (CN) (Fig. 14.4) using the SCS equation:

$$\text{S} = 254 \times (100 / \text{CN} - 1) \qquad (14.14)$$

Runoff curves are specified by numbers (CN), which vary from 0 (no runoff) to 100 (all runoff). The SCS handbook provides a list of runoff CNs for various hydrological soil groups and soil cover complexes. The value of CN for some default soils can be found in Table 14.2. Figure 14.5 shows how daily runoff depends on rainfall and curve number.

There are some approaches to adjust CN for soil cover by the crop foliage and/or straw mulch (Chapman *et al.*, 1993) and slope of the land (Williams,

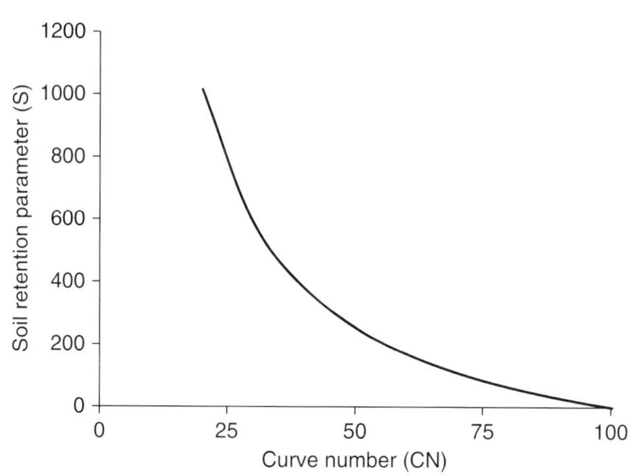

Fig. 14.4. Soil retention parameter as a function of soil curve number for calculation of runoff.

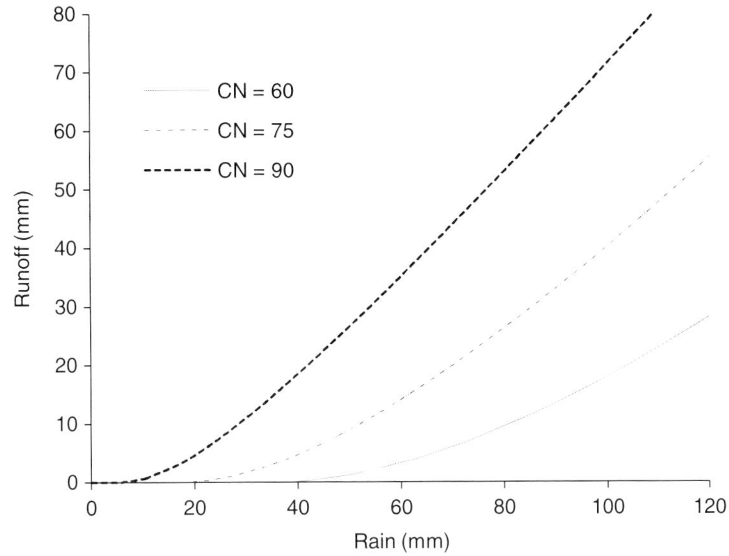

Fig. 14.5. Daily runoff as a function of daily rain for three curve numbers (CN).

1991), but they are not applied in the examples presented here to keep the method simple.

Soil saturation can greatly increase runoff. Once the soil is saturated, it is usually not possible for additional water to enter the soil. Therefore, when the soil becomes saturated, all additional RAIN is assumed to be dispersed by runoff under both rainfed and irrigated conditions.

Soil evaporation (SEVP)

To calculate soil evaporation, potential evapotranspiration from a wet surface for the environmental conditions on each day needs to be determined. Then, potential soil evaporation is multiplied by the fraction of the soil surface that is exposed to incident solar radiation, that is, not shaded by the crop. Additionally, adjustments in potential soil evaporation need to be made if the top soil layer is not actually wet.

Potential soil evaporation

The first step in calculating soil evaporation is to account for the effect of environmental conditions on potential evaporation from a wet soil surface. Here a simplified Penman equation is used to account for potential evaporation, which is calculated from the slope of saturated vapor pressure curve versus temperature (DELT), incident daily solar radiation (SRAD), and soil albedo (SALB).

The DELT term is calculated from mean daily temperature (TMP) using the following equation:

$$\text{DELT} = \text{EXP}(21.255 - 5304 / (273 + \text{TMP})) \times (5304 / ((273 + \text{TMP})^2)) \quad (14.15)$$

Solar radiation incident to the soil (SRAD) needs to be discounted to account for the interception of the radiation by the crop canopy. This is done using the exponential Beer–Bouguer–Lambert equation (Eqn 10.1) based on the canopy extinction coefficient for global solar radiation (KET, ~0.5) and the leaf area intercepting the solar radiation.

The leaf area intercepting solar radiation is different from the LAI used in the calculation of crop growth. During grain filling period, some leaves fall on the soil and cover the soil. Further, some senesced leaves remain attached to the plant that do not participate in dry matter production but intercept solar radiation and reduce interception of solar radiation by the soil surface. Therefore, the LAI required in the calculation of soil radiation penetration to the soil surface for estimating soil evaporation (ETLAI) is defined as the total leaf area produced by the crop. Before seed growth LAI and ETLAI are the same, but after seed growth ETLAI is kept constant at the value of LAI at beginning seed growth (BSGLAI) because senesced leaves, attached to the plant or fallen on the soil surface, still shade the soil, intercept solar radiation, and reduce soil evaporation:

$$\begin{aligned} \text{ETLAI} &= \text{LAI} \quad && \textit{if} \quad \text{CTU} < \text{tuBSG} \\ \text{ETLAI} &= \text{BSGLAI} \quad && \textit{if} \quad \text{CTU} \geq \text{tuBSG} \end{aligned} \quad (14.16)$$

where CTU is cumulative temperature unit (Chapter 6) and tuBSG is cumulative temperature unit from sowing to beginning seed growth.

Finally, the albedo, or reflectance of solar radiation from the soil surface, must be taken into account. SALB is a soil input and commonly has a value close to 0.12 (Table 14.2).

Consequently, potential soil evaporation from a bare, wet soil each day (EOS, mm day^{-1}) is calculated as:

$$\text{EOS} = \text{SRAD} \times (1 - \text{SALB}) \times \text{EXP}(-\text{KET} \times \text{ETLAI}) \times \text{DELT} / (\text{DELT} + 0.68) \quad (14.17)$$

$$\text{EOS} = \text{EOS} \times 239 / 583 \quad (14.18)$$

The number "239/583" in Eqn 14.18 is used to convert EOS from energy unit to millimeters of water evaporated.

In Eqn 14.17, high values of ETLAI can result in very low values of EOS. Since some radiation always provides energy for soil evaporation, a minimum value of EOS is used. Generally, it is assumed the minimum EOS is 1.5 mm (Amir and Sinclair, 1991). Figure 14.6 represents EOS as a function of ETLAI for three levels of potential evapotranspiration.

Actual soil evaporation

Actual soil evaporation (SEVP, mm day^{-1}) is calculated using a two-stage model (Amir and Sinclair, 1991). Stage I evaporation occurs when the top soil layer contains water (ATSW1 > 1) and water is freely evaporated as described in Eqns 14.17 and 14.18. An additional definition of Stage I evaporation is that the total soil profile is not dry (FTSW > 0.5). Under these conditions, Stage I SEVP is equal to EOS. Stage II evaporation occurs after the soil in the top layer dries to the point where evaporation is no longer equivalent to that of a wet surface. At this point, Stage II is initiated. In Stage II, the potential rate of soil evaporation is also equal to EOS, but it is decreased substantially as a function of the square root of time since the start of Stage II (DYSE, days). The calculation of soil evaporation returns to Stage I only when rainfall and/or irrigation of greater than 10 mm occurs (Fig. 14.7).

$$\text{SEVP} = \text{EOS} \quad \text{for Stage I}$$
$$\text{SEVP} = \text{EOS} \times ((\text{DYSE} + 1)^{0.5} - \text{DYSE}^{0.5}) \quad \text{for Stage II} \qquad (14.19)$$

Plant transpiration (TR)

Analysis by Tanner and Sinclair (1983) indicated that crop dry matter production and transpiration are linked, as both processes are dependent on gas diffusion through stomata. Therefore, daily transpiration (TR, mm day^{-1}) is calculated

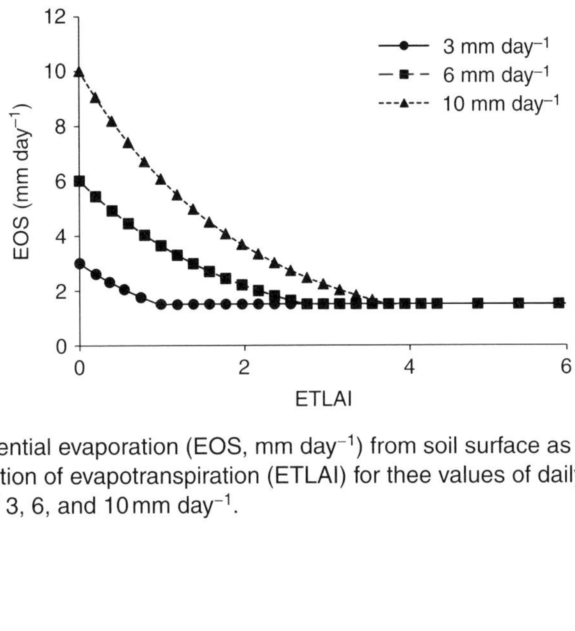

Fig. 14.6. Potential evaporation (EOS, mm day^{-1}) from soil surface as a function of LAI for calculation of evapotranspiration (ETLAI) for thee values of daily potential evaporation of 3, 6, and 10 mm day^{-1}.

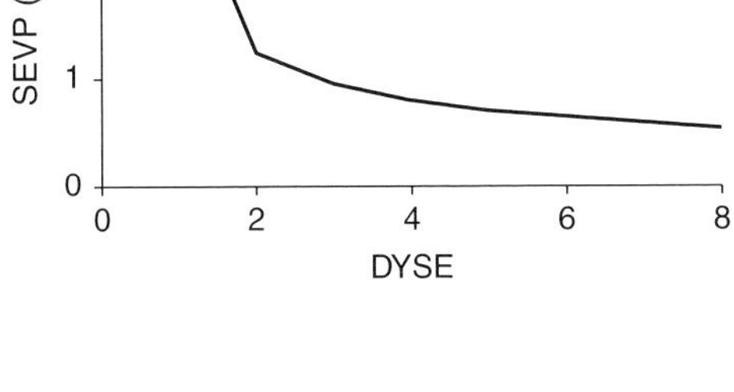

Fig. 14.7. Soil evaporation (SEVP) as a function of days since last rainfall and/or irrigation event (DYSE). Potential soil evaporation of 3 mm was assumed for all the days.

from daily dry matter production (DDMP, g m^{-2} day^{-1}), effective daily vapor pressure deficit for transpiration (VPD, kPa), and a transpiration efficiency coefficient (TEC, Pa).

$$TR = (DDMP \times VPD) / TEC \tag{14.20}$$

Note that in Eqn 14.20 VPD is in kPa but TEC is in Pa and this results in conversion of transpiration unit from g m^{-2} day^{-1} to kg m^{-2} day^{-1} or mm day^{-1}.

TEC depends on crop characteristics and is mainly dependent on the crop photosynthetic pathway (C_3 versus C_4) and biochemical composition of plant tissues. C_4 plants and plants with a high proportion of carbohydrate in their tissues have higher values of TEC, and hence transpire less water per unit of produced dry matter. For more details about TEC and its impact on crop production under water-limited conditions refer to Tanner and Sinclair (1983) and Sinclair (1994, 2010). TEC for some field crops can be found in Table 14.1.

TEC is also sensitive to CO_2 concentration. This should be considered for simulation of crop production under elevated CO_2 concentration. Doubling CO_2 concentration generally results in 10–60% increase in TEC. For examples of the studies, refer to Sinclair and Rawlins (1993), Reyenga *et al.* (2001), and Ludwig and Asseng (2006).

Vapor pressure deficit is the difference or deficit between the partial pressure of water in the air when it is saturated and the actual partial pressure of water in the air. Weighted VPD for transpiration through the daily cycle is computed as a fraction (VPDF) of the difference between saturation vapor pressure calculated from maximum (VPTMAX) and minimum (VPTMIN) daily temperatures as outlined by Tanner and Sinclair (1983). VPDF commonly has values in the range of 0.65 to 0.75.

$$VPD = VPDF \times (VPTMAX - VPTMIN) \tag{14.21}$$

Saturation vapor pressure (e°(T), kPa) is related to temperature and can be found for any temperature (T, °C):

$$e^{o}(T) = 0.6108 \exp\left[\frac{17.27T}{T + 237.3}\right] \tag{14.22}$$

Direct measures of dewpoint temperature, dry and wet bulb temperatures, or relative humidity data allow direct computation of vapor pressure deficit. However, there is still a challenge in estimating the weighted VPD for transpiration over the daily cycle. It is likely that an estimate of VPDF will be needed in estimating effective VPD.

Since transpiration is a tightly linked consequence of stomata opening for photosynthesis, it is physiologically appropriate to have transpiration rate calculated from daily crop growth. The linkage between these two variables is illustrated in Fig. 14.8 for a range of daily VPDs.

When the crop rooting layer is thinner than the soil top layer all transpirational water is withdrawn from the top layer. However, when the thickness of the rooting layer is greater than the top layer a procedure is required to calculate the

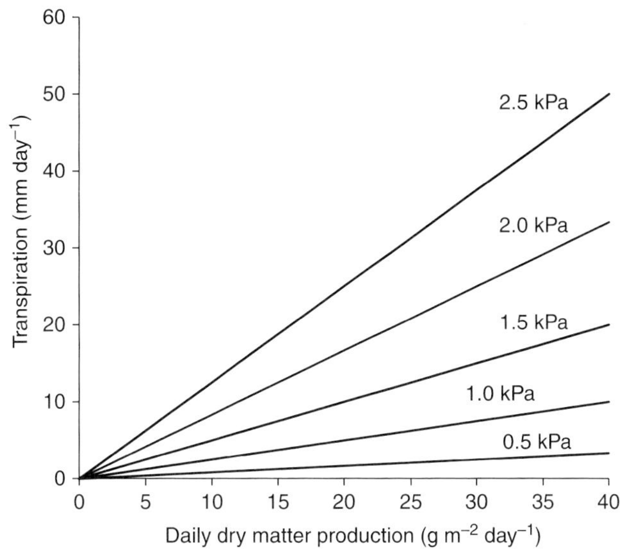

Fig. 14.8. Daily transpiration as a function of daily dry matter production as influenced by daily vapor pressure deficit.

fraction (RT1) of transpirational water that is withdrawn from the top layer. This value of RT1 depends on the soil water content of the top layer (FTSW1):

$$\begin{aligned} RT1 &= 1 && \textit{if} \quad FTSW1 > WSSG \\ RT1 &= FTSW1 / WSSG && \textit{if} \quad FTSW1 \leq WSSG \end{aligned} \qquad (14.23)$$

where WSSG is the threshold FTSW when crop transpiration starts to decline (discussed in Chapter 15). The amount of transpirational water withdrawn from the top layer (TR1) is then computed as the product of TR and the factor (RT1):

$$TR1 = TR \times RT1 \qquad (14.24)$$

Required Parameters and Inputs

Predicting soil water balance using the submodel of this chapter requires several soil inputs and crop parameters. Required soil inputs are: volumetric soil water contents at saturation (SAT), drained upper limit (DUL), volumetric extractable soil water content (EXTR), drainage factor (DRAINF), curve number (CN), and soil albedo (SALB). The depth of the top layer (DEP1) and soil depth (SOLDEP) need to be known. Estimates of these inputs for some default soils are presented in Tables 13.1 and 14.2. These estimates can be used if measured values are not available, especially for educational purposes.

Plant parameters that are needed are: the rate of root growth depth (GRTD), maximum effective rooting depth for water extraction (MEED), and transpiration efficiency coefficient (TEC). Estimates of these parameters are provided in Table 14.1 for some crops. These estimates can be employed if users do not have appropriate values for their crops or cultivars. Crop parameters that are required to model response of plant processes to water deficit and excess are discussed in Chapter 15.

There are a few other inputs/parameters that were fixed in the model. They are: extinction coefficient for total global solar radiation (KET, 0.5), minimum soil evaporation when the soil is covered by crop (EOSMIN, 1.5 mm day^{-1}), and amount of rainfall and/or irrigation required to wet the top soil layer to return soil evaporation from Stage II to Stage I (WETWAT, 10 mm).

Actual transpirable soil water that exists in both the top layer and the rooting layer at the initiation of the simulation is required. That is, what is the soil water content when the simulations are begun? To facilitate defining initial soil water, two variables are defined and need to be inputted by the user; they are called moisture availability indices (MAI and MAI1). These variables have values between 0 and 1. A value of 0 indicates that soil water is at the lower limit and in the same way a value of 1 indicates that soil water is at the drained upper limit. MAI is used for total soil profile and MAI1 for the top layer. The crop rooting layer at the time of emergence has a thickness not greater than the top layer, so MAI1 applies for root layer, too. Therefore, the initial actual transpirable soil water in the top layer (ATSW1) and root layer (ATSW) are estimated using the thickness of the top layer (DEP1) and the rooting depth at crop emergence (DEPORT) and MAI1:

$$\text{ATSW1} = \text{DEP1} \times \text{EXTR} \times \text{MAI1} \quad (14.25)$$

$$\text{ATSW} = \text{DEPORT} \times \text{EXTR} \times \text{MAI} \quad (14.26)$$

To account for the increasing availability of soil water as a result of root growth, it is also necessary to calculate the available soil water below the rooting layer (WSTORG). The value of WSTORG is dependent on the initial amount of water stored in the total soil depth (IPATSW). That is:

$$\text{IPATSW} = \text{SOLDEP} \times \text{EXTR} \times \text{MAI} \quad (14.27)$$

where SOLDEP is the soil depth, EXTR is the volumetric extractable soil water, and MAI is an input to the model.

The difference between IPATSW and ATSW is the initial available soil water stored below the rooting layer (WSTORG) for possible crop exploitation due to root growth during later stages. The initial value of WSTORG is also calculated as:

$$\text{WSTORG} = \text{IPATSW} - \text{ATSW} \quad (14.28)$$

Programming

Computer codes for the soil water submodel are prepared in VBA (Box 14.1). For a full list of names of variables and definitions used in the submodel refer to the text or Appendix III. In Chapter 16, this submodel will be integrated in a model of water-limited production.

Box 14.1. Program of soil water submodel as written in VBA in Excel. For the names of variables refer to the text or Appendix III.

```
SoilWater:
'------------------------------ Parameters and Initials
  If iniSW = 0 Then
      DEPORT = Sheet5.[B36]
      MEED = Sheet5.[B37]
      GRTDP = Sheet5.[B38]
      tuBRG = tuEMR
      tuTRG = tuBSG
      TEC = Sheet5.[B39]
      WSSG = Sheet5.[B40]
      WSSL = Sheet5.[B41]
      WSSD = Sheet5.[B42]

      SOLDEP = Sheet7.[b7]
      DEP1 = Sheet7.[b8]
      SALB = Sheet7.[b9]
      CN = Sheet7.[b10]
      DRAINF = Sheet7.[b11]
      SAT = Sheet7.[b12]
      DUL = Sheet7.[b13]
      EXTR = Sheet7.[b14]
      CLL = DUL - EXTR

      MAI1 = Sheet7.[b16]
      MAI = Sheet7.[b17]

      IPATSW = SOLDEP * EXTR * MAI

      ATSW = DEPORT * EXTR * MAI1
      TTSW = DEPORT * EXTR
      FTSW = ATSW / TTSW
      WSTORG = IPATSW - ATSW

      ATSW1 = DEP1 * EXTR * MAI1
      TTSW1 = DEP1 * EXTR
      FTSW1 = ATSW1 / TTSW1

      WLL1 = DEP1 * CLL
      WAT1 = WLL1 + ATSW1
      WSAT1 = DEP1 * SAT

      EOSMIN = 1.5:    WETWAT = 10:    KET = 0.5:      CALB = 0.23:

      DYSE = 1:        CTR = 0:        CE = 0:         CRAIN = 0:
      CRUNOF = 0:      CIRGW = 0:      IRGNO = 0:      iniSW = 1
  End If
'------------------------------ Irrigation
  If water = 1 And FTSW <= IRGLVL And CTU < tuTSG Then
    IRGW = (TTSW - ATSW)
    IRGNO = IRGNO + 1
```

Continued

Box 14.1. Continued.

```
  Else
    IRGW = 0
  End If

  CIRGW = CIRGW + IRGW

'------------------------------ Drainage
  If ATSW1 <= TTSW1 Then
    DRAIN1 = 0
  ElseIf ATSW1 > TTSW1 Then
    DRAIN1 = (ATSW1 – TTSW1) * DRAINF
  End If

  If ATSW <= TTSW Then
    DRAIN = 0
  ElseIf ATSW > TTSW Then
    DRAIN = (ATSW – TTSW) * DRAINF
  End If

  WSTORG = WSTORG + DRAIN – EWAT
  If WSTORG < 0 Then WSTORG = 0

'------------------------------ Water exploitation by root growth
  GRTD = GRTDP       'mm per day
  If CTU < tuBRG Then GRTD = 0
  If CTU > tuTRG Then GRTD = 0
  If DDMP = 0 Then GRTD = 0
  If DEPORT >= SOLDEP Then GRTD = 0
  If DEPORT >= MEED Then GRTD = 0
  If WSTORG = 0 Then GRTD = 0
  DEPORT = DEPORT + GRTD

  EWAT = GRTD * EXTR
  If EWAT > WSTORG Then EWAT = WSTORG

'------------------------------ Runoff
  RUNOF = 0
  If water = 2 And RAIN > 0.01 Then
    S = 254 * (100 / CN – 1)
    SWER = 0.15 * ((WSAT1 – WAT1) / (WSAT1 – WLL1))
    If SWER < 0 Then SWER = 0
    If (RAIN – SWER * S) > 0 Then
      RUNOF = (RAIN – SWER * S) ^ 2 / (RAIN + (1 – SWER) * S)
    Else
      RUNOF = 0
    End If
  End If

  If (WAT1 – DRAIN1) > WSAT1 Then
    RUNOF = RUNOF + (WAT1 – DRAIN1 – WSAT1)
  End If
```

Continued

Box 14.1. Continued.

```
  CRAIN = CRAIN + RAIN
  CRUNOF = CRUNOF + RUNOF
'------------------------------ LAI for soil evaporation
  If CTU <= tuBSG Then ETLAI = LAI Else ETLAI = BSGLAI
'------------------------------ Potential ET
  TD = 0.6 * TMAX + 0.4 * TMIN
  ALBEDO = CALB * (1 – Exp(–KET * ETLAI)) + SALB * Exp(–KET * ETLAI)
  EEQ = SRAD * (0.004876 – 0.004374 * ALBEDO) * (TD + 29)
  PET = EEQ * 1.1
  If TMAX > 34 Then PET = EEQ * ((TMAX – 34) * 0.05 + 1.1)
  If TMAX < 5 Then PET = EEQ * 0.01 * Exp(0.18 * (TMAX + 20))
'------------------------------ Soil evaporation
  EOS = PET * Exp(–KET * ETLAI)
  If PET > EOSMIN And EOS < EOSMIN Then EOS = EOSMIN

  SEVP = EOS
  If (RAIN + IRGW) > WETWAT Then DYSE = 1
  If ATSW1 < 1 or DYSE > 1 Or FTSW < 0.5 Then
    SEVP = EOS * ((DYSE + 1) ^ 0.5 – DYSE ^ 0.5)
    DYSE = DYSE + 1
  End If

  CE = CE + SEVP
'------------------------------ Plant transpiration
  VPTMIN = 0.6108 * Exp(17.27 * TMIN / (TMIN + 237.3))
  VPTMAX = 0.6108 * Exp(17.27 * TMAX / (TMAX + 237.3))
  VPD = VPDF * (VPTMAX – VPTMIN)
  TR = DDMP * VPD / TEC          'VPD in kPa, TEC in Pa
  If TR < 0 Then TR = 0

  CTR = CTR + TR

  If DEPORT <= DEP1 Then
    TR1 = TR
  ElseIf DEPORT > DEP1 Then
    If FTSW1 > WSSG Then RT1 = 1 Else RT1 = FTSW1 / WSSG
    TR1 = TR * RT1
  End If
'------------------------------ Updating
  ATSW1 = ATSW1 + RAIN + IRGW – DRAIN1 – RUNOF – TR1 – SEVP
  If ATSW1 < 0 Then ATSW1 = 0
  FTSW1 = ATSW1 / TTSW1
  WAT1 = WLL1 + ATSW1

  ATSW = ATSW + RAIN + IRGW + EWAT – DRAIN – RUNOF – TR – SEVP
  If ATSW < 0 Then ATSW = 0
  TTSW = DEPORT * EXTR
  FTSW = ATSW / TTSW
```

Continued

Box 14.1. Continued.

```
'------------------------------ Water-stress-factors
  If FTSW > WSSL Then WSFL = 1 Else WSFL = FTSW / WSSL
  If FTSW > WSSG Then WSFG = 1 Else WSFG = FTSW / WSSG
  WSFD = (1 – WSFG) * WSSD + 1

  If WAT1 > (0.95 * WSAT1) Then
    WSFG = 0:   WSFL = 0:   WSFD = 0
  End If
Return
```

Additional Notes

Priestley and Taylor method for calculation of potential evaporation

A simplified Penman equation in combination with the exponential Beer–Bouguer–Lambert equation (to account for fraction uncovered soil) was used to calculate potential evaporation from bare, wet soil surface. Another method that is commonly used is the Priestley and Taylor method (Priestley and Taylor, 1972) as modified and described by Ritchie (1998). This method also needs daily maximum and minimum temperature and solar radiation. Potential evaporation (EOS) from bare, wet soil surface is obtained from potential evapotranspiration (PET, mm day^{-1}) and the fraction of soil that is not covered by the crop. As mentioned before, the fraction of uncovered soil is calculated using ETLAI and KET based on exponential Beer–Bouguer–Lambert equation.

$$\text{EOS} = \text{PET} \times \text{EXP}(-\text{KET} \times \text{ETLAI}) \qquad (14.29)$$

Potential evapotranspiration (PET, mm day^{-1}) is calculated as the equilibrium evaporation (EEQ, mm day^{-1}) multiplied by 1.1 to account for the effect of unsaturated air. The multiplier is increased above 1.1 to allow for advection (TMAX > 34) and is reduced to account for the influence of frozen soil on evaporation and cold temperatures on stomatal closure (TMAX < 5) when necessary (Fig. 14.9).

$$\begin{aligned}
\text{PET} &= \text{EEQ} \times 1.1 && \textit{if} \quad 5 \le \text{TMAX} \le 34 \\
\text{PET} &= \text{EEQ}\ ((\text{TMAX} - 34) \times 0.05 + 1.1) && \textit{if} \quad \text{TMAX} > 34 \\
\text{PET} &= \text{EEQ} \times 0.01 \times \text{EXP}\ (0.18 \times (\text{TMAX} + 20)) && \textit{if} \quad \text{TMAX} < 5 \qquad (14.30)
\end{aligned}$$

EEQ is obtained from surface (crop plus soil) albedo (ALBEDO), average air temperature during the day (TD, °C), and daily solar radiation (SRAD, MJ m^{-2} day^{-1}):

$$\text{EEQ} = \text{SRAD} \times (0.004876 - 0.004374 \times \text{ALBEDO}) \times (\text{TD} + 29) \qquad (14.31)$$

TD is computed using a higher weight for daily maximum temperature (TMAX, °C) and a lower weight for daily minimum temperature (TMIN, °C):

$$\text{TD} = 0.6 \times \text{TMAX} + 0.4 \times \text{TMIN} \qquad (14.32)$$

Surface albedo (ALBEDO) depends on the proportion of the field surface that is covered by crop or soil and the albedos of the crop (CALB) and the soil (SALB). CALB is fairly constant at a value of 0.23.

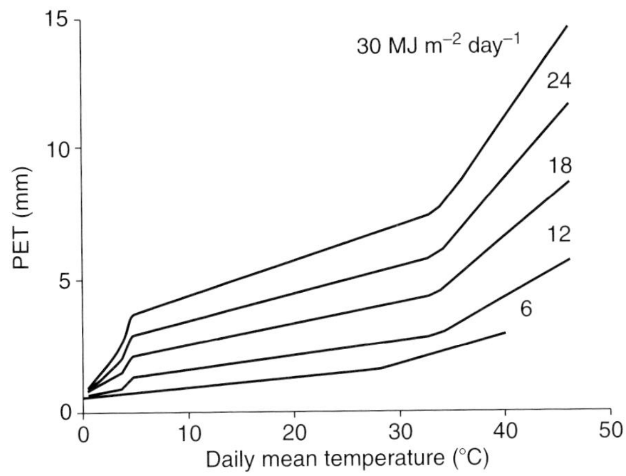

Fig. 14.9. Potential evapotranspiration (PET, mm) as a function of mean daily temperature and daily solar radiation (MJ m^{-2} day^{-1}) calculated using the Priestley and Taylor method.

Accounting for the effect of soil water content on runoff

In Eqn 14.13, a constant value of 0.2 was used in the calculation of RUNOF. However, this term can be allowed to vary as a function of the total soil water in the top soil layer (WAT1, mm). The value of WAT1 is the sum of available soil water (ATSW1) and water unavailable for direct use by plants. The unavailable soil water WLL1 is equivalent to the thickness of the top layer (DEP1, mm) and volumetric soil water content at the lower limit (LL, mm^3 mm^{-3}):

$$\text{WLL1} = \text{DEP1} \times \text{LL} \tag{14.33}$$

Total soil water is then calculated as:

$$\text{WAT1} = \text{WLL1} + \text{ATSW1} \tag{14.34}$$

Having defined WLL1 and WAT1 it is now possible to replace the 0.2 constant by a coefficient (SWER), that is dependent on soil water content. The value of SWER is estimated from the following equation (Ritchie, 1998):

$$\text{SWER} = 0.15 \times ((\text{WSAT1} - \text{WAT1}) / (\text{WSAT1} - \text{WLL1})) \tag{14.35}$$

Therefore, SWER is adjusted for the soil water in the layer between the lower limit (WLL1, mm) and saturation (WSAT1, mm).

Crop cover and runoff

Crop cover can result in decreased runoff. Chapman *et al.* (1993) used a simple method to simulate this effect. In this method, first crop cover (COVER, %) is obtained based on Beer–Bouguer–Lambert Law (Chapter 10; Eq. 10.2) using ETLAI and KET.

$$COVER = (1 - EXP(-KSEVP \times ETLAI)) \times 100 \quad (14.36)$$

Then, the curve number is reduced proportional to crop cover but the reduction of CN is limited to 20:

$$CNC = CN - \min(COVER \times 0.25, 20) \quad (14.37)$$

where CNC is CN number adjusted (reduced) for the crop cover.

Mulch straw and soil water balance

The simple soil water of this chapter can be extended to include the effect of mulch residue as a result of no-tillage management leaving straw on the soil surface. Straw mulch can directly affect runoff and soil evaporation. O'Leary and Connor (1996) used simple methods to consider these effects, which are used here, too.

Equations 14.36 and 14.37 can be used to explain the effect of straw mulch on runoff, if the fraction of the soil surface covered by straw mulch (SMCVR, %) is available. If not available, SMCVR is obtainable from stubble weight (STBLW, t ha^{-1}) (Fig. 14.10):

$$SMCVR = (1 - \exp(-0.8 \times STBLW)) \times 100 \quad (14.38)$$

Soil coverage due to straw mulch results in lower potential soil evaporation (Fig. 14.11). This effect can be modeled by reducing potential soil evaporation as a function of stubble weight. The reduced potential evaporation (EOSM, mm day^{-1}) is obtained as:

$$EOSM = EOS \times (1.5 - 0.2 \times LN(100 \times STBLW)) \quad (14.39)$$

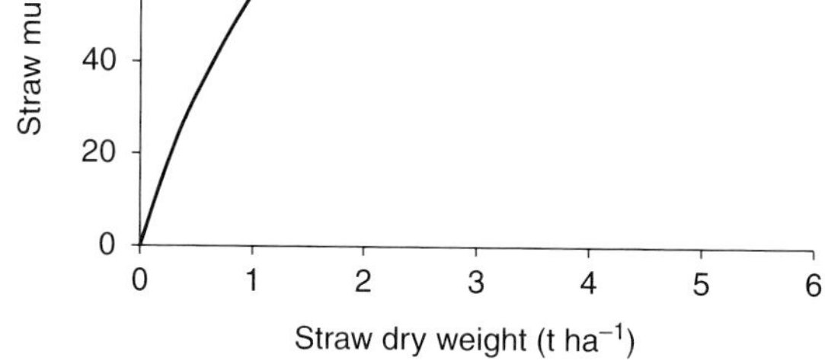

Fig. 14.10. Straw mulch cover as a function of straw dry weight.

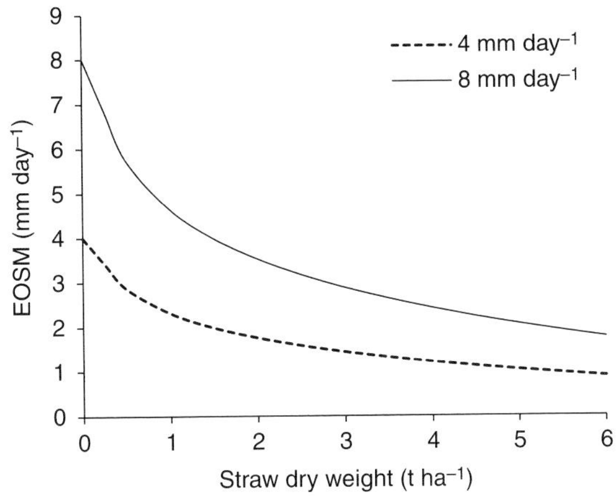

Fig. 14.11. Potential soil evaporation as influenced by straw mulch (EOSM) as a function of straw dry weight for two values of potential evaporation of 4 and 8 mm day^{-1}.

Exercises

1. Try to gather or obtain soil inputs and crop parameters required for calculation of soil water balance for your area and crops. Compare these values with those presented in Tables 13.1, 14.1, and 14.2.

2. Complete the table below by calculating potential evapotranspiration. Units and definitions are the same as presented in the text.

TMAX	TMIN	SRAD	EEQ	PET
0	–20	5		
10	–5	8		
20	12	16		
30	10	24		
40	25	33		
50	30	36		

15 Plant Responses to Soil Water Deficit and Excess

Modeling plant responses to soil water deficit and excess is challenging because nearly all of the physiological processes discussed in Chapters 6 to 11 – phenology, leaf area development and dry matter production and distribution – are involved. Therefore, crop models dealing with soil water deficits and excess require adjustments in many of the relationships established previously. Often simple stress factors are included to quantify the effect of water stress on plant processes. Usually several stress factors are calculated for different processes that have different sensitivities to water deficit. These stress factors are scalar factors that vary between 0 and 1 and adjust the rate of the processes for the effect of water deficit. A factor of 1 means no water deficit effect and a factor of 0 means a complete stop in the process. The stress factors are computed as a function of the degree of water deficit.

Defining Crop Stress Level

A critical issue is to quantify the stress level to which the crop is being subjected. Progress in developing quantitative response functions to soil water deficits was initially slow because many studies have attempted to characterize water deficits with thermodynamic variables (Sinclair and Ludlow, 1985). These thermodynamic variables have not related directly to physiological processes such as leaf gas exchange or leaf expansion. Later, it was discovered that many physiological responses could be well described by conservative functions based on available soil water (e.g. Ritchie, 1981; Lecoeur and Sinclair, 1996; Ray and Sinclair, 1997).

Saseendran *et al.* (2008) reviewed the simulation of plant water stress and its integration with crop growth and development processes in many agricultural models. They concluded that the models, in general, used the ratio of actual to potential transpiration or evapotranspiration to represent the degree of water deficit stress. For example, CropSyst model (Stockle *et al.*, 2003) uses the ratio of actual to potential transpiration. DSSAT models (Jones *et al.*, 2003) use the ratio of potential uptake to potential transpiration. APSIM (Keating *et al.*, 2003) utilizes both the ratio of supply to demand for soil water and fraction available (transpirable) soil water.

In this book, water stress factors are calculated as functions of the fraction of transpirable soil water (FTSW) that remains in the soil on each day. The calculation of the water balance in Chapter 14 tracks the changes in the actual transpirable soil water (ATSW). FTSW is calculated as the ratio of ATSW to the total transpirable soil water that can be stored in each of the two soil layers (TTSW). TTSW for any soil depth of interest was previously defined in Chapter 13 as the difference between the soil water content at drained upper limit (DUL) and the soil water content at crop lower limit (LL) when transpiration of the drought-stressed plants decreased to 10% or less of that of well-watered plants.

As indicated in Chapter 14, total transpirable soil water in the root zone layer (TTSW, mm) is obtained each day as the product of current crop root depth (DEPORT, mm) and EXTR:

$$\text{TTSW} = \text{DEPORT} \times \text{EXTR} \qquad (15.1)$$

Fraction transpirable soil water in the root layer (FTSW) is computed from ATSW and TTSW in the root layer.

$$\text{FTSW} = \text{ATSW} / \text{TTSW} \qquad (15.2)$$

Sinclair and Ludlow (1985) outlined how physiological processes to water deficit could be expressed as functions of FTSW. Simple so-called dry-down experiments can be used to quantify the response of different plant processes to FTSW (e.g. Ray and Sinclair, 1997).

Dry-down Experiments

In dry-down experiments, plants are first grown in pots under well-watered conditions until there is sufficient leaf area to result in readily measurable variation in plant transpiration rates, i.e. five to ten leaves on main stem, depending on crop species. The pots are then sealed to prevent water loss from the soil so that all water loss is via transpiration. The pots are divided between two watering regimes: a well-watered control and a water-deficit regime. Water deficits are imposed by simply not rewatering the pots. The plants are weighed daily to track changes in transpiration rate and to calculate soil water content. Other processes of interest such as leaf expansion and symbiotic nitrogen fixation rates are also measured daily during the soil dry down.

At the end of the experiment, relative or normalized rate of the plant process in the stressed plants is calculated and evaluated versus FTSW calculated on each day of the dry down for each pot. Figure 15.1 shows normalized (relative) transpiration rate in sorghum as a function of FTSW (Gholipoor *et al.*, 2012). The response of normalized plant process rate to FTSW can be described using a 2-segment linear regression; one sloping line that describes the decrease in the process with soil drying and a horizontal line that represents maximum rate of the process (=1). Where these lines intersect is the FTSW threshold at which the rate of the process in stressed plants starts to decline. The threshold is a crop parameter that is needed to model the response of the process to soil drying. As illustrated in Fig. 15.1, transpiration rate (dry matter production) usually remains unaffected until FTSW is less than 0.3 to 0.4. After that transpiration rate decreases linearly with decrease in FTSW and by definition reaches zero when FTSW=0. For the objective of this book, threshold FTSWs are inputted to the model for leaf area development and transpiration (dry matter production). Threshold FTSW for nitrogen fixation in legume crops will be needed for nitrogen-limited situations (Part IV of this book). For specific examples of dry-down experiments refer to Lecoeur and Sinclair (1996), Serraj *et al.* (1999), and Devi *et al.* (2009).

The use of FTSW has led to fairly consistent response functions to soil dehydration across a range of conditions. Transpiration was shown to be unaffected by soil drying until FTSW decreased to 0.25 to 0.35 in several grain legumes (Sinclair and Ludlow, 1986). In addition, a similar response pattern has been found for leaf area development (e.g. Sinclair, 1986; Sinclair *et al.*, 1987; Muchow and Sinclair, 1991; Soltani *et al.*, 1999). In these cases, leaf area expansion started to decline at about the same or slightly higher FTSW than transpiration. Sadras and Milroy (1996) reviewed literature for the FTSW threshold and stated that average threshold was 0.56 for tissue expansion and 0.4 for gas exchange using different experimental procedures. Table 15.1 includes threshold FTSW for leaf area expansion and transpiration in some field crops.

Sadras and Milroy (1996) concluded that the threshold FTSW can be affected by evaporative demand, root distribution, and soil texture and soil bulk density. Ray and Sinclair (1997) examined the effect of pot size in soybean and maize and indicated that regardless of pot size, the overriding factor determining transpirational response to drought stress was soil water content.

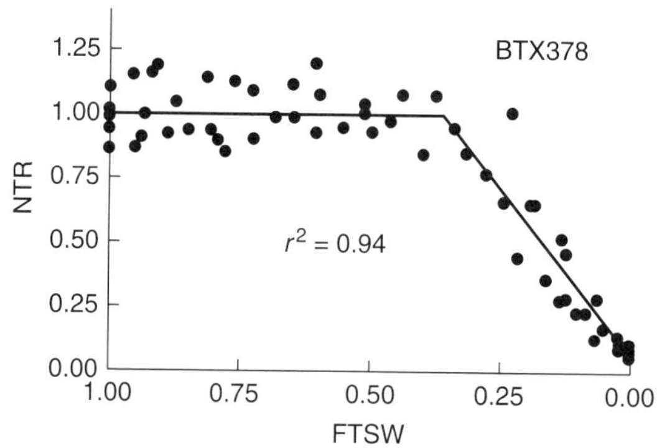

Fig. 15.1. Normalized transpiration rate (NTR) versus fraction of transpirable soil water (FTSW) for sorghum genotype BTX378 (Gholipoor *et al.*, 2012). The solid line is the result of the 2-segment linear regression with a FTSW threshold of 0.364.

Table 15.1. Threshold FTSW for leaf area development (WSSL) and growth (WSSG), and a coefficient of phenological development response to drought (WSSD) in some grain crops (Sinclair, 1986; Amir and Sinclair, 1991; Hammer *et al.*, 1995; Sadras and Milroy, 1996; Soltani *et al.*, 1999).

Crop	WSSL	WSSG	WSSD
Wheat	0.40	0.30	0.40
Barley	0.40	0.30	
Rice	0.60	0.60	
Maize	0.35	0.25	
Sorghum	0.35	0.25	
Soybean	0.31	0..25	
Peanut	0.35	0.35	
Canola	0.40	0.30	
Sunflower	0.35	0.35	
Dry bean	0.40	0.30	
Chickpea	0.40	0.30	0.40

Sinclair (2005) provided a theoretical basis for explaining the consistency in response of relative transpiration to FTSW observed over a wide range of conditions. Predicted response of relative transpiration using the derivation was consistent with experimental derivation. In the theoretical analysis, there was a broad range over which there was only a small change in relative transpiration when the soil was relatively wet. The decrease in relative transpiration became much greater when the soil dried to FTSW values of about one-third. The threshold FTSW was the soil water content where soil water potential started to decrease substantially. Further decease in soil water resulted in greater decreases in soil water potential and relative transpiration. The results of the analysis further indicated that the response was, to a large extent, independent of root length density, leaf water potential, transpiration rate, and soil depth. The response of relative transpiration to soil drying as a function of FTSW was also found to be nearly independent of soil texture.

Modeling Effects of Water Deficit

The impact of a water-deficit stress factor (WSF) is considered in four processes:

1. Growth, or specifically transpiration/dry matter accumulation (WSFG).
2. Leaf area development (WSFL).
3. Phenological development (WSFD).
4. Nitrogen accumulation (WSFN), discussed in Chapter 17.

Growth and transpiration

As shown in Figs 15.1 and 15.2, the water-deficit stress factor for growth (WSFG) is equal to 1 when FTSW is higher than the FTSW threshold of

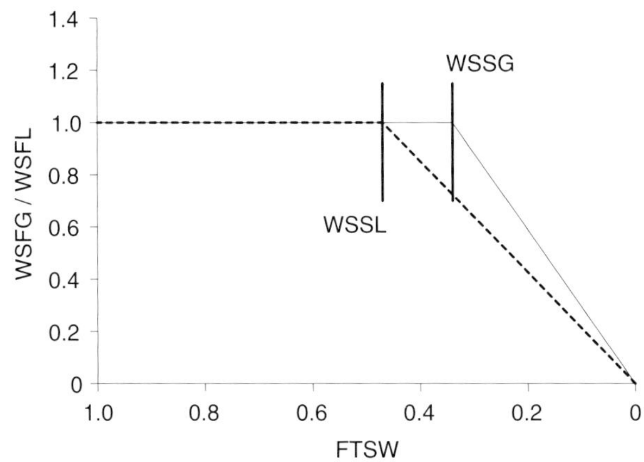

Fig. 15.2. Water-deficit stress factor for leaf area development (WSFL) and dry matter production (WSFG) as a function of fraction transpirable soil water (FTSW).

WSSG. Decreases in FTSW below the threshold result in a linear decrease in WSFG from 1 and reach 0 at FTSW = 0.

$$\begin{aligned} &WSFG = 1 && \textit{if} \quad FTSW \geq WSSG \\ &WSFG = FTSW / WSSG && \textit{if} \quad FTSW < WSSG \end{aligned} \qquad (15.3)$$

Therefore, the daily growth is decreased as a result of a WSFG value less than 1. This calculation is handled daily by multiplying radiation use efficiency (RUE, Eqn 10.4, Chapter 10) by WSFG.

$$RUE = RUE \times WSFG \qquad (15.4)$$

The decrease in RUE results directly in a decrease in daily dry matter production and hence total crop transpiration rate as described in Chapter 14 (Eqn 14.20).

Leaf area

The water-deficit stress factor for leaf area development (WSFL) is calculated in a manner very similar to that used for growth. WSFL is equal to 1 when FTSW is higher than threshold FTSW for leaf development (WSSL). However, with further decrease in FTSW below the threshold, the stress factor decreases linearly from 1 and reaches 0 at FTSW = 0.

$$\begin{aligned} &WSFL = 1 && \textit{if} \quad FTSW \geq WSSL \\ &WSFL = FTSW / WSSL && \textit{if} \quad FTSW < WSSL \end{aligned} \qquad (15.5)$$

Figure 15.2 shows how WSFL is changed with FTSW as described by Eqn 15.5.

Application of WSFL in calculating leaf area development is more complicated than growth. Daily increase in leaf area can be calculated based on daily increase in the number of leaves/nodes on the main stem (Eqn 9.1, Chapter 9) or daily increase in leaf area index (Eqn 9.5, Chapter 9). Each of these can be adjusted for development of water deficit in the soil.

$$\text{INODE} = \text{INODE} \times \text{WSFL} \tag{15.6}$$

$$\text{GLAI} = \text{GLAI} \times \text{WSFL} \tag{15.7}$$

Selection of the responsive rate variable to use in the model depends on their comparative sensitivity to water deficit. Some species respond to water deficit by a slower rate of leaf appearance, but some others keep a constant rate of leaf appearance and decrease leaf expansion and leaf size. For the first group, Eqn 15.6 must be applied and for the second group, Eqn 15.7 is used.

Development

Water deficit may hasten or delay phenological development. This depends on crop species. McMaster *et al.* (2009) reviewed the effect of drought on phenological development (Table 15.2). From their review it seems that acceleration of development rates is more common. They also presented several hypotheses to justify the effect of water deficit on phenological development rate. Amir and Sinclair (1991) simulated the effect of water deficit on phenological development in wheat by adding 6°C to calculated daily temperature unit (Chapter 6) on each day when FTSW was less than 0.2.

Here, water stress factor for phenological development (WSFD) is related to FTSW via WSFG (Soltani *et al.*, 1999). It is assumed that when water is not limited for

Table 15.2. The effect of water deficit on phenological development (McMaster *et al.*, 2009). Flower initiation is the appearance of the inflorescence/flower primordium, flowering is the appearance of flower or anthesis, duration of grain filling is from pollination to physiological maturity, and physiological maturity is when maximum seed weight is attained. Symbols – and + indicate later and earlier occurrence of the event under water deficits. Symbol 0 indicates no response to water deficit and question marks indicate conflicting or uncertain responses.

Crop	Flower initiation	Flowering	Duration of grain filling	Physiological maturity
Wheat	0?	+	+	+
Barley	0?	+	+	+
Maize		–	+	+
Sorghum		–	+	+
Soybean			+	+
Peanut		–	+	+
Sunflower	0		+	+
Dry bean	0	0/+	+	+
Chickpea		+	+	+

dry matter production and stomata are opened (FTSW > WSSG), development is not influenced by water deficit. However, with development of soil deficit and closing of stomata, phenological development is hastened. The maximum expected ratio of development in water-stressed plant to well-watered plant will be required for calculation of WSFD. This ratio is indicated by WSSD.

$$\text{WSFD} = (1 - \text{WSFG}) \times \text{WSSD} + 1 \qquad (15.8)$$

According to Eqn 15.8, maximum increase in development rate will be (WSSD × 100)%. The ratio of leaf temperature in a water-stressed plant to a well-watered plant might be a measure of WSSD.

For example, if WSSD is 0.4, the maximum value of WSFD at WSFG = 0 is equal to 1.4 (= 1 × 0.4 + 1). The variation in WSFD over the range of FTSW is illustrated in Fig. 15.3. Acceleration in development rate is incorporated by multiplying WSFD by daily temperature unit calculated in each day after emergence.

$$\text{DTU} = \text{DTU} \times \text{WSFD} \qquad (15.9)$$

For those crops in which development rates are decreased due to water deficit, Eqn 15.8 can still be used with negative values for WSSD. For example, if WSSD is −0.4, then WSFD will be 0.6 when FTSW and hence WSFG reach 0. That is, these values define a maximum 40% decrease in phenological development due to drought. However, the scientific basis and a procedure to measure WSSD need to be sought. Table 15.3 includes sample calculations of WSFD.

Flooding Effects

Flooding is a situation when soil water content is close to or at saturation content. Heavy rains or irrigations may result in flooding, especially if soil drainage is poor or there is a restricting subsurface layer in the soil. Furthermore, a high water table close to the soil surface may cause flooding. Flooding can impede different physiological processes in plants. However, quantitative data for the effect of flooding based on experimental observations are scarce.

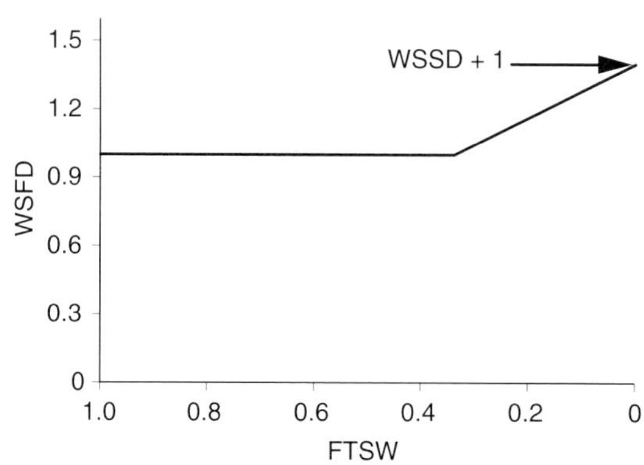

Fig. 15.3. Water stress factor for phenological development (WSFD) as a function of fraction transpirable soil water (FTSW).

Table 15.3. Sample calculation of water stress factor for phenological development (WSFD) as a function of water stress factor for growth (WSDG) with two values of WSSD of 0.4 or –0.4.

WSFG	WSFD with WSSD = 0.4	WSFD with WSSD = –0.4
1.0	(1 – 1.0) × 0.4 +1 = 1.00	(1 – 1.0) × –0.4 +1 = 1.00
0.7	(1 – 0.7) × 0.4 +1 = 1.12	(1 – 0.7) × –0.4 +1 = 0.88
0.5	(1 – 0.5) × 0.4 +1 = 1.20	(1 – 0.5) × –0.4 +1 = 0.80
0.2	(1 – 0.2) × 0.4 +1 = 1.32	(1 – 0.2) × –0.4 +1 = 0.68
0.0	(1 – 0.0) × 0.4 +1 = 1.40	(1 – 0.0) × –0.4 +1 = 0.60

Therefore, a very simple procedure is used here. On any day when soil water in the root layer is 95% of soil water at saturation, all water stress factors (WSFL, WSFG, WSFD, and WSFN) are set to 0.

Crop Termination Due to Water-deficit Stress

A combination of very low FTSW and high vapor pressure deficit can result in early crop termination. This phenomenon is important for crop simulation under water deficit (Sinclair and Muchow, 2001). Sinclair and Amir (1996) in their wheat simulation model included crop termination due to water deficit. They assumed that various combinations of FTSW, VPD, and duration resulted in crop termination (Table 15.4). Similar combinations can be applied for other crops. The model for water-limited conditions of this book (Chapter 16) does not include crop termination but this can be included easily if needed.

Similarly, crop termination due to flooding can be modeled. For example, crop terminates after given consecutive number of days of saturated soil water.

Table 15.4. Combinations of very low fraction transpirable soil water (FTSW) and high vapor pressure deficit (VPD), which result in crop termination due to severe drought (Sinclair and Amir, 1996).

FTSW	VPD (kPa)	Duration (day)
< 0.10	> 2.20	3
< 0.02	> 2.20	1
< 0.00	> 1.75	1

Exercise

1. A crop has threshold FTSWs of 0.3 for WSSG and 0.4 for WSSL. Calculate water stress factors (WSFG and WSFL) for the crop if FTSW in the soil is (a) 0.1, (b) 0.2, (c) 0.3, and (d) 0.5.

16 A Model for Water-limited Conditions

In Chapter 14, the elements of a soil water submodel were developed. To undertake simulations for water-limited conditions, this submodel needs to be integrated with the model of potential production as presented in Chapter 12. For this objective, simulation of different crop processes including phenology, crop leaf area, and dry matter production need to be adjusted for possible water deficit/excess stress as discussed in Chapter 15.

Model Structure

Model structure is the same as the potential production model (Chapter 12). The model is written in Visual Basic for Application (VBA) in Excel. The entire model is a subroutine (macro) in Excel. It includes a main part and submodels. The submodels are called by the main part as needed. That is, submodels are subroutines within Excel's macro subroutine structure (Box 16.1; please see Chapter 12).

The three water-stress factors as described in Chapter 15 are added to each of three key crop submodels:

- *"Phenology"* as developed in Chapter 6;
- *"CropLAI"* as developed in Chapter 9; and
- *"DMProduction"* as developed in Chapter 10.

In addition to simulating the plant responses to water-limited conditions, it is necessary to track dynamic changes in soil water content. Therefore, an additional soil water submodel is included using the relationships described in Chapter 14. To facilitate the application of both potential-production and water-limited models, an auxiliary variable was added named *"water"* that allows combining of the two models. Thus, it is not necessary to run two separate models to compare potential and water-limited conditions. The variable *"water"* can have one of the three values below:

Box 16.1. Main part of the crop model. Submodels are called by the main part when necessary.

```
'--------------------------- Main program
  GoSub ManagInputs
  GoSub InitialsHeaders
  GoSub FindSowingDate
  Do Until MAT = 1
    GoSub Weather
    GoSub Phenology
    GoSub CropLAI
    GoSub DMProduction
    GoSub DMDistribution
    If water = 1 Or water = 2 Then GoSub SoilWater
    GoSub DailyPrintOut
  Loop
  GoSub SummaryPrintOut
  Exit Sub
'--------------------------- End of main program
```

0 For potential production situation so that soil water submodel is bypassed.
1 For water-limited conditions with irrigation application.
2 For water-limited conditions without irrigation application (rainfed).

The flow diagram of the water-limited model and the sequences of events are the same as presented in Fig. 12.2 for the potential production model, except that the soil water submodel is called daily if "*water*" has a value of 1 or 2 (Box 16.1). For more information refer to Chapter 12.

Box 16.2 includes the model program to simulate water-limited conditions. Parameter estimates in this program belong to a wheat cultivar (cv. Tajan). The complete list of the variables can be found in Appendix III.

Structure of Excel File Containing the Model

Similar to the potential production model (Chapter 12), the Excel file containing the model has several sheets and a module (model program macro). The program uses Excel's sheets for inputs and outputs as indicated in Fig. 12.3 in Chapter 12.

The sheets in the Excel file are again the same as in the potential production model, except that a new sheet has been added. This new sheet is called "Soils" and includes soil inputs required to simulate water-limited conditions (Fig. 16.1). In addition, changes have been made in the "Run", "Outputs" and "Figures" sheets that were necessary in the model to switch from simulations of potential production to water-limited conditions.

Box 16.2. Program to simulate crop development, growth, and yield under water-limited conditions. The model can be requested from the authors or can be downloaded from the book's website (https://sites.google.com/site/CropModeling).

```
Sub wlm()
'-------------------------- A simple crop model for water-limited conditions
'-------------------------- Main program
  GoSub ManagInputs
  GoSub InitialsHeaders
  GoSub FindSowingDate
  Do Until MAT = 1
    GoSub Weather
    GoSub Phenology
    GoSub CropLAI
    GoSub DMProduction
    GoSub DMDistribution
    If water = 1 Or water = 2 Then GoSub SoilWater
    GoSub DailyPrintOut
  Loop
  GoSub SummaryPrintOut
  Exit Sub
'-------------------------- End of main program

ManagInputs:
  pyear = Sheet1.[b7]
  pdoy = Sheet1.[b8]
  PDEN = Sheet1.[b9]

  water = Sheet1.[b11]
  VPDF = Sheet1.[b12]
  IRGLVL = Sheet1.[b13]
Return

Weather:
  Row = Row + 1
  Yr = Sheet2.Range("A" & Row)
  DOY = Sheet2.Range("B" & Row)
  SRAD = Sheet2.Range("C" & Row)
  TMAX = Sheet2.Range("D" & Row)
  TMIN = Sheet2.Range("E" & Row)
  RAIN = Sheet2.Range("F" & Row)
  TMP = (TMAX + TMIN) / 2
Return

Phenology:
'-------------------------- Parameters and Initials
  If iniPheno = 0 Then
    TBD = Sheet5.[b7]
    TP1D = Sheet5.[b8]
    TP2D = Sheet5.[b9]
    TCD = Sheet5.[b10]
```

Continued

Box 16.2. Continued.

```
    tuSOWEMR = Sheet5.[b11]
    tuEMRTLM = Sheet5.[b12]
    tuTLMBSG = Sheet5.[b13]
    tuBSGTSG = Sheet5.[b14]
    tuTSGMAT = Sheet5.[b15]

    tuEMR = tuSOWEMR
    tuTLM = tuEMR + tuEMRTLM
    tuBSG = tuTLM + tuTLMBSG
    tuTSG = tuBSG + tuBSGTSG
    tuMAT = tuTSG + tuTSGMAT

    DAP = 0:   CTU = 0:   WSFD = 1:
    iniPheno = 1
End If
'--------------------------- Temperature unit calculation
  If TMP <= TBD Or TMP >= TCD Then
    tempfun = 0
  ElseIf TMP > TBD And TMP < TP1D Then
    tempfun = (TMP – TBD) / (TP1D – TBD)
  ElseIf TMP > TP2D And TMP < TCD Then
    tempfun = (TCD – TMP) / (TCD – TP2D)
  ElseIf TMP >= TP1D And TMP <= TP2D Then
    tempfun = 1
  End If

  DTU = (TP1D – TBD) * tempfun
  If CTU > tuEMR Then DTU = DTU * WSFD
  CTU = CTU + DTU
  DAP = DAP + 1

  If CTU < tuEMR Then DTEMR = DAP + 1 'Saving days to EMR
  If CTU < tuTLM Then DTTLM = DAP + 1 'Saving days to TLM
  If CTU < tuBSG Then DTBSG = DAP + 1 'Saving days to BSG
  If CTU < tuTSG Then DTTSG = DAP + 1 'Saving days to TSG
  If CTU < tuMAT Then DTMAT = DAP + 1 'Saving days to MAT

  If CTU > tuMAT Then MAT = 1
Return

CropLAI:
'--------------------------- LAI initials and pars
  If iniLAI = 0 Then
    PHYL = Sheet5.[b17]
    PLACON = Sheet5.[b18]
    PLAPOW = Sheet5.[b19]
    SLA = Sheet5.[b20]
    MSNN = 1:   PLA2 = 0:   PLA1 = 0:   LAI = 0:
    MXLAI = 0:   WSFL = 1:   iniLAI = 1
  End If
```

Continued

Box 16.2. Continued.

```
'--------------------------- Yesterday LAI to intercept PAR today
  LAI = LAI + GLAI – DLAI
  If LAI < 0 Then LAI = 0
  If LAI > MXLAI Then MXLAI = LAI          'Saving maximum LAI
'--------------------------- Daily increase and decrease in LAI today
  If CTU <= tuEMR Then
    GLAI = 0:     DLAI = 0
  ElseIf CTU > tuEMR And CTU <= tuTLM Then
    INODE = DTU / PHYL
    MSNN = MSNN + INODE
    PLA2 = PLACON * MSNN ^ PLAPOW
    GLAI = ((PLA2 – PLA1) * PDEN / 10000) * WSFL
    PLA1 = PLA2
    DLAI = 0
  ElseIf CTU > tuTLM And CTU <= tuBSG Then
    GLAI = GLF * SLA
    BSGLAI = LAI          'Saving LAI at BSG
    DLAI = 0
  ElseIf CTU > tuBSG Then
    GLAI = 0
    DLAI = DTU / (tuMAT – tuBSG) * BSGLAI
  End If
Return
DMProduction:
'--------------------------- Parameters and Initials
  If iniDMP = 0 Then
    TBRUE = Sheet5.[b22]
    TP1RUE = Sheet5.[b23]
    TP2RUE = Sheet5.[b24]
    TCRUE = Sheet5.[b25]
    KPAR = Sheet5.[b26]
    IRUE = Sheet5.[b27]
    WSFG = 1:   iniDMP = 1:
  End If
'--------------------------- Adjustment of RUE
  If TMP <= TBRUE Or TMP >= TCRUE Then
    TCFRUE = 0
  ElseIf TMP > TBRUE And TMP < TP1RUE Then
    TCFRUE = (TMP – TBRUE) / (TP1RUE – TBRUE)
  ElseIf TMP > TP2RUE And TMP < TCRUE Then
    TCFRUE = (TCRUE – TMP) / (TCRUE – TP2RUE)
  ElseIf TMP >= TP1RUE And TMP <= TP2RUE Then
    TCFRUE = 1
  End If
  RUE = IRUE * TCFRUE * WSFG
```

Continued

Box 16.2. Continued.

```
'--------------------------- Daily dry matter production
  FINT = 1 – Exp(–KPAR * LAI)
  DDMP = SRAD * 0.48 * FINT * RUE
Return

DMDistribution:
'--------------------------- Parameters and Initials
  If iniDMD = 0 Then
    FLF1A = Sheet5.[b29]
    FLF1B = Sheet5.[b30]
    WTOPL = Sheet5.[b31]
    FLF2 = Sheet5.[b32]
    FRTRL = Sheet5.[b33]
    GCC = Sheet5.[b34]

    WLF = 0.5:   WST = 0.5:   WVEG = WLF + WST:
    WGRN = 0:   iniDMD = 1:
  End If

'--------------------------- Biomass partitioning and yield formation
  If CTU <= tuEMR Or CTU > tuTSG Then
    DDMP = 0:   GLF = 0:   GST = 0:   TRANSL = 0:   SGR = 0
  ElseIf CTU > tuEMR And CTU <= tuTLM Then
    If WTOP < WTOPL Then FLF1 = FLF1A Else FLF1 = FLF1B
    GLF = FLF1 * DDMP
    GST = DDMP – GLF
    SGR = 0
  ElseIf CTU > tuTLM And CTU <= tuBSG Then
    GLF = FLF2 * DDMP
    GST = DDMP – GLF
    SGR = 0
    BSGDM = WTOP      'Saving WTOP at BSG
  ElseIf CTU > tuBSG And CTU <= tuTSG Then
    GLF = 0:
    GST = 0:
    TRLDM = BSGDM * FRTRL
    TRANSL = DTU / (tuTSG – tuBSG) * TRLDM
    SGR = (DDMP + TRANSL) * GCC
  End If

  WLF = WLF + GLF
  WST = WST + GST
  WGRN = WGRN + SGR
  WVEG = WVEG + DDMP – (SGR / GCC)
  WTOP = WVEG + WGRN
Return

SoilWater:
'--------------------------- Parameters and Initials
  If iniSW = 0 Then
```

Continued

Box 16.2. Continued.

```
        DEPORT = Sheet5.[B36]
        MEED = Sheet5.[B37]
        GRTDP = Sheet5.[B38]
        tuBRG = tuEMR
        tuTRG = tuBSG
        TEC = Sheet5.[B39]
        WSSG = Sheet5.[B40]
        WSSL = Sheet5.[B41]
        WSSD = Sheet5.[B42]

        SOLDEP = Sheet7.[b7]
        DEP1 = Sheet7.[b8]
        SALB = Sheet7.[b9]
        CN = Sheet7.[b10]
        DRAINF = Sheet7.[b11]
        SAT = Sheet7.[b12]
        DUL = Sheet7.[b13]
        EXTR = Sheet7.[b14]
        CLL = DUL - EXTR

        MAI1 = Sheet7.[b16]
        MAI = Sheet7.[b17]

        IPATSW = SOLDEP * EXTR * MAI

        ATSW = DEPORT * EXTR * MAI1
        TTSW = DEPORT * EXTR
        FTSW = ATSW / TTSW
        WSTORG = IPATSW - ATSW

        ATSW1 = DEP1 * EXTR * MAI1
        TTSW1 = DEP1 * EXTR
        FTSW1 = ATSW1 / TTSW1

        WLL1 = DEP1 * CLL
        WAT1 = WLL1 + ATSW1
        WSAT1 = DEP1 * SAT

        EOSMIN = 1.5:   WETWAT = 10:   KET = 0.5:    CALB = 0.23:
        DYSE = 1:       CTR = 0:       CE = 0:       CRAIN = 0:
        CRUNOF = 0:     CIRGW = 0:     IRGNO = 0:    iniSW = 1
  End If

'--------------------------- Irrigation
  If water = 1 And FTSW <= IRGLVL And CTU < tuTSG Then
    IRGW = (TTSW - ATSW)
    IRGNO = IRGNO + 1
  Else
    IRGW = 0
  End If

  CIRGW = CIRGW + IRGW
```

Continued

Box 16.2. Continued.

```
'--------------------------- Drainage
  If ATSW1 <= TTSW1 Then
    DRAIN1 = 0
  ElseIf ATSW1 > TTSW1 Then
    DRAIN1 = (ATSW1 - TTSW1) * DRAINF
  End If

  If ATSW <= TTSW Then
    DRAIN = 0
  ElseIf ATSW > TTSW Then
    DRAIN = (ATSW - TTSW) * DRAINF
  End If

  WSTORG = WSTORG + DRAIN - EWAT
  If WSTORG < 0 Then WSTORG = 0

'--------------------------- Water exploitation by root growth
  GRTD = GRTDP          'mm per day
  If CTU < tuBRG Then GRTD = 0
  If CTU > tuTRG Then GRTD = 0
  If DDMP = 0 Then GRTD = 0
  If DEPORT >= SOLDEP Then GRTD = 0
  If DEPORT >= MEED Then GRTD = 0
  If WSTORG = 0 Then GRTD = 0
  DEPORT = DEPORT + GRTD

  EWAT = GRTD * EXTR
  If EWAT > WSTORG Then EWAT = WSTORG

'--------------------------- Runoff
  RUNOF = 0
  If water = 2 And RAIN > 0.01 Then
    S = 254 * (100 / CN - 1)
    SWER = 0.15 * ((WSAT1 - WAT1) / (WSAT1 - WLL1))
    If SWER < 0 Then SWER = 0
    If (RAIN - SWER * S) > 0 Then
      RUNOF = (RAIN - SWER * S) ^ 2 / (RAIN + (1 - SWER) * S)
    Else
      RUNOF = 0
    End If
  End If

  If (WAT1 - DRAIN1) > WSAT1 Then
    RUNOF = RUNOF + (WAT1 - DRAIN1 - WSAT1)
  End If

  CRAIN = CRAIN + RAIN
  CRUNOF = CRUNOF + RUNOF

'--------------------------- LAI for soil evaporation
  If CTU <= tuBSG Then ETLAI = LAI Else ETLAI = BSGLAI
```

Continued

Box 16.2. Continued.

```
'--------------------------- Potential ET
  TD = 0.6 * TMAX + 0.4 * TMIN
  ALBEDO = CALB * (1 - Exp(-KET * ETLAI)) + SALB * Exp(-KET * ETLAI)
  EEQ = SRAD * (0.004876 - 0.004374 * ALBEDO) * (TD + 29)
  PET = EEQ * 1.1
  If TMAX > 34 Then PET = EEQ * ((TMAX - 34) * 0.05 + 1.1)
  If TMAX < 5 Then PET = EEQ * 0.01 * Exp(0.18 * (TMAX + 20))

'--------------------------- Soil evaporation
  EOS = PET * Exp(-KET * ETLAI)
  If PET > EOSMIN And EOS < EOSMIN Then EOS = EOSMIN

  SEVP = EOS
  If (RAIN + IRGW) > WETWAT Then DYSE = 1
  If ATSW1 < 1 or DYSE > 1 Or FTSW < 0.5 Then
    SEVP = EOS * ((DYSE + 1) ^ 0.5 - DYSE ^ 0.5)
    DYSE = DYSE + 1
  End If

  CE = CE + SEVP

'--------------------------- Plant transpiration
  VPTMIN = 0.6108 * Exp(17.27 * TMIN / (TMIN + 237.3))
  VPTMAX = 0.6108 * Exp(17.27 * TMAX / (TMAX + 237.3))
  VPD = VPDF * (VPTMAX - VPTMIN)
  TR = DDMP * VPD / TEC          'VPD in kPa, TEC in Pa
  If TR < 0 Then TR = 0

  CTR = CTR + TR

  If DEPORT <= DEP1 Then
    TR1 = TR
  ElseIf DEPORT > DEP1 Then
    If FTSW1 > WSSG Then RT1 = 1 Else RT1 = FTSW1 / WSSG
    TR1 = TR * RT1
  End If

'--------------------------- Updating
  ATSW1 = ATSW1 + RAIN + IRGW - DRAIN1 - RUNOF - TR1 - SEVP
  If ATSW1 < 0 Then ATSW1 = 0
  FTSW1 = ATSW1 / TTSW1
  WAT1 = WLL1 + ATSW1

  ATSW = ATSW + RAIN + IRGW + EWAT - DRAIN - RUNOF - TR - SEVP
  If ATSW < 0 Then ATSW = 0
  TTSW = DEPORT * EXTR
  FTSW = ATSW / TTSW

'--------------------------- Water-stress-factors
  If FTSW > WSSL Then WSFL = 1 Else WSFL = FTSW / WSSL
  If FTSW > WSSG Then WSFG = 1 Else WSFG = FTSW / WSSG
  WSFD = (1 - WSFG) * WSSD + 1
```

Continued

Box 16.2. Continued.

```
  If WAT1 > (0.95 * WSAT1) Then
    WSFG = 0:   WSFL = 0:   WSFD = 0:   WSFN = 0
  End If
Return

FindSowingDate:
  Row = 10
  Do
    Row = Row + 1
    Yr = Sheet2.Range("A" & Row)
    DOY = Sheet2.Range("B" & Row)
    SRAD = Sheet2.Range("C" & Row)
    TMAX = Sheet2.Range("D" & Row)
    TMIN = Sheet2.Range("E" & Row)
    RAIN = Sheet2.Range("F" & Row)
  Loop Until Yr = pyear And DOY = pdoy
Return

InitialsHeaders:
'--------------------------- Initials
  MAT = 0
  iniPheno = 0
  iniLAI = 0
  iniDMP = 0
  iniDMD = 0
  iniSW = 0
  iniPNB = 0
  iniSNB = 0
'--------------------------- Headers
  Sheet4.Cells(2, 1) = "Year"
  Sheet4.Cells(2, 2) = "DOY"
  Sheet4.Cells(2, 3) = "DAP"
  Sheet4.Cells(2, 4) = "TMP"
  Sheet4.Cells(2, 5) = "DTU"
  Sheet4.Cells(2, 6) = "CTU"
  Sheet4.Cells(2, 7) = "MSNN"
  Sheet4.Cells(2, 8) = "GLAI"
  Sheet4.Cells(2, 9) = "DLAI"
  Sheet4.Cells(2, 10) = "LAI"
  Sheet4.Cells(2, 11) = "TCFRUE"
  Sheet4.Cells(2, 12) = "FINT"
  Sheet4.Cells(2, 13) = "DDMP"
  Sheet4.Cells(2, 14) = "GLF"
  Sheet4.Cells(2, 15) = "GST"
  Sheet4.Cells(2, 16) = "SGR"
  Sheet4.Cells(2, 17) = "WLF"
  Sheet4.Cells(2, 18) = "WST"
  Sheet4.Cells(2, 19) = "WVEG"
```

Continued

Box 16.2. Continued.

```
  Sheet4.Cells(2, 20) = "WGRN"
  Sheet4.Cells(2, 21) = "WTOP"
  If water = 1 Or water = 2 Then
  Sheet4.Cells(2, 22) = "DEPORT"
  Sheet4.Cells(2, 23) = "RAIN"
  Sheet4.Cells(2, 24) = "IRGW"
  Sheet4.Cells(2, 25) = "RUNOF"
  Sheet4.Cells(2, 26) = "PET"
  Sheet4.Cells(2, 27) = "SEVP"
  Sheet4.Cells(2, 28) = "TR"
  Sheet4.Cells(2, 29) = "DRAIN1"
  Sheet4.Cells(2, 30) = "ATSW"
  Sheet4.Cells(2, 31) = "FTSW"
  Sheet4.Cells(2, 32) = "CRAIN"
  Sheet4.Cells(2, 33) = "CIRGW"
  Sheet4.Cells(2, 34) = "IRGNO"
  Sheet4.Cells(2, 35) = "CRUNOF"
  Sheet4.Cells(2, 36) = "CE"
  Sheet4.Cells(2, 37) = "CTR"
  Sheet4.Cells(2, 38) = "WSTORG"
  End If
Return

SummaryPrintOut:
  Sheet1.[g8] = DTEMR
  Sheet1.[g9] = DTTLM
  Sheet1.[g10] = DTBSG
  Sheet1.[g11] = DTTSG
  Sheet1.[g12] = DTMAT
  Sheet1.[g15] = MXLAI
  Sheet1.[g16] = BSGLAI
  Sheet1.[g17] = BSGDM
  Sheet1.[G20] = WTOP
  Sheet1.[G21] = WGRN
  Sheet1.[G22] = WGRN / WTOP * 100

  If water = 1 Or water = 2 Then
    Sheet1.[G25] = IPATSW
    Sheet1.[G26] = CRAIN
    Sheet1.[G27] = CIRGW
    Sheet1.[G28] = IRGNO

    Sheet1.[G30] = ATSW
    Sheet1.[G31] = CRUNOF
    Sheet1.[G32] = CE
    Sheet1.[G33] = CTR
    Sheet1.[G34] = WSTORG
```

Continued

Box 16.2. Continued.

```
    Sheet1.[G36] = CE + CTR
    Sheet1.[G37] = CE / (CE + CTR)
  End If
Return

DailyPrintOut:
  Sheet4.Cells(DAP + 2, 1) = Yr
  Sheet4.Cells(DAP + 2, 2) = DOY
  Sheet4.Cells(DAP + 2, 3) = DAP
  Sheet4.Cells(DAP + 2, 4) = TMP
  Sheet4.Cells(DAP + 2, 5) = DTU
  Sheet4.Cells(DAP + 2, 6) = CTU
  Sheet4.Cells(DAP + 2, 7) = MSNN
  Sheet4.Cells(DAP + 2, 8) = GLAI
  Sheet4.Cells(DAP + 2, 9) = DLAI
  Sheet4.Cells(DAP + 2, 10) = LAI
  Sheet4.Cells(DAP + 2, 11) = TCFRUE
  Sheet4.Cells(DAP + 2, 12) = FINT
  Sheet4.Cells(DAP + 2, 13) = DDMP
  Sheet4.Cells(DAP + 2, 14) = GLF
  Sheet4.Cells(DAP + 2, 15) = GST
  Sheet4.Cells(DAP + 2, 16) = SGR
  Sheet4.Cells(DAP + 2, 17) = WLF
  Sheet4.Cells(DAP + 2, 18) = WST
  Sheet4.Cells(DAP + 2, 19) = WVEG
  Sheet4.Cells(DAP + 2, 20) = WGRN
  Sheet4.Cells(DAP + 2, 21) = WTOP
  If water = 1 Or water = 2 Then
  Sheet4.Cells(DAP + 2, 22) = DEPORT
  Sheet4.Cells(DAP + 2, 23) = RAIN
  Sheet4.Cells(DAP + 2, 24) = IRGW
  Sheet4.Cells(DAP + 2, 25) = RUNOF
  Sheet4.Cells(DAP + 2, 26) = PET
  Sheet4.Cells(DAP + 2, 27) = SEVP
  Sheet4.Cells(DAP + 2, 28) = TR
  Sheet4.Cells(DAP + 2, 29) = DRAIN1
  Sheet4.Cells(DAP + 2, 30) = ATSW
  Sheet4.Cells(DAP + 2, 31) = FTSW
  Sheet4.Cells(DAP + 2, 32) = CRAIN
  Sheet4.Cells(DAP + 2, 33) = CIRGW
  Sheet4.Cells(DAP + 2, 34) = IRGNO
  Sheet4.Cells(DAP + 2, 35) = CRUNOF
  Sheet4.Cells(DAP + 2, 36) = CE
  Sheet4.Cells(DAP + 2, 37) = CTR
  Sheet4.Cells(DAP + 2, 38) = WSTORG
  End If
Return

End Sub '---------------------------
```

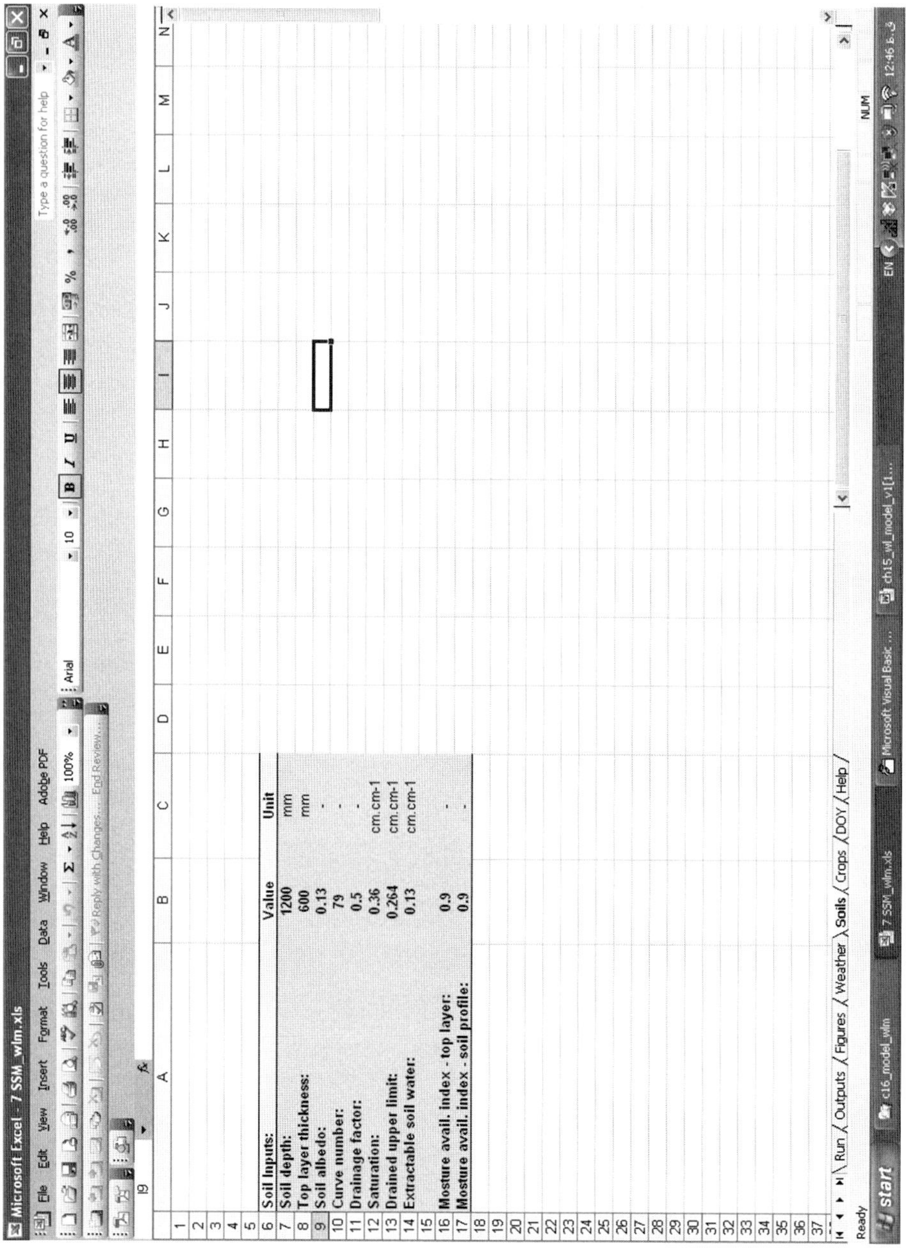

Fig. 16.1. Appearance of the "Soils" sheet in the Excel file containing the model.

In the "Run" sheet, the following additional management inputs are needed:

- "*water*";
- vapor pressure deficit factor; and
- irrigation level.

If "*water*" has a value of 1, then irrigation level must be specified (see Chapter 14, section on irrigation).

In the "Run" sheet summary outputs are also expanded to include important soil water outputs. The new summary outputs to this sheet are:

- available soil water at sowing;
- cumulative rainfall during growing season;
- cumulative irrigation;
- irrigation number;
- available soil water at maturity;
- cumulative runoff;
- cumulative soil evaporation;
- cumulative plant transpiration;
- cumulative drainage;
- total evapotranspiration; and
- evaporation/evapotranspiration (E/ET) ratio.

The sheets "Outputs" and "Figures" are also expanded to include soil water outputs and their related figures. The content of the "Outputs" sheet can be used for further analyses and more figures might be included in the "Figures" sheet.

Appendix I gives a step-by-step practical guide that can be used for troubleshooting if model predictions are not in agreement with observed data.

Sample Runs of the Model

Sample simulations are presented to show how the water-limited model can be applied. In this example, the model is used to simulate wheat growth and yield for irrigated and rainfed conditions of Gorgan in northeast Iran. The weather data for growing season 2005/06 is used. The sowing date was mid-December 2005 and the sowing density was 300 plants m^{-2}. Under irrigated conditions, irrigation was decided to be done by the model by applying water when FTSW dropped below 0.5. Irrigation amount will also be calculated by the model to return the soil to the drained upper limit. All other conditions including soil water at sowing time were the same for the irrigated and rainfed simulations.

A summary of simulation results is presented in Table 16.1. Days to different phenological stages occurred sooner under rainfed conditions, except for days to emergence. Days to beginning seed growth was 124 days for irrigation conditions and 123 days for rainfed conditions. Seed filling duration was 4 days shorter under rainfed conditions (32 versus 28 days) due to acceleration in development rate under the water-deficit conditions that developed in the rainfed simulation.

Table 16.1. Summary of crop characteristics simulated for wheat in Gorgan during growing season of 2005/06 under irrigated and rainfed conditions.

	Irrigated	Rainfed
Phenology:		
Days to emergence	11	11
Days to termination leaf growth	95	94
Days to beginning seed growth	124	123
Days to termination seed growth	156	151
Days to maturity	168	163
Growth:		
Maximum LAI	5.26	5.16
Crop mass at beginning seed growth	729	722
Yield:		
Crop total mass (g m^{-2})	1212	973
Grain yield (g m^{-2})	635	402
Harvest index (%)	52	41
Soil water:		
Available soil water at sowing (mm)	140	140
Cumulative rainfall during season (mm)	302	302
Cumulative irrigation (mm)	142	0
Irrigation number	3	0
Available soil water at maturity (mm)	68	34
Cumulative run-off (mm)	0	65
Cumulative soil evaporation (mm)	182	128
Cumulative transpiration (mm)	190	149
Cumulative drainage (mm)	144	66

Maximum LAI and crop mass at the beginning of seed growth were nearly identical for both conditions, less than 2% difference. Maximum LAI was 5.26 and 5.16 under irrigated and rainfed conditions, respectively (Table 16.1). However, crop characteristics were significantly different at maturity. Final crop mass was 25% greater under irrigated conditions (1212 versus 973 g m^{-2}). A look at changes in daily dry matter production (Fig. 16.2a) and cumulative crop mass (Fig. 16.2b) indicate that these two variables were nearly identical until about 125 days after sowing, but after this time, the results for the irrigated and rainfed conditions diverge. More dry matter production was predicted under irrigated conditions.

For grain yield the difference was greater than crop mass. Grain yield decreased from 635 g m^{-2} under irrigated conditions to 402 g m^{-2} under rainfed conditions, which means a 37% decrease without irrigation in this particular case. The water limitation also resulted in lower harvest index for the rainfed crop: 41% for rainfed and 52% for irrigated conditions.

To better understand the reason(s) for the yield difference, important soil water balance components can be evaluated. Figure 16.3 presents daily variation in simulated FTSW. Defined conditions for irrigation resulted in three irrigations with cumulative irrigation water of 142 mm (Table 16.1). As a result of defined conditions for irrigation, these irrigations took place at 7, 82, and

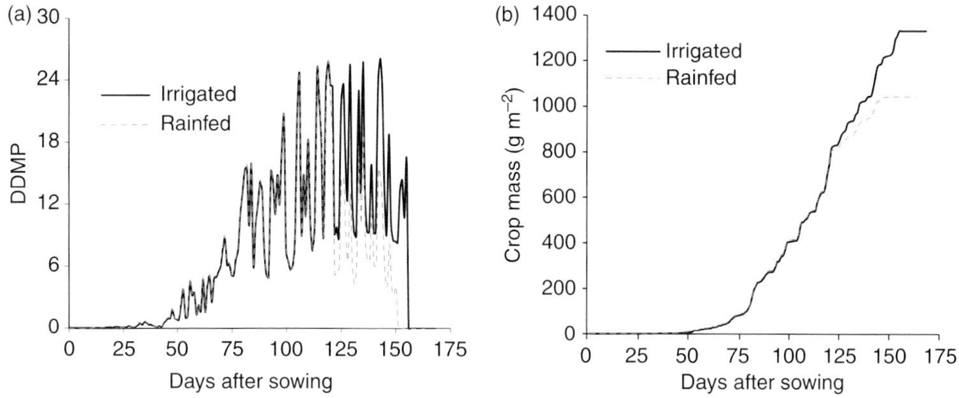

Fig. 16.2. Simulated changes in (a) daily dry matter production (DDMP, g m^{-2} day^{-1}) and (b) cumulative crop mass (g m^{-2}) for wheat in Gorgan during growing season of 2005/06 under irrigated and rainfed conditions.

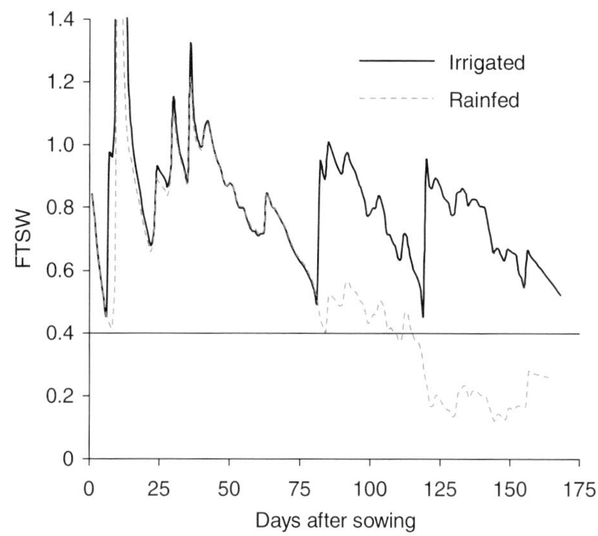

Fig. 16.3. Simulated changes in fraction transpirable soil water (FTSW) for wheat in Gorgan during growing season of 2005/06 under irrigated and rainfed conditions. Horizontal line marks FTSW of 0.4, below which crop growth is negatively affected.

120 days after sowing with 14, 58, and 70 mm, respectively (data not shown). These irrigations can be found in Fig. 16.3 as an abrupt increase in FTSW. Under rainfed conditions, FTSW falls below 0.4 after about 115 days after sowing and accounts for the decrease in daily dry matter production observed under rainfed conditions in Fig. 16.2a.

Cumulative soil evaporation for both conditions is the same until 82 days after sowing, i.e. until the time of the second irrigation (Fig. 16.4a). After this time, cumulative soil evaporation under irrigated conditions continues to

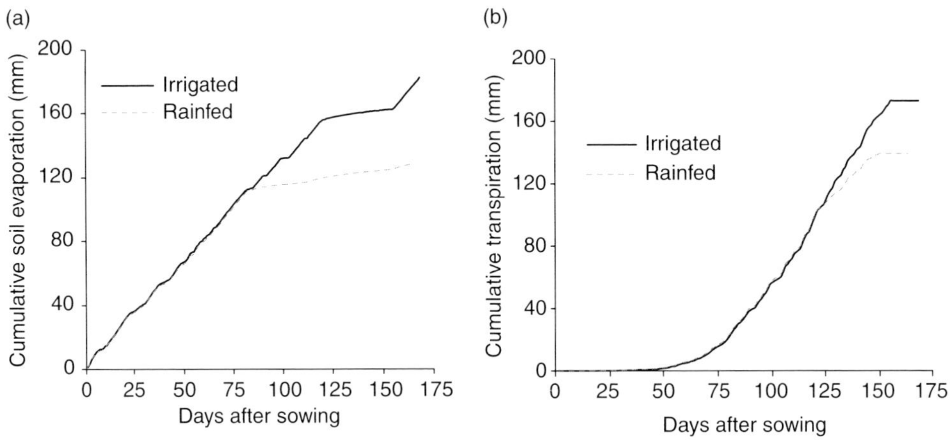

Fig. 16.4. Simulated changes in (a) cumulative soil evaporation and (b) plant transpiration for wheat in Gorgan during growing season of 2005/06 under irrigated and rainfed conditions.

increase due to soil wetness as a result of the irrigation, but under rainfed conditions cumulative evaporation increase is very slow. The trend in cumulative transpiration (Fig. 16.4b) is similar to that of cumulative crop mass due to the linkage between these two that was discussed in Chapter 14. Total soil evaporation and plant transpiration were 182 and 190 mm for the irrigated crop, respectively (Table 16.1). For the rainfed crop, soil evaporation and transpiration were 128 and 149 mm, respectively. Cumulative evapotranspiration was 372 mm for the irrigated simulation and 277 mm for the rainfed simulation (74% of the irrigated crop) (Table 16.1).

Runoff was assumed to be zero for irrigated conditions (Chapter 14), however, under rainfed conditions 65 mm runoff was predicted. Cumulative drainage was greater under irrigated (144 mm) than the rainfed (66 mm) conditions.

Exercises

1. Try to parameterize the water-limited model of this chapter for your crops/cultivars. Crop parameter estimates and soil inputs presented in previous chapters can be used as default values, but precise estimates of required temperature units are necessary.
2. Use the model of Exercise 1 to estimate your crops'/cultivars' potential yields at your location under irrigated conditions. Calculate how much irrigation water and how many irrigations are needed. Then, simulate the same crops/cultivars under rainfed conditions. Compare the predicted yields from irrigated and rainfed conditions. Explain how these two are similar or different. Why?
3. Prepare a list of applications for the water-limited model. Some ideas are presented in Chapter 4.

4. Use the model of Exercise 1 to simulate the impact of higher temperatures on your crops/cultivars as done in Chapter 12, but pay attention to the effect of higher temperatures on water use and soil water balance components.
5. Repeat simulations of Exercise 4 for higher concentrations of CO_2. Do not forget to correct radiation use efficiency and transpiration efficiency coefficient for higher CO_2 concentrations.
6. Try to use the model of Exercise 1 for other applications you have listed in Exercise 3.

17 Plant Nitrogen Budget

Nitrogen (N) and water are the two key resources that often limit crop production. N is necessary for crop production since it is an essential component of the proteins, nucleic acids, and many other components of cells. Without N for inclusion in these compounds they cannot be synthesized, and there can be no plant growth. Much of the history of advances in crop production is very closely tied to improved N management (Sinclair and de Wit, 1976; Godwin and Jones, 1991; Sinclair, 2004; Sinclair and Weiss, 2010).

Ironically, plants are surrounded by an atmosphere that is 79% dinitrogen (N_2). However, only legumes have evolved special structures and biochemistry to access the atmospheric N_2. Also, many soils contain large quantities of N in the form of organic matter, but often the organic matter is not readily broken down for uptake by plants. Hence, N fertilizers are the most widely used fertilizers and their application has become crucial in high-yield cropping systems. Unfortunately, N from fertilizers and from other sources is ephemeral and can be readily released to the environment through volatilization, denitrification, and nitrate leaching. These releases all can have harmful effects on the environment and human health.

Understanding the processes that govern N fluxes, particularly N uptake and distribution in crops, is considered important with respect to both environmental concerns and the quantity and quality of crop products (Gastal and Lemaire, 2002). Thus, modeling N dynamics in crops and soil has always been important for crop/soil modelers. Crop models are increasingly used to understand N limitation in crop production (Sinclair, 1986), to evaluate N-related traits for yield improvement (Sinclair *et al.*, 2003), and to optimize use of N in cropping systems (Robertson *et al.*, 2005).

The objective of this part of the book is to extend the water-limited model presented in Chapter 16 to cover N-limited conditions. An important part of this effort is to model N accumulation and distribution in crops.

Crop N uptake is a principal process in the N cycle. N uptake and/or fixation vary depending on crop yield and quality. Values of seasonal N accumulation can be as low as near 0 g m^{-2} to high values of 25 g m^{-2}, and even much higher accumulation by high-protein legumes. This chapter deals with a submodel for N accumulation and distribution in plants in legumes and non-legume crops. In Chapter 18, a simple soil N submodel will be presented. Chapter 19 will indicate how the submodels are merged with the water-limited model of Chapter 16 to form an N-limited model.

N and Crop Production

N plays many roles in plants in different scales from biochemical level to ecosystem level. This section indicates how N is important for crop production by affecting major crop processes as described in Chapters 6 to 11.

N and phenological development

N generally has negligible effect on crop phenological development under usual crop production situations. Only under severe N deficit might crop phenological development be impeded. Therefore, N will be ignored as a variable influencing development under practical production situations.

N for leaf area development

Leaves require large amounts of N as a critical component of the enzymes that carry on photosynthesis. About three-quarters of leaf N is connected with photosynthesis (Sinclair and Horie, 1989). In fact, the most abundant protein on earth is ribulose bisphosphate carboxylase, which is the enzyme that captures carbon dioxide as the first step in photosynthesis. The partitioning of N among leaves is critical because there is a trade-off between leaf N content (area basis) and leaf area development (Sinclair and de Wit, 1976; Sinclair and Horie, 1989).

Nitrogen deficit limits crop leaf area development, although dry matter accumulation by the leaves is affected by a lesser degree and the result is increased specific leaf weight. Large variation in specific leaf weight of two-fold has been reported under N shortage (Grindlay, 1997). On the other hand, as large amounts of N are invested in crop leaves, leaf area expansion is considered as a key determinant of crop N demand (Grindlay, 1997).

Sinclair and Horie (1989) analyzed the effects of N supply rate (from soil or biological fixation) and leaf N content on leaf area growth (Fig. 17.1). They selected N supplies of 0.05, 0.10, 0.20, and 0.40 g N m^{-2} day^{-1} to cover a broad range of N availability. They examined the influence of leaf N content (g N m^{-2} leaf) over a range from 0 to 3 g N m^{-2}. For example, leaf N content often is 1.5 and 2.5 g N m^{-2} in wheat and soybean, respectively, during vegetative growth.

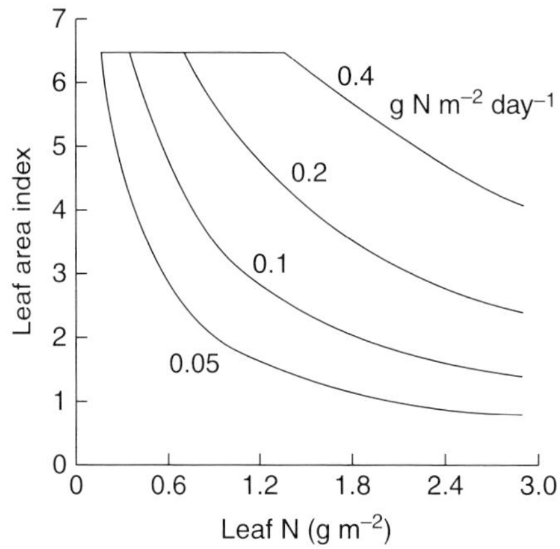

Fig. 17.1. Predicted LAI after 40 days of growth in response to four N supply rates for leaf growth plotted as a function of leaf N content per unit area (Sinclair and Horie, 1989).

Figure 17.1 shows how crop leaf area might be limited by low N supply and/or high leaf N content. According to Fig. 17.1, with an N supply rate of 0.2 g N m^{-2} day^{-1}, a soybean crop develops an LAI of 3, but the same value will be 4.5 for a wheat crop with lower leaf N content. For a crop with leaf N content of 2.4 g N m^{-2}, decrease in N supply from 0.4 to 0.05 g N m^{-2} day^{-1} results in decline in LAI development from 5 to 1.

N and dry matter production

Many studies have shown that there is a close relationship between carbon exchange rate of the leaves and their N contents (e.g. Fig. 17.2; Sinclair and Horie, 1989). Sinclair and Horie (1989) analyzed the effects of N on crop mass accumulation by elucidating quantitative relationships among leaf N content, CO_2 assimilation rate, and crop radiation use efficiency in soybean, rice, and maize. They developed relationships predicting crop radiation use efficiency (RUE; Chapter 10) for each of the crops as a function of leaf N content (Fig. 17.3). Therefore, crop growth expressed via RUE is directly related to leaf N content.

In the species examined by Sinclair and Horie (1989), RUE showed great sensitivity to leaf N at the low rates of leaf N (Fig. 17.3). At high leaf N levels, RUE was nearly maximal and little response to further increases in leaf N was predicted. Consequently, they concluded that accumulation of very high leaf N results in no advantage in RUE, while decreased leaf N results in substantial reductions in the potential for crop mass accumulation. RUE within each species was nearly constant at high leaf N contents, but decreased appreciably at low leaf N contents.

Shiraiwa and Sinclair (1993) conducted field experiments on soybean and found that canopies with expanding leaves had a fairly uniform leaf N in the top 1.5 to 2.0 LAI. Below this top zone there was a linear decline in leaf N with increased cumulative LAI. In contrast, mature canopies with fully expanded leaves had a continuous linear decrease in leaf N with cumulative LAI from the

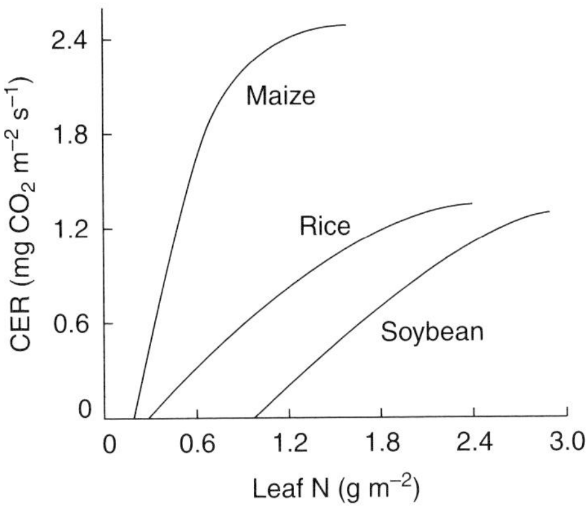

Fig. 17.2. Relationships between leaf carbon dioxide exchange rates (CER) at light saturation and leaf N content in maize, rice and soybean (Sinclair and Horie, 1989).

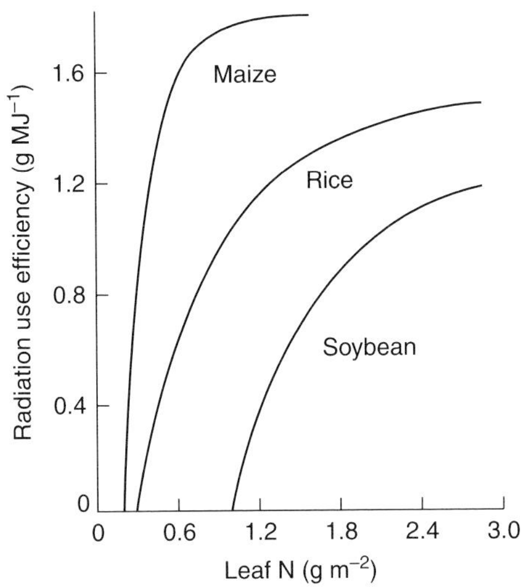

Fig. 17.3. Radiation use efficiency of maize, rice, and soybean as a function of leaf N content per unit area (Sinclair and Horie, 1989).

top of the canopy. Leaf N of the leaves at the top of the mature canopies was substantially greater than in those canopies with expanding leaves. Neither plant density nor N fertility substantially altered the linear decline in leaf N with cumulative LAI. Their results indicated that leaf N at the canopy top was such that the leaves approached maximum photosynthetic rates, a phenomenon that was called leaf N optimization by Grindlay (1997).

Sinclair and Shiraiwa (1993) evaluated the effect of non-uniform leaf N distribution on RUE in soybean and indicated that non-uniform leaf N distribution enhances RUE. A non-uniform distribution of leaf N has the possibility of increasing RUE if more of the N is present in the top leaves and less in the bottom leaves. The advantage of the non-uniform N distribution results because the top leaves intercept much more of the incident solar radiation than the bottom leaves and, therefore, the top leaves are responsible for much of the photosynthetic activity (Sinclair and Shiraiwa, 1993; Hammer and Wright, 1994).

Leaves as temporary N storage for grain

An important aspect of plant N budget is the role of crop leaves as storage of N for translocation to growing seeds. Sinclair (2004) stated it is important for yield increase to increase storage N in vegetative tissue, especially in greater leaf area, for later transfer to developing seeds. Sinclair and Sheehy (1999) pointed out that one approach to achieving a large amount of leaf area was the development of erect leaves to allow small amounts of light penetration to the bottom leaves to sustain these leaves. The retention of these additional leaves increases plant N storage for translocation to the seeds during grain fill. Although increases in leaf erectness as a result of Green Revolution breeding are often traditionally ascribed to better capture of light, there is little support for this hypothesis and increased N storage is the major advantage to increase crop yield.

Figure 17.4 shows how grain yield and LAI via its role as an N reservoir are related in wheat in a temperate, sub-humid climate. In the figure, it has been assumed that 75 or 100% of grain N come from mobilization from leaves and stems. Stem dry weight at anthesis has been obtained from a leaf area–stem mass relationship. Total N available for mobilization from leaf and stem has been calculated from leaf and stem N content at anthesis and maturity. It is obvious in this example that an LAI of at least 6 is required for high yields of 600 to 800 g m^{-2}.

Seed growth need for N results in leaf senescence

With the beginning of seed growth, the plant N budget changes dramatically. Generally, N accumulation by the crop after beginning of seed growth constitutes a negligible or small portion of total N accumulation (e.g. Kichey *et al.*, 2007). Therefore, seeds are nearly completely dependent on pre-seed N storage in leaves and stems. N mobilization from leaves and stems towards the grain results in the "self-destruction" phenomenon described by Sinclair and

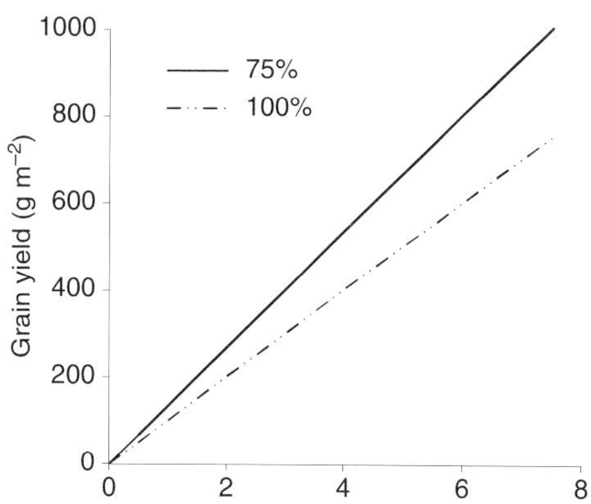

Fig. 17.4. Dependency of grain yield on LAI at anthesis in wheat with two scenarios: that 75% of grain N comes from translocation from leaves and stems and the remaining from soil uptake after anthesis; or 100% of grain N comes from translocation from leaves and stems (A. Soltani and T.R. Sinclair, unpublished).

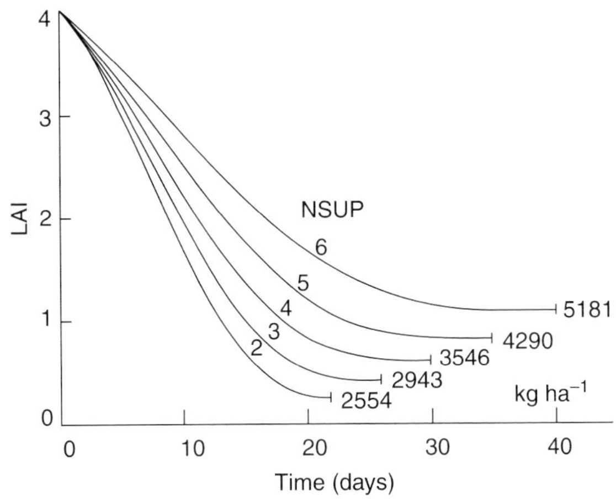

Fig. 17.5. Simulated changes in LAI with time after beginning seed growth for soybean crops with an initial LAI of 4 and various rates of root N supplies (NSUP, kg ha^{-1} day^{-1}). Final grain yields are presented in the bottom, right portion of the figure (Sinclair and de Wit, 1976).

de Wit (1976). Figure 17.5 presents the impact of seed growth on N depletion from leaves and hence leaf senescence in soybean under five scenarios of N supply by crop roots. A similar effect has been observed in other grain crops and many crop models relate leaf senescence to N mobilization (e.g. Jones *et al.*, 2003; Sinclair *et al.*, 2003).

Background and Basics of Plant N Budget

A range of approaches has been used to calculate crop N demand, N accumulation and distribution, and the effects of N deficiency on crop growth and yield (Sinclair, 1986; Boote *et al.*, 1998; Robertson *et al.*, 2002a; Sinclair *et al.*, 2003; Stockle *et al.*, 2003). In general, N demand is computed based on crop mass growth and N concentration in the plant or plant tissues. The N concentration in various tissues can be allowed to vary in the model as a function of temperature unit or crop mass.

Accumulated N is partitioned to plant tissues in proportion to their individual demands or based on partitioning coefficients. If demand exceeds supply, there is a deficiency, which affects crop growth. During the grain growth period, N demand of growing grains is computed by multiplying grain growth rate and anticipated grain N concentration.

The effects of N limitation are accounted for by calculating a number of N stress factors (similar to water stress factors; Chapter 15) from actual and optimal N concentration in the plant. One approach suggested by Lemaire *et al.* (2007) is based on critical N uptake curves as a function of biomass accumulation. They found that there were different but fairly stable curves for C_3 and C_4 species. However, they concluded that if crop simulation models are to capture the genotypic and environmental control of crop N dynamics in a physiologically functional manner, a more detailed structure to account for N in the plant is needed and plant growth has to be considered as the sum of the metabolic (e.g. leaves) and the structural (e.g. stems) compartments, each with its own demand for metabolic and structural N.

Jamieson and Semenov (2000) proposed an appealing method to simulate N accumulation and distribution by wheat. This method was then successfully applied to soybean (Sinclair *et al.*, 2003), sorghum (Hammer *et al.*, 2010), and chickpea (Soltani and Sinclair, 2011). In this approach N demand is set by the need to maintain target N concentration in new leaves (area basis), allowing for leaves to become thicker if N available is not sufficient for expansion. Stems (and similar organs like leaf sheaths or petioles) act as a reservoir for extra N between minimum and target N concentration.

This approach will be used in this chapter. The approach mimics decreasing N concentration of the entire canopy during seed fill as a well-defined function of total shoot mass, which is consistent with observations (e.g. Ney *et al.*, 1997; Stockle and Debaeke, 1997; Gastal and Lemaire, 2002). In addition, the crop parameters needed to define N accumulation and distribution with this method are limited to a few relatively conservative parameters, i.e. leaf and stem N concentration during growth and at senescence, and the target grain N concentration. These parameters can be readily measured on tissue samples and require no model calibration. The approach can be used to evaluate the potential influence of key plant and environmental variables on N accumulation by major grain crops (Sinclair *et al.*, 2003).

An overview of the approach is shown in Box 17.1. Plant N balance is simulated separately before and after seed growth. Sinks for N and developmental intervals during which these sinks are active, are identical to those of dry

Box 17.1. Overview of the approach that is used to simulate plant N budget.

N accumulation and distribution during vegetative growth:

- Daily N demand is obtained from daily development in leaf area and daily growth in stem weight and their target N concentrations.
- The demand is adjusted for maximum capacity of N accumulation rate, flooding condition, and soil available N for crop uptake.
- N is distributed to leaves and stems according to their demands.
- Under limited N conditions, first stem N is diluted until its minimum, second leaf area expansion is stopped, and finally leaves are killed to provide N for stem growth at its minimum N content.

N accumulation and distribution during seed growth:

- All seeds' demand for N is supplied by N mobilization from leaves and stems.
- Fraction N mobilized from leaves and stems is proportional to their relative mobilizable N.
- N mobilization from leaves results in leaf senescence.

matter described in Chapter 11. These sinks are: (i) leaves; (ii) "stems"; and (iii) grains or other storage organs. The stem component includes the actual stems and any other tissue other than leaf blades and grains. Due to biological N_2 fixation by legume crops, a slightly different plant N submodel will be required for legumes. Thus, the procedure is described for non-legume crops and then modification will be discussed for legume crops.

Plant N Budget During Vegetative Growth

Before seed growth, daily demand for N accumulation (NUP, g N m^{-2} day^{-1}) is computed based on N requirements of leaves and stems. The demand for leaves is obtained by multiplying daily increase in LAI (GLAI, m^2 m^{-2} day^{-1}; Chapter 9) by N content per unit leaf area (SLNG, g N m^{-2}). SLNG is called specific leaf N in green leaves. The N demand for stems is calculated by multiplying daily increase in stem dry matter (GST, g m^{-2} day^{-1}; Chapter 11) by N content per unit stem weight (SNCG, g N g^{-1}). The total demand, i.e. NUP, is the sum of the leaf and stem demand:

$$\text{NUP} = (\text{GLAI} \times \text{SLNG}) + (\text{GST} \times \text{SNCG}) \qquad (17.1)$$

Both SLNG and SNCG are target N contents of leaves and stems during vegetative growth when N is not limited. A constant value is used for both SLNG and SNCG in the model.

A third term might be added to Eqn 17.1 to calculate daily N demand as a result of N deficiencies in the plant that developed early in the season. The implementation of this demand assumes that such "make up" N accumulation is physiologically possible for the crop species late in the growing season. In this case, this component (NSTDF, g m^{-2}) is obtained from current stem weight (WST, g m^{-2}) and N accumulation (NST, g N m^{-2}) and target stem N content (SNCG):

$$\text{NSTDF} = \min(\text{WST} \times \text{SNCG} - \text{NST}, 0) \qquad (17.2)$$

Daily NUP may not be met by supply for several reasons:

- There is evidence that crops can process daily only a maximum amount of accumulated N. Therefore, NUP has a maximum rate (MXNUP, g N m^{-2} day^{-1}). Thus, under conditions where demand is higher than MXNUP, actual rate of N accumulation is limited to this value.
- Under the hypoxic conditions of soil saturation NUP is inhibited. Therefore, NUP is set to 0 under flooding conditions whenever soil water in the nitrogen layer (WAT1) is at least 95% of water saturation.
- NUP is limited to the amount of soil N available for crop uptake (SNAVL, g N m^{-2}; Chapter 18). Therefore, if NUP is calculated to be greater than SNAVL, it is adjusted at SNAVL. SNAVL is calculated by another submodel which is described in the next chapter (Chapter 18).

Figure 17.6 indicates the factors that govern potential and actual NUP as described above.

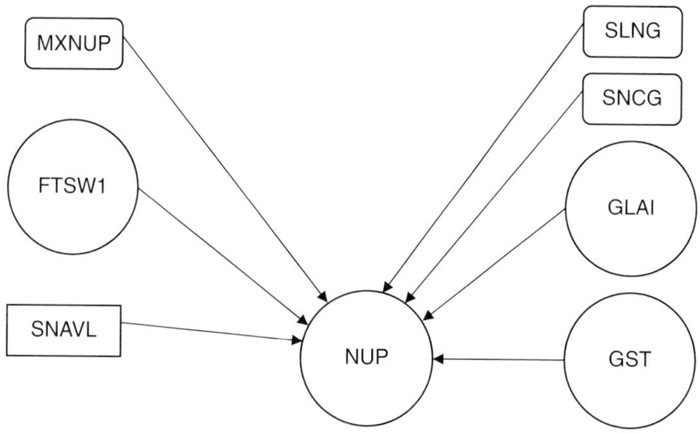

Fig. 17.6. Relational diagram indicating factors affecting potential (right side) and actual (left side) N accumulation rate (NUP, g m^{-2} day^{-1}) by crops during vegetative growth. SLNG is the specific leaf N in green leaves (g N m^{-2}), SNCG the N concentration in green stem (g g^{-1}), GLAI the daily increase (growth) in leaf area index (m^2 m^{-2} day^{-1}), GST the daily growth in stem weight (g m^{-2} day^{-1}), MXNUP the maximum daily rate of N accumulation (g N m^{-2} day^{-1}), FTSW1 the fraction transpirable soil water in the top layer, and SNAVL the soil available N for crop uptake (g N m^{-2}).

At times when NUP does not fully meet the demand by the crop for N during vegetative growth, the model responds in the following sequence.

1. Initially, leaf area development is sustained with adequate N, and the N deficiency results in a decrease in stem N concentration to allow leaf area development at its target N concentration. The stem N concentration is allowed to decrease until it reaches a minimum (SNCS, g N g^{-1}).
2. With continued deficiency in NUP, and after the stems have reached their minimum N concentration, leaf area development is inhibited. Therefore, continued stem growth at minimum N concentration (to maintain SNCS) receives priority and the development of leaf area is limited only to the N available in excess after accounting for stem growth.
3. Under very low NUP when there is no excess N for leaf area development, leaves are senesced to provide N to sustain stem growth at SNCS. Thus, LAI decreases under very limited NUP as a result of this leaf senescence.

To calculate the amount of N available from the senesced leaves for stem growth, it is necessary to know the difference in leaf concentration between the green leaves (SLNG) and the senesced leaves (SLNS, g N m^{-2} leaf). Therefore, the mobilizable N from the leaves is (SLNG – SLNS).

Figure 17.7 shows how daily N increase for leaves (INLF, g N m^{-2} day^{-1}) and stems (INST, g N m^{-2} day^{-1}) are determined each day based on above described criteria. It also indicates how daily N decrease from leaves (XNLF, g N m^{-2} day^{-1}) and stems (XNST, g N m^{-2} day^{-1}) are obtained.

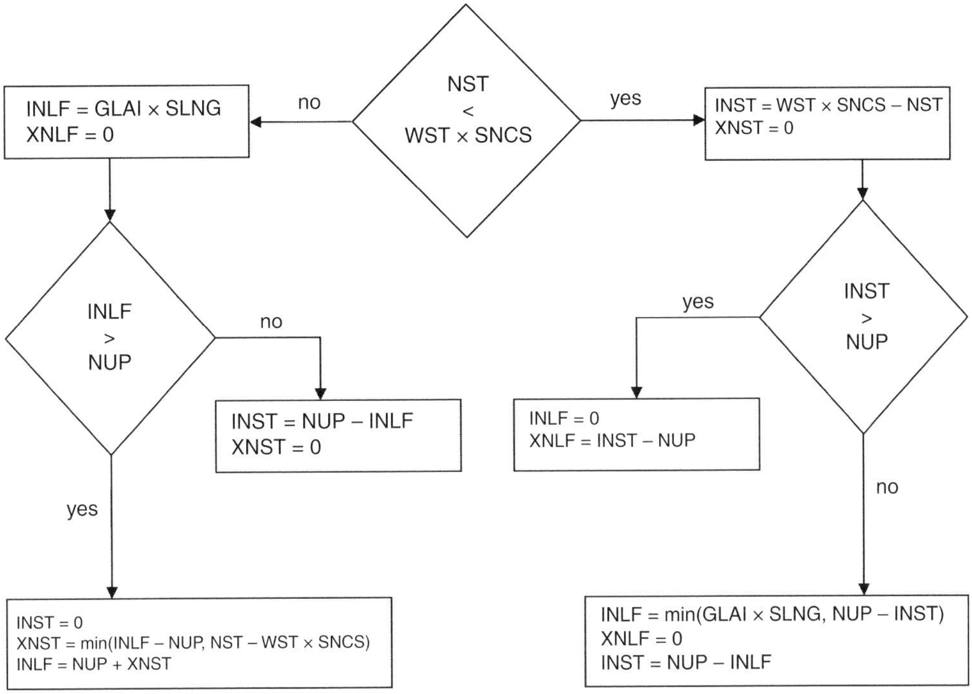

Fig. 17.7. Flow diagram showing how daily accumulated N is distributed between leaves and stems. NST is accumulated N in stems (g N m^{-2}), WST the cumulative stem weight (g m^{-2}), SNCS the minimum N concentration in stems (g g^{-1}), INST the daily increase in stem N (g N m^{-2} day^{-1}), XNST the daily rate of N extraction from stems (g N m^{-2} day^{-1}), INLF the daily rate of N accumulation in leaves (g N m^{-2} day^{-1}), XNLF the daily rate of N extraction from leaves (g N m^{-2} day^{-1}), NUP the daily rate of N accumulation (g N m^{-2} day^{-1}), GLAI the daily increase in leaf area index (m^2 m^{-2} day^{-1}), and SLNG the specific leaf N in green leaves (g N m^{-2}). NST is for yesterday and all other variables are for today.

Plant N Budget During Seed Growth

After beginning seed growth, seeds become the prime sink for N and the N balance shifts to a focus on N mobilization to the seeds. Daily N demand by the seeds (INGRN, g m^{-2} day^{-1}) is calculated as the product of seed growth rate (SGR, g m^{-2} day^{-1}; Chapter 11) and grain N concentration (GNC, g N g^{-1}).

$$INGRN = SGR \times GNC \tag{17.3}$$

GNC is a crop parameter and is assumed constant throughout seed growth.

Daily N demand (NUP) is then set equal to INGRN, as it was assumed in Chapter 11 that grains are the only sink during seed growth. All the limitations described above for limiting NUP during the vegetative stage apply in seed growth.

In addition to the previous limitations on NUP, there needs to be excess photosynthate after meeting the needs of seed growth to support NUP. NUP is set equal to 0 when daily dry matter production (DDMP, Chapter 10) by the crop does not exceed SGR. Due to this restriction, as a practical matter in the current structure of the model the value of NUP is often 0 during grain fill. This is the case because it was assumed in the calculation of SGR in Chapter 11 that all daily dry matter production is distributed to the growing seeds.

Rather than assigning all photosynthate to SGR, an alternative in calculating SGR would be to base it on daily increase in harvest index (Chapter 11), and then checking to determine if there is surplus photosynthate to support NUP. However, this alternative approach to calculating NUP during seed fill still results in a negligible or a small part of total N accumulation by the crop for both legumes and non-legumes during seed fill (e.g. Sinclair and Amir, 1992; Sinclair, 2004) due to limited photosynthate availability. While this approach has been used by Sinclair *et al.* (2003) and Soltani and Sinclair (2011), the simpler approach is used here since it commonly gives satisfactory results.

Therefore in the N submodel presented here, all daily N demand (INGRN) for growing seeds is obtained as a result of N mobilization from the leaves and stems. Total mobilizable N available in the plant at the beginning of seed fill (TRLN, g N m^{-2}) is calculated in the N submodel. Not all N in leaves and stems can be mobilized because some N is in structural components of these tissues and is not accessible for breakdown. The potential N available for translocation to the growing grain is calculated as the sum of the mobilizable N in the leaves and in the stems. The mobilizable N in the leaves is a product of LAI multiplied by the N concentration difference between green leaves and senesced leaves (SLNS and SNCS). The mobilizable N from the stems is calculated as the difference between the total N in the stems (NST, g N m^{-2}) and the amount of N that remains in senesced stems. Senesced stem N is equal to stem dry weight (WST, g m^{-2}) multiplied by the senesced stem N concentration (SNCS, g N g^{-1}).

$$\mathrm{TRLN} = \mathrm{LAI} \times (\mathrm{SLNG} - \mathrm{SLNS}) + (\mathrm{NST} - \mathrm{WST} \times \mathrm{SNCS}) \quad (17.4)$$

Each day the N demand of the seeds (INGRN) is satisfied by removing N from TRLN. To track the physiological impact of the loss in N from the vegetative tissue, specifically the leaves, it is necessary to calculate each day the amount of N remaining in the leaves. The proportion of the daily N transfer from the leaves (FXLF) is equal to the proportion of the total translocatable N (TRLN) that is in the leaves. That is:

$$\mathrm{FXLF} = \mathrm{LAI} \times (\mathrm{SLNG} - \mathrm{SLNS}) / \mathrm{TRLN} \quad (17.5)$$

Using FXLF, daily decrease in leaf N (XNLF, g N m^{-2} day^{-1}) and stem N (XNST, g N m^{-2} day^{-1}) are computed:

$$\mathrm{XNLF} = (\mathrm{SGR} \times \mathrm{GNC} - \mathrm{NUP}) \times \mathrm{FXLF} \quad (17.6)$$

$$\mathrm{XNST} = (\mathrm{SGR} \times \mathrm{GNC} - \mathrm{NUP}) \times (1 - \mathrm{FXLF}) \quad (17.7)$$

Crop models are different with respect to the effect of the decrease of N in the leaves on LAI or RUE. Some models incorporate effects of N on both RUE and

LAI (e.g. Porter *et al.*, 1993). Some others only account for the effect of N on LAI (e.g. Jamieson and Semenov, 2000; Sinclair *et al.*, 2003) or RUE (e.g. Hansen *et al.*, 1991). Olesen *et al.* (2002) made a comparison of the models and indicated that all the models could reproduce observed development of LAI and dry matter over time. Further, the observed response of grain yield to increasing rates of N application was also reproduced by most of them.

Here, it is assumed that the transfer of N from the leaves to the grain results in a loss in leaf area. That is, it is assumed that leaf N per unit leaf area (SLNG) remains constant, so there is no effect of N translocation on RUE. The main reason for this assumption as explained by Jamieson and Semenov (2000) is that changes in leaf N concentration are not uniform throughout the canopy (Shiraiwa and Sinclair, 1993; Sinclair and Shiraiwa, 1993; Bindraban, 1999; van Oosterom *et al.*, 2010). Leaf N concentration in top leaves in the canopy remains optimal or near optimal, which intercept most of the incident photosynthetic active radiation (PAR) during seed fill. Sinclair and Amir (1992) and Sinclair and Muchow (1995) demonstrated by a sensitivity test that grain yield prediction was not sensitive to the proportion of leaf N mobilization that goes to leaf senescence or decline in leaf N concentration and hence RUE. Jamieson and Semenov (2000) stated that probably any error in the assumption of constant leaf N concentration (hence a slight overestimation of RUE) will be compensated by an overestimate in leaf senescence.

As shown by Sinclair and de Wit (1976), translocation of N from leaves to support seed growth results in a self-destruction phenomenon leading to a loss in crop productivity. Accelerated transfer of N from vegetative tissue under limited N conditions results in earlier maturity than expected based on temperature unit. This earlier exhaustion of TRLN lowers grain yield. It is assumed the crop has reached termination seed growth when LAI decreases below 0.05. Thus, limited N conditions can result in early termination of seed growth.

Daily changes in leaf area can be calculated from N mobilization from leaves. N mobilization from leaves results in leaf senescence and hence a decrease in radiation interception and crop mass production. Daily decrease in LAI (DLAI, $m^2\ m^{-2}\ day^{-1}$) can be obtained from XNLF and the translocatable N per unit leaf (SLNG – SLNS).

$$DLAI = XNLF / (SLNG - SLNS) \qquad (17.8)$$

In the model that accounts for plant N budget, linear decrease in LAI during grain filling period defined by Eqn 9.7 (Crop LAI submodel; Chapter 9) is replaced with Eqn 17.8.

Plant N Budget in Legumes

The fact that legumes are able to fix atmospheric N_2 can minimize or eliminate the need to simulate the soil N budget (Sinclair *et al.*, 2003). Thus, plant N balance is further simplified for grain legume crops. It is assumed that N demand can be fully met by either N uptake from the soil or N_2 fixation, or a combination of both. Consequently, simulation of soil N balance (Chapter 18) can be bypassed for grain legume crops.

However, there are specific circumstances where daily demand for N may not be fully met by daily N accumulation rate during vegetative growth from emergence to beginning seed growth. These circumstances include the following:

- N_2 fixation in grain legumes is not active until a certain amount of temperature unit (or biological days) after sowing has elapsed. During this period, N accumulation is dependent on soil N availability. Thus, soil limited N availability can result in decreased N accumulation. The amount of soil N available for crop uptake is an input for the legume model.
- A maximum N uptake rate (MXNUP) is also assumed for legumes, so if daily demand is greater than MXNUP, then the N supply to the plant is limited to the value of MXNUP.
- Soil water deficit can have a large negative influence on grain legume N accumulation via N_2 fixation activity. In soybean, for example, it has been reported that N_2 fixation is highly sensitive to soil water deficit and decreased rates are initiated earlier with soil drying than all other physiological processes (Serraj *et al.*, 1999). This sensitivity is accounted for in the model by calculating a water stress factor for N fixation (WSFN). WSFN is obtained from FTSW using a threshold that specifies the sensitivity of N_2 fixation to water deficit (WSSN) in the crop/genotype under consideration. WSFN is calculated in a similar manner that was described for growth and leaf area development in Chapter 15.
- N accumulation is sensitive to flooding conditions (Sinclair *et al.*, 2003). N accumulation rate in the model is set equal to 0 whenever WAT1 is at least 95% of saturation.

When N accumulation rate is not sufficient to support fully the N requirements for new leaf area development and new stem growth, adjustments are required in the N distribution within the plant. The same sequence of processes described previously for non-legume crops is also invoked for legumes.

During the seed growth period, from beginning to termination seed growth, no N accumulation from the soil and/or N_2 fixation is simulated (i.e. NUP = 0) because it is assumed that all daily dry matter production goes to the grains. The process of N accumulation and resultant leaf senescence in legumes is again identical to that described for non-legumes.

Although it is not used here, nitrogen fixation by grain legumes has been simulated during grain filling for the brief periods when there is excess photosynthate available to support fixation (Sinclair *et al.*, 2003; Soltani and Sinclair, 2011). The simple approach outlined by Sinclair *et al.* (2003) assumed that during seed growth the daily amount of N accumulation rate can be calculated based on observation that N_2 fixation rate is closely correlated with vegetative mass. It is assumed that the capacity for N accumulation during the final stage of vegetative growth carries over to the seed growth period. Therefore, the N_2 fixation coefficient to do this calculation can be determined by calculating the ratio of N accumulation rate and vegetative mass during the days immediately preceding seed growth. Then, the calculated coefficient can be multiplied by the vegetative mass during seed fill to calculate accumulation of N during seed filling period when photosynthate supply exceeds that required by the growing seeds. If the calculated rate of N accumulation during seed filling is greater than N demand by growing seeds, the excess N is distributed to the stems.

Parameterization

As mentioned above, readily measurable crop parameters are required for simulation of plant N balance by the approach used here. The required crop parameters are: leaf N concentration per unit area in green leaves (SLNG) during vegetative growth and in senesced leaves (SLNS), stem N concentration per unit weight in green stems (SNCG) during vegetative growth and in senesced stems (SNCS), grain N content (GNC), and maximum N accumulation rate (MXNUP). Table 17.1 includes estimates of these parameters in some field crops.

The concentration parameters can be measured directly in the field. MXNUP can be found by examining the N accumulation curve for the maximum slope in the curve. Maximum rate of N accumulation is generally between 0.2 and 0.6 g N m^{-2} day^{-1} (Viets, 1965; Sinclair and Amir, 1992; Sinclair *et al.*, 2003). For an example of the determination of the parameters, refer to Soltani *et al.* (2006e).

SLNS and SNCS indicate minimum leaf and stem N content that is structural and is not available for mobilization. The minimum values of N have

Table 17.1. Estimates of crop parameters that govern plant N accumulation and distribution in some crops (Sinclair and Amir, 1992; Sinclair and Muchow, 1995; Robertson *et al.*, 2002a; Sinclair *et al.*, 2003; Hammer *et al.*, 2010; Soltani and Sinclair, 2011).

Crop	SLNG	SLNS	SNCG	SNCS	GNC	MXNUP
Wheat	1.50	0.40	0.015	0.005	0.015	0.25
Barley					0.016	
Rice					0.023	
Maize	1.35	0.55			0.011	
Sorghum	1.06	0.55			0.016	
Soybean	2.50	0.80	0.020	0.005	0.065	0.60
Peanut	2.20	1.00	0.020	0.010	0.048	0.50
Canola					0.010	
Sunflower					0.024	
Dry bean	1.80	0.55	0.020	0.008	0.035	0.50
Chickpea	2.30	0.78	0.025	0.0078	0.043	0.45

SLNG: Specific leaf N in green leaves (g N m^{-2})
SLNS: Specific leaf N in senesced leaves (g N m^{-2})
SNCG: N content in green stems (g N g^{-1})
SNCS: N content in senesced stems (g N g^{-1})
GNC: Grain N content (g N g^{-1})
MXNUP: Maximum N uptake rate (g N m^{-2} day^{-1})

another application in some crop models to calculate N remaining in crop residues after harvest.

Programming

Accounting for plant N budget makes it necessary to prepare three submodels in the program code. The first submodel is to calculate the plant N balance in non-legume crops as described in above sections. Box 17.2 represents this submodel. This submodel needs to be accompanied by a soil N submodel, which will be developed in the next chapter.

The second submodel is for grain legume crops in which the plant N submodel is somewhat different because there is no need for a soil N submodel. Box 17.3 includes this submodel.

A third submodel can be developed in which increase in crop LAI (GLAI; Eqns 9.5 and 9.6, Chapter 9) is limited to available N for leaf expansion (INLF) and leaf senescence is predicted from N mobilization from leaves, Eqn 17.8. Box 17.4 includes this submodel, which is identical to the crop LAI submodel developed in Chapter 9, except for two changes. In the former submodel (Box 9.2), GLAI was not limited to available N and LAI senescence was simulated by linear decrease in crop LAI from its value at the beginning of seed growth to zero at maturity.

Box 17.2. Program of plant N submodel for non-legume crops. For names of variables refer to the text or Appendix III.

```
NonLegumPlantN:
  If iniPNB = 0 Then
     SLNG = Sheet5.[B44]
     SLNS = Sheet5.[B45]
     SNCG = Sheet5.[B46]
     SNCS = Sheet5.[B47]
     GNC = Sheet5.[B48]
     MXNUP = Sheet5.[B49]

     NST = WST * SNCG:    NLF = LAI * SLNG:
     CNUP = NST + NLF:    NGRN = 0:          iniPNB = 1:
  End If

  If CTU <= tuEMR Or CTU > tuTSG Then
       NUP = 0:      XNLF = 0:     XNST = 0:
       INLF = 0:     INST = 0:     INGRN = 0:
  ElseIf CTU > tuEMR And CTU < tuBSG Then
       INGRN = 0
       NSTDF = (WST * SNCG) – NST
       If NSTDF < 0 Then NSTDF = 0
       NUP = (GST * SNCG) + (GLAI * SLNG)   '+ NSTDF      '<----- Inactive!
       If NUP > MXNUP Then NUP = MXNUP
```

Continued

Box 17.2. Continued.

```
        If NUP < 0 Then NUP = 0
        If DDMP = 0 Then NUP = 0
        If NUP > SNAVL Then NUP = SNAVL

        If NST <= (WST * SNCS) Then
            INST = WST * SNCS – NST
            XNST = 0
            If INST >= NUP Then
                INLF = 0
                XNLF = INST – NUP
            ElseIf INST < NUP Then
                INLF = GLAI * SLNG
                If INLF > (NUP – INST) Then INLF = NUP – INST
                INST = NUP – INLF
                XNLF = 0
            End If
        ElseIf NST > (WST * SNCS) Then
            INLF = GLAI * SLNG
            XNLF = 0
            If INLF >= NUP Then
                INST = 0
                XNST = INLF – NUP
                If XNST > (NST – WST * SNCS) Then XNST = NST – WST * SNCS
                INLF = NUP + XNST
            ElseIf INLF < NUP Then
                INST = NUP – INLF
                XNST = 0
            End If
        End If

    ElseIf CTU >= tuBSG And CTU <= tuTSG Then
        INGRN = SGR * GNC
        NUP = INGRN
        If FTSW1 > 1 Then NUP = 0
        If DDMP <= (SGR / GCC) Then NUP = 0
        If DDMP = 0 Then NUP = 0
        If NUP > SNAVL Then NUP = SNAVL

        If NUP > (SGR * GNC) Then
            'N is excess of seed needs
            INLF = 0
            INST = NUP – SGR * GNC
            XNLF = 0
            XNST = 0
        ElseIf NUP <= (SGR * GNC) Then
            'Need to transfer N from vegetative tissue
            INLF = 0
            INST = 0
```

Continued

Box 17.2. Continued.

```
            XNLF = (SGR * GNC – NUP) * FXLF
            XNST = (SGR * GNC – NUP) * (1 – FXLF)
      End If
   End If

   NST = NST + INST – XNST
   NLF = NLF + INLF – XNLF
   NVEG = NLF + NST
   NGRN = NGRN + INGRN
   CNUP = CNUP + NUP

   TRLN = LAI * (SLNG – SLNS) + (NST – WST * SNCS)
   FXLF = LAI * (SLNG – SLNS) / (TRLN + 0.000000000001)
   If FXLF > 1 Then FXLF = 1
   If FXLF < 0 Then FXLF = 0
Return
```

Box 17.3. Program of plant N submodel for legume crops. For names of variables refer to the text or Appendix III.

```
LegumPlantN:
  If iniPNB = 0 Then
    SLNG = Sheet5.[B44]
    SLNS = Sheet5.[B45]
    SNCG = Sheet5.[B46]
    SNCS = Sheet5.[B47]
    GNC = Sheet5.[B48]
    MXNUP = Sheet5.[B49]
    tuBNF = Sheet5.[B51]
    WSSN = Sheet5.[B52]
    INSOL = Sheet5.[B53]

    NST = WST *SNCG:    NLF = LAI * SLNG:    WSFN = 1:
    CNUP = NST + NLF:   NGRN = 0:            iniPNB = 1:
  End If

  If CTU <= tuEMR Or CTU > tuTSG Then
      NUP = 0:  XNLF = 0:  XNST = 0:  INLF = 0:  INST = 0:  INGRN = 0:
  ElseIf CTU > tuEMR And CTU < tuBSG Then
      INGRN = 0
      NSTDF = (WST * SNCG) – NST
      If NSTDF < 0 Then NSTDF = 0
      NUP = (GST * SNCG) + (GLAI * SLNG)   '+ NSTDF    '<----- Inactive!
      If CTU < tuBNF And CNUP > INSOL Then NUP = 0
      If NUP > MXNUP Then NUP = MXNUP
      NFC = NFC * 3 / 4 + NUP / WVEG * (1 / 4)     'from Sinclair et al. 2003
      NUP = NUP * WSFN
```

Continued

Box 17.3. Continued.

```
    If NUP < 0 Then NUP = 0
    If FTSW > 1 Then NUP = 0
    If DDMP = 0 Then NUP = 0

    If NST <= (WST * SNCS) Then
        INST = WST * SNCS - NST
        XNST = 0
        If INST >= NUP Then
          INLF = 0
          XNLF = INST - NUP
        ElseIf INST < NUP Then
          INLF = GLAI * SLNG
          If INLF > (NUP - INST) Then INLF = NUP - INST
          INST = NUP - INLF
          XNLF = 0
        End If
    ElseIf NST > (WST * SNCS) Then
        INLF = GLAI * SLNG
        XNLF = 0
        If INLF >= NUP Then
          INST = 0
          XNST = INLF - NUP
          If XNST > (NST - WST * SNCS) Then XNST = (NST - WST *
             SNCS)
          INLF = NUP + XNST
        ElseIf INLF < NUP Then
          INST = NUP - INLF
          XNST = 0
        End If
    End If

ElseIf CTU >= tuBSG And CTU <= tuTSG Then
    INGRN = SGR * GNC
    NUP = INGRN
    PDNF = NFC * WVEG
    If PDNF > NUP Then PDNF = NUP
    DNF = PDNF * WSFN
    If DNF < 0 Then DNF = 0
    If DDMP <= (SGR / GCC) Then DNF = 0
    If DDMP = 0 Then DNF = 0
    NUP = DNF

    If NUP > (SGR * GNC) Then
        'N is excess of seed needs
        INLF = 0
        INST = NUP - SGR * GNC
        XNLF = 0
        XNST = 0
```

Continued

Box 17.3. Continued.

```
      ElseIf NUP <= (SGR * GNC) Then
         'Need to transfer N from vegetative tissue
         INLF = 0
         INST = 0
         XNLF = (SGR * GNC – NUP) * FXLF
         XNST = (SGR * GNC – NUP) * (1 – FXLF)
      End If
   End If

   NST = NST + INST – XNST
   NLF = NLF + INLF – XNLF
   NVEG = NLF + NST
   NGRN = NGRN + SGR * GNC
   CNUP = CNUP + NUP

   TRLN = LAI * (SLNG – SLNS) + (NST – WST * SNCS)
   FXLF = LAI * (SLNG – SLNS) / (TRLN + 0.000000000001)
   If FXLF > 1 Then FXLF = 1
   If FXLF < 0 Then FXLF = 0
Return
```

Box 17.4. Program of crop LAI submodel that accounts for the effect of N on leaf area development and senescence. Former crop LAI submodel (Chapter 9) is replaced by this submodel in crop model for N-limited conditions. For names of variables refer to the text or Appendix III.

```
CropLAIN:
'---------------------------- LAI initials and pars
  If iniLAI = 0 Then
    PHYL = Sheet5.[b17]
    PLACON = Sheet5.[b18]
    PLAPOW = Sheet5.[b19]
    SLA = Sheet5.[b20]

    MSNN = 1:   PLA2 = 0:   PLA1 = 0:   LAI = 0:
    MXLAI = 0:  WSFL = 1:   SLNG = 2:   iniLAI = 1:
  End If

'---------------------------- Yesterday LAI to intercept PAR today
  If GLAI > (INLF / SLNG) Then GLAI = (INLF / SLNG)
  LAI = LAI + GLAI – DLAI
  If LAI < 0 Then LAI = 0
  If LAI > MXLAI Then MXLAI = LAI          'Saving maximum LAI

'---------------------------- Daily increase and decrease in LAI today
  If CTU <= tuEMR Then
    GLAI = 0:
  ElseIf CTU > tuEMR And CTU <= tuTLM Then
    INODE = DTU / PHYL
    MSNN = MSNN + INODE
```

Continued

Box 17.4. Continued.

```
    PLA2 = PLACON * MSNN ^ PLAPOW
    GLAI = ((PLA2 – PLA1) * PDEN / 10000) * WSFL
    PLA1 = PLA2
  ElseIf CTU > tuTLM And CTU <= tuBSG Then
    GLAI = GLF * SLA
    BSGLAI = LAI                  'Saving LAI at BSG
  ElseIf CTU > tuBSG Then
    GLAI = 0
  End If
  DLAI = XNLF / (SLNG – SLNS)
Return
```

Exercises

1. Examine N parameter estimates of Table 17.1. What are crop differences and similarities with respect to the parameters? Are the differences related to any crop classifications?
2. Try to find N parameters for your cultivars/crops at your location.
3. Assume a crop has a LAI of 5 at beginning seed growth. SLNG and SLNS are 1.5 and 0.5 g N m^{-2}, respectively. If GNC of the crop is 0.015 g N g^{-1} and N stored in leaves is the only source of N for grain filling, how much will be the grain yield? What will be the grain yield if GNC is 0.020 g N g^{-1}?
4. In Exercise 3, if daily SGR is 25 g m^{-2}, how long will be the grain filling period?
5. With a SLN of 2.5 g N m^{-2} and MXNUP of 0.5 g N m^{-2}, how much will be the maximum daily increase in LAI (GLAI) if all daily N uptake goes for LAI growth?

18 Soil Nitrogen Balance

The ability to simulate nitrogen (N) dynamics in the soil is crucial for simulating crop growth for several reasons. First, N is the nutrient that is required in the greatest amounts by plants, and it generally limits crop growth and yield more often than any other nutrient. Second, N is used as a fertilizer and the costs related to N application account for an appreciable part of field expenses. Third, N release to the environment as nitrate into waterways or gaseous forms into the atmosphere can be important environmental menaces. Many environmental concerns such as eutrophication of surface waters, acid rains, hypoxia, and global warming are related to environmental burdens of N.

N processes in the soil are important components of agricultural systems and reasonable predictions of these processes are essential for crop models. Modeling N dynamics in the soil has been attempted since the 1960s, but N dynamics in the soil is still considered one of the least understood areas in crop production (Ma and Shaffer, 2001). There are many models that predict N dynamics within agricultural soils. Ma and Shaffer (2001) reviewed nine US carbon and N models and McGechan and Wu (2001) reviewed four European models.

A soil N submodel is needed in crop growth models primarily to provide estimates of available soil N for crop uptake. This objective is less rigorous than an objective of fully simulating N release to the environment. Consequently, a fairly simple soil N submodel is usually sufficient for providing input to crop growth models. The model presented in this chapter is based on the simple soil N submodel successfully used by Sinclair and Amir (1992), Sinclair and Muchow (1995), and Sinclair *et al.* (1997) in their wheat, maize, and sorghum models.

Soil N Balance

Similar to soil water balance, soil N balance in any soil layer is the result of processes that add N to the soil and those that remove it from the soil. Thus, soluble N in the soil solution (NSOL, g N m^{-2}) can be defined as:

$$\text{NSOL} = \text{soil N balance} = \text{N inputs} - \text{N losses} \tag{18.1}$$

Various processes add and remove N from the soil solution. N is added to the soil by mineralization of organic matter (NMIN, g N m^{-2} day^{-1}) and by fertilizer application (NFERT, g N m^{-2} day^{-1}). On the other hand, N volatilization (NVOL, g N m^{-2} day^{-1}), leaching (NLEACH, g N m^{-2} day^{-1}), denitrification (NDNIT, g N m^{-2} day^{-1}) and crop uptake (NUP, g N m^{-2} day^{-1}) are the principal processes that remove N from the soil solution. Approaches to calculate each of these processes are presented in this chapter.

Equation 18.1 can be rewritten to include explicitly each of the terms contributing to the soil N balance.

$$\text{NSOL}_i = \text{NSOL}_{i-1} + \text{NMIN} + \text{NFERT} - \text{NVOL} - \text{NLEACH} - \text{NDNIT} - \text{NUP} \tag{18.2}$$

A relational diagram of the interaction among the processes in the submodel is shown in Fig. 18.1. Box 18.1 presents a summary of the approaches used in the submodel.

In addition to the processes included in Eqn 18.2, there are other processes that participate in the dynamic soil N balance but these are usually quite small. N can be added to the soil due to atmospheric reduction of N_2 as a result of lightning activity and biological fixation by free-living organisms. N removal can occur in runoff when soil particles to which N might be tightly attached is eroded from a field. In addition, high levels of carbon (organic matter) in the soil encourage microbial growth, which will consume N and sequester it in the microbes. Of course, when the microbes die they contribute to the soil organic matter and the N can be returned to the soil solution. To account for this microbial activity, net mineralization of N into the soil solution is considered in this chapter.

In this book, soil N is only tracked in the soil top layer (see Chapter 14). The reason for limiting the soil N submodel to a top soil layer is because this is the layer where nearly all the dynamic processes of the soil N balance occur. That is, fertilizer is added to this top layer, and by definition virtually all of the soil organic matter is in this top layer, and the uptake of soil N must necessarily be from this top layer. Soil organic matter decreases exponentially by soil depth, so lower layers in the soil usually have very low amounts of organic matter (Fig. 18.2; Franzluebbers, 2010). The actual thickness of this top layer is defined by the model user and will usually have a thickness between 200 and 600 mm. In addition, the user will need to input soil N characteristics/inputs, such as total N and initial N in soil solution, related to the top layer. If the soil characteristics or the model objective require, a model that accounts for several soil layers may be required. A multi-layer soil version of the model can be found on the book's website.

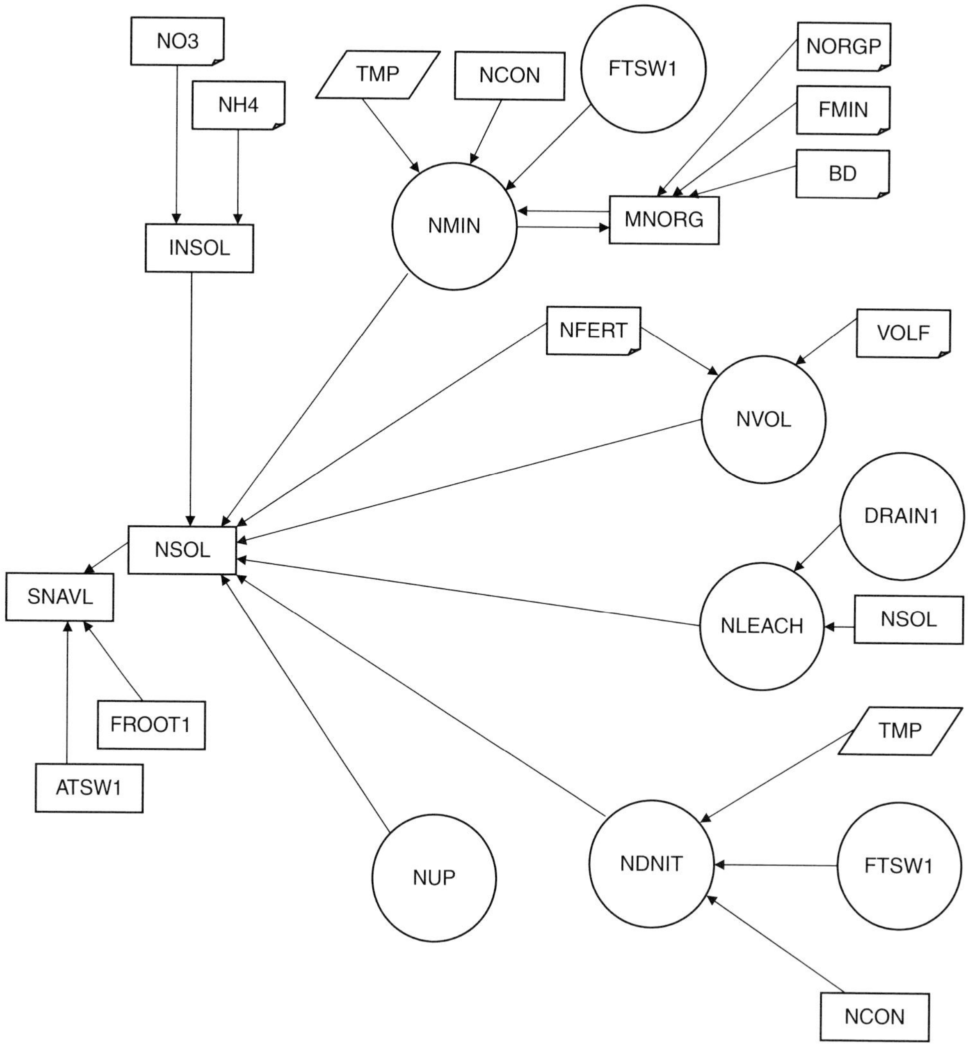

Fig. 18.1. Relational diagram of soil N submodel. SNAVL is the soil available N (g N m^{-2}), NSOL the total soil soluble N (g N m^{-2}), INSOL the initial total soil soluble N (g N m^{-2}), FROOT1 the ratio of current crop rooting depth to the top layer thickness (0–1), ATSW1 the actual transpirable soil water in the top layer (mm), NH4 the NH_4^+ content in soil solution (mg kg^{-1} or ppm), NO3 the NO_3^- content in soil solution (mg kg^{-1} or ppm), NMIN the daily rate of mineralization of organic N (g N m^{-2} day^{-1}), TMP the average daily temperature (°C), NCON the concentration of N in the soil solution (g N g^{-1} water), FTSW1 the fraction of transpirable soil water, MNORG the potentially mineralizable soil N (g N m^{-2}), NORGP the soil organic N (%), FMIN the fraction soil organic N available to mineralization (g g^{-1}), BD the soil bulk density (g cm^{-3}), NVOL the daily rate of N volatilization (g N m^{-2} day^{-1}), NFERT the rate of N applied as fertilizer (g N m^{-2}), VOLF the fraction of N fertilizer volatilized (g g^{-1}), NLEACH the daily rate of N leaching from the top layer (g N m^{-2} day^{-1}), DRAIN1 the daily rate of drainage from the top layer (mm day^{-1}), NDNIT the daily rate of N denitrification (g N m^{-2} day^{-1}), and NUP the daily rate of N uptake by the crop (g N m^{-2} day^{-1}).

Box 18.1. Overview of the approach that is used to simulate soil N balance.

N inputs from mineralization and fertilization are considered:

- N mineralization is obtained from soil organic and soluble N, soil water, and temperature.
- N from fertilizer application should be known.

N losses from volatilization, leaching, denitrification, and crop uptake are considered:

- N volatilization is obtained as a fraction of applied N.
- N leaching from drainage from top layer and N concentration.
- N denitrification from soil N concentration, soil water, and temperature.
- N uptake (as described in Chapter 17).

Available N in Soil Solution

Not all the N in the soil solution is readily available to crops. Three factors determine the amount of soil N which is accessible to the crop (SNAVL, g N m^{-2}). First, part of the soil solution is held very tightly by soil particles and roots do not have direct access to this part of the soil solution. Therefore, the N in this fraction of the soil solution is not available to the plants. Only the available soil water in the soil top layer holds the N in soil solution available to the crop. To facilitate the presentation, it will be assumed that the depth of top soil layer for the N submodel corresponds to the depth of the top layer for soil evaporation. In the use of the model, this assumption is not required. Therefore, in the remainder of the chapter the available water for calculating N availability is ATSW1 (mm; Chapter 14).

To track the amount of N in the accessible soil solution it is convenient to base calculations on the concentration of N in the soil solution (NCON, g N g^{-1} water). The N concentration is equal to the amount of N in the soil solution (NSOL) divided by the amount of soil water per unit horizontal area in the soil layer of interest (WAT1, mm; Chapter 14).

$$NCON = NSOL / (WAT1 \times 1000) \qquad (18.3)$$

The number 1000 in Eqn 18.3 converts WAT1 from mm to g. Of course, Eqn 18.3 can be used reversibly to calculate NSOL from NCON.

A second factor limiting accessibility of the soil N solution is the inability of plants to extract N from the solution at very low concentrations. Commonly, little N is taken up by plants when NCON is less than 0.000001 g N g^{-1} water (1 mg N l^{-1}).

The third limiting factor that might limit access to soil solution N is a root system that does not fully occupy the top soil layer. This is almost always only a possibility early in the crop growth season when roots are developing. The fraction of the top layer occupied by roots is simply taken as the ratio of the current depth of rooting (DEPORT, mm; Chapter 14) divided by the depth of the top soil layer for the N soil solution (DEP1, mm; Chapter 14). Since DEPORT continues

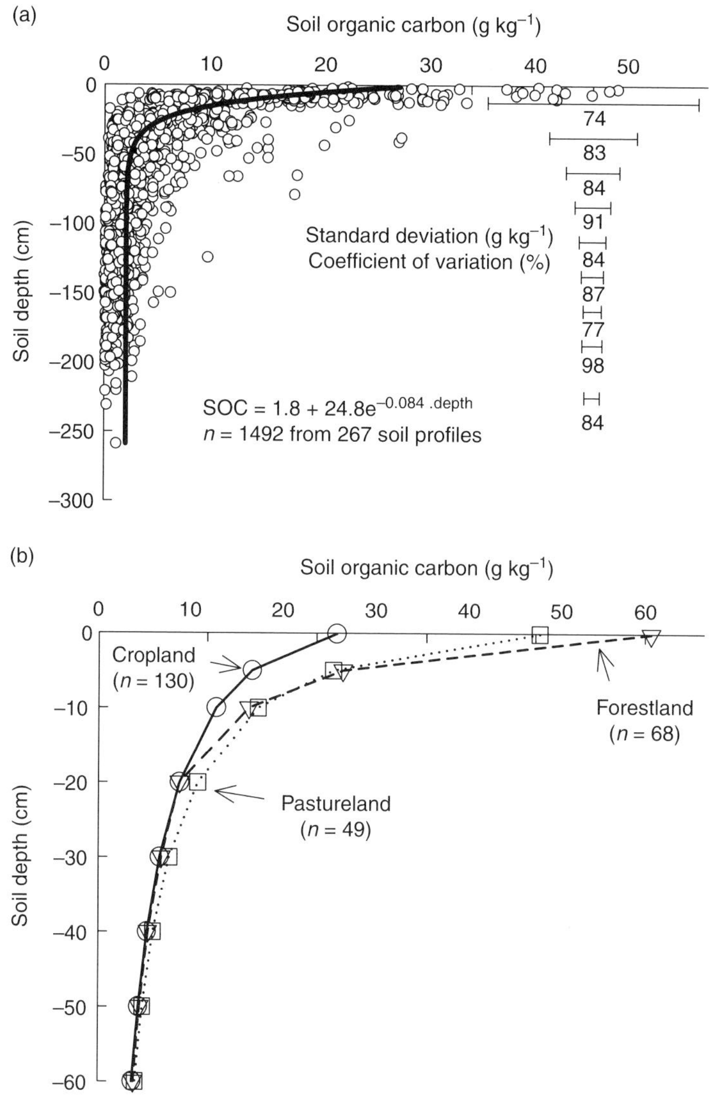

Fig. 18.2. (a) Soil organic carbon depth distribution in 267 soil profiles in Georgia, USA and (b) the distribution when data averaged within a land use category (Franzluebbers, 2010).

to grow beyond the depth of the top soil layer, it is necessary to set the maximum value of this ratio at 1.0. This is expressed in the following equation.

$$\text{FROOT1} = \min(\text{DEPORT} / \text{DEP1},\ 1) \tag{18.4}$$

According to Eqn 18.4, if crop rooting depth is equal to or greater than the top layer depth (DEP1), all available soil N will be accessible to the crop. For example, a soil with DEP1 of 600 mm and crops with rooting depths of 200 mm or 1000 mm will have access to 0.3 (30%) or 1.0 (100%) of soil available N, respectively.

Taken together, the equation defining the amount of N available to the crop from the soil solution is defined in the following equation.

$$SNAVL = (NCON - 0.000001) \times ATSW1 \times 1000 \times FROOT1 \quad (18.5)$$

N Inputs

Mineralization

Most agricultural soils include several thousand kilograms of organic N per hectare. However, only a fraction of the N in this organic matter becomes available to the crop each year through mineralization depending on the composition of the organic matter (Stanford and Smith, 1972; Campbell *et al.*, 1995). In natural environments, mineralization is the single most important process that releases N to the soil for plant use. In croplands, mineralization can also be a very important source of N.

Mineralization of organic N to ammonium (NH_4^+) and the subsequent transformation to nitrate (NO_3^-) is modeled as one transformation. This transformation is dependent on microbial activity and generally occurs readily in well-drained soils. The transformation of NH_4^+ to NO_3^- occurs very rapidly, so ammonium constitutes only a small portion of N in the soil solution. Net mineralization (mineralization – immobilization) is simulated using a simple approach (Stanford and Smith, 1972; Watts and Hanks, 1978) as presented by Sinclair and Amir (1992) and Sinclair and Muchow (1995).

Daily net mineralization (NMIN, g N m^{-2} day^{-1}) is obtained as a function of potentially mineralizable soil N (MNORG, g N m^{-2}) and the suitability of soil temperature (KN) and water (RN) for mineralization.

$$NMIN = MNORG \times KN \times RN \quad (18.6)$$

KN describes the sensitivity of mineralization to soil temperature (TMPS, °C) and is described by an exponential function (Watts and Hanks, 1978; Fig. 18.3a). For simplicity in presentation, soil temperature is assumed equal to the air temperature.

$$KN = 1 - Exp(-KNMIN) / 168 \quad (18.7)$$

$$KNMIN = 24 \times Exp(17.753 - 6350.5 / (TMPS + 273)) \quad 18.8)$$

RN describes the sensitivity of mineralization rate to soil moisture conditions. Fraction transpirable soil water in the top layer (FTSW1) is used as an indicator of soil moisture status (Watts and Hanks, 1978; Fig. 18.3b).

$$RN = 1.111 \times FTSW1 \quad \textit{if} \quad FTSW1 < 0.9 \quad (18.9)$$

$$RN = 10 - 10 \times FTSW1 \quad \textit{if} \quad FTSW1 \geq 0.9 \quad (18.10)$$

Since mineralization is a biological activity, it is not surprising that high concentrations of N in the soil solution result in a feedback that inhibits mineralization. In the approach used by Sinclair and Amir (1992), NMIN is decreased linearly as a function of N concentration in soil solution (NCON) so that NMIN reaches 0 as NCON increases to 0.0002 g g^{-1} (= 200 mg N l^{-1}) (Fig. 18.4).

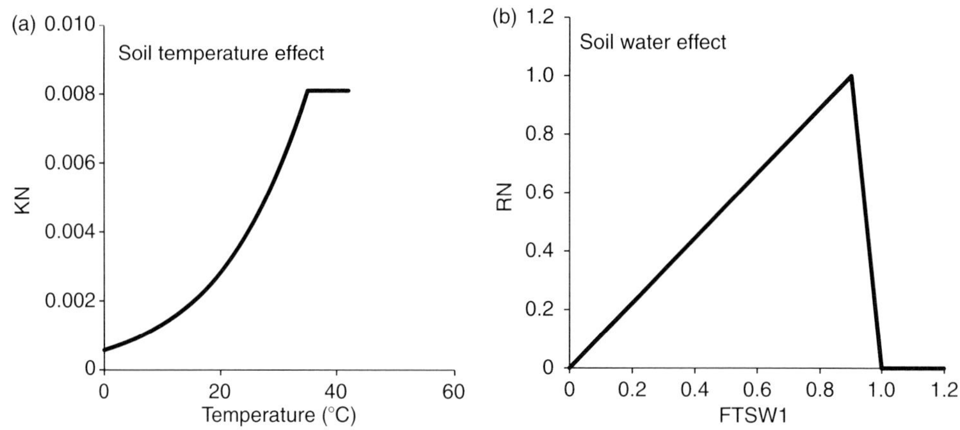

Fig. 18.3. (a) Sensitivity of N mineralization to soil temperature (KN) and (b) fraction transpirable soil water in the top layer (FTSW1) (RN).

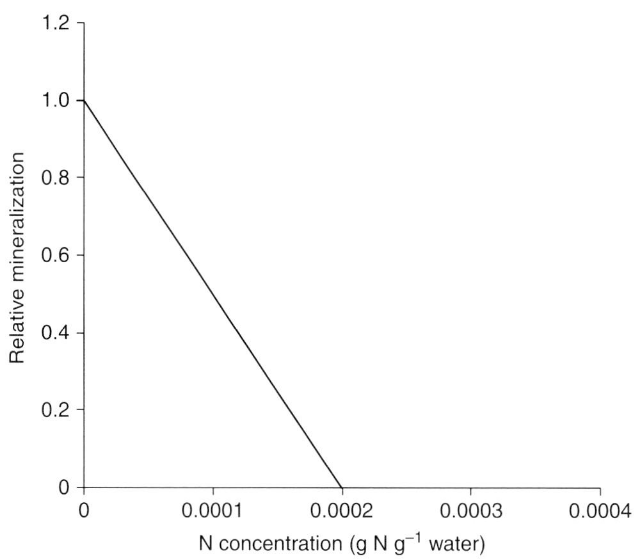

Fig. 18.4. Relative N mineralization as a function of N concentration in soil solution.

$$\text{NMIN} = \text{NMIN} \times \max(0, (0.0002 - \text{NCON}) / 0.0002) \tag{18.11}$$

Cumulative net mineralization (CNMIN) is tracked in the model by summing the daily values of NMIN. Cumulative mineralization up to today (CNMIN_i, g N m^{-2}) is obtained by adding current NMIN to cumulative NMIN of the previous day (CNMIN_{i-1}, g N m^{-2}).

$$\text{CNMIN}_i = \text{CNMIN}_{i-1} + \text{NMIN} \tag{18.12}$$

Fertilizer application

The submodel also accounts for N application to the soil of inorganic fertilizers. (Fertilization with organic fertilizers or manures can be included in the mineralization calculations above by increasing MNORG appropriately.) It is assumed that fertilizer is immediately solubilized in the soil solution of the top soil layer. While this assumption is not immediately true, the high solubility of N and its rapid transport in the soil solution allows the assumption to quickly become appropriate. The soil N submodel allows up to ten fertilization applications but this can be increased if required. Amount of net N in each application (NFERT) needs to be defined as N, not in the form of NO_3^- or NH_4^+. Thus, equivalent N in the fertilizer must be inputted into the model along with the time of application as days after sowing.

N Losses

Volatilization

Volatilization of N is a complicated process depending on weather and soil conditions, as well as cultural management such as type of N fertilizer and the time and method of application. For a review of existing approaches refer to Ma and Shaffer (2001) and McGechan and Wu (2001). Fortunately, a simple, empirical method can be used to compute volatilization rate with sufficient accuracy for most simulations of crop growth.

Following any fertilizer application, N loss through volatilization (NVOL, g N m^{-2} day^{-1}) is calculated as a single pulse as a fraction (VOLF, g g^{-1}) of applied N (NFERT, g N m^{-2}). That is:

$$\mathrm{NVOL} = \mathrm{VOLF} \times \mathrm{NFERT} \tag{18.13}$$

Therefore, VOLF must be inputted into the model for each N fertilization event. Table 18.1 includes VOLF for a number of N fertilizers under three conditions

Table 18.1. N volatilization factor (%) as influenced by environment, fertilizer type, and management (Delgado *et al.*, 2010; originally from Meisinger and Randall, 1991).

		Environment		
N fertilizer	Application	Humid	Sub-humid	Dry
Urea	Surface applied	10	15	25
Urea	Incorporated	2	3	5
$(NH_4)_2SO_4$	Surface applied	4	8	15
$(NH_4)_2SO_4$	Incorporated	1	1	2
NH_4NO_3	Surface applied	2	4	10
NH_4NO_3	Incorporated	0	5	1
Anhydrous-NH_3	Incorporated	1	2	3

of humid, sub-humid, or dry environmental conditions and type of application (Delgado *et al.*, 2010; originally from Meisinger and Randall, 1991). For example, if N fertilizer is surface applied as urea just before a rainfall event, VOLF of 0.1 (10%) or 0.15 (15%) can be used.

Cumulative N volatilization is also calculated by the submodel.

Leaching

N can be carried below the active root zone with the water moving downward in the soil. In the model, leaching is calculated from the top layer, and assumed to be lost to any major recovery by the crop. The amount of N loss through leaching (NLEACH, g N m^{-2} day^{-1}) is obtained as a function of soil soluble N (NSOL, g N m^{-2}), the amount of drained water from the top layer (DRAIN1, Chapter 14), and total soil water in the top layer (WAT1, Chapter 14).

$$\text{NLEACH} = \text{NSOL} \times (\text{DRAIN1} / (\text{WAT1} + \text{DRAIN1})) \qquad (18.14)$$

According to Eqn 18.14, N leached from the top layer is proportional to the fraction of water that leaves the layer, as specified in the second term of the equation. However, NLEACH is set to 0 if N concentration in the top layer is very low, i.e. less than 0.000001 g N g^{-1} water (1 mg N l^{-1}).

The above calculation might overestimate N loss via leaching because part of the N leached to below the top N layer may be taken up by the crop. However, it will give an indication of the N loss through leaching. Inclusion of bypass water might offset this overestimation. Bypassed water is the fraction of soil water that is retained in small soil pores and does not participate in the leaching process, i.e. bypassed (Corwin *et al.*, 1991; Stockle *et al.*, 2003). Therefore, the fraction of soil water bypassed and its solute content is not mixed with new incoming water during infiltration. If the bypass coefficient (BC) is known for the top layer, a better estimate of NLEACH will be obtained as:

$$\text{NLEACH} = \text{NSOL} \times (\text{DRAIN1} \times (1 - \text{BC})) / (\text{WAT1} + \text{DRAIN1})) \qquad (18.15)$$

For more information about bypass and a method of obtaining BC, refer to Corwin *et al.* (1991).

Denitrification

Saturation of the soil with water results in the activation of microbes that thrive in anaerobic conditions. One set of microbes consume nitrate resulting in N denitrification. Denitrification can result in major losses of soil, especially if the soil is saturated frequently by heavy rains or irrigations. It is assumed that denitrification occurs whenever the water content of the top layer exceeds the well-drained soil water content, i.e. FTSW1 exceeds 1.

When FTSW1 > 1, N loss through denitrification (NDNIT, g N g^{-1} water day^{-1}) is obtained by an exponential function of N concentration (NCON) in the top layer, but a maximum soluble N concentration of 0.0004 g N g^{-1} water (400 mg N l^{-1}) is used (Reddy, 1976). That is, if NCON in the top layer is greater than 0.0004 g N g^{-1} water, NCON will be set to 0.0004 g N g^{-1} water (Fig. 18.5).

$$\text{NDNIT} = \min(\text{NCON}, 0.0004) \times (1 - \text{Exp}(-\text{KDNIT})) \tag{18.16}$$

The exponent, KDNIT, is calculated as a function of soil temperature (Fig. 18.6; Sinclair and Muchow, 1995) based on data collected for eight soils (Stanford *et al.*, 1975). Again, air temperature is used as substitute for soil temperature.

$$\text{KDNIT} = 6 \times \text{Exp}(0.07735 \times \text{TMPS} - 6.593) \tag{18.17}$$

The value of NDNIT calculated in Eqn 18.16 is in g N g^{-1} water per day; this term needs to be converted to units of ground area. Therefore, NDNIT needs to be multiplied by the amount of water in the top layer (WAT1).

$$\text{NDNIT} = \text{NDNIT} \times \text{WAT1} \times 1000 \tag{18.18}$$

The number 1000 in Eqn 18.18 converts WAT1 from mm to g. Therefore, NDNIT finally has the units of g N m^{-2} day^{-1}. However, porous soils may drain rapidly so that FTSW decreases to less than 1.0 within the first day. In this case, the value of NDNIT needs to be multiplied by the fraction of the day when FTSW is greater than 1.0. For example, Sinclair and Muchow (1995) multiplied NDNIT in their simulations by 0.25 to reflect the fact that they anticipated the soil to have the high soil water content for only about 6 h.

Cumulative NDNIT is computed by the submodel.

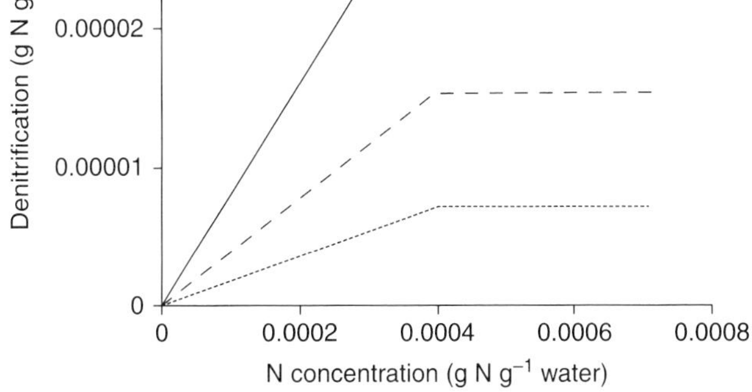

Fig. 18.5. N denitrification as a function of soil N concentration under three temperatures of 10, 20, and 30°C.

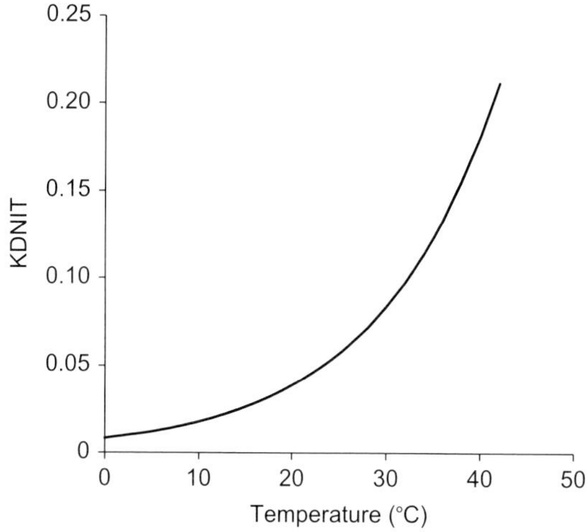

Fig. 18.6. Sensitivity (KDNIT) of N denitrification to temperature.

Crop N uptake

A major component in soil N balance is daily N uptake by the crop (NUP, g N m^{-2} day^{-1}; Chapter 17). Discussion of the calculation of NUP and the various factors that influence NUP was discussed in detail in Chapter 17.

Required Inputs

The soil N submodel as described in this chapter requires a few simple soil inputs. The depth of the top layer (DEP1), initial soil soluble N (NSOL), and initial soil organic N available for mineralization (NORG), as well as the time, amount, and volatilization fraction of each N application are required.

NSOL and NORG at the beginning of the simulation can be determined internally in the submodel from other soil inputs. These inputs are usually available from the results of routine soil tests:

- soil coarse fraction (FG, %);
- soil bulk density (BD, g cm^{-3});
- soil organic N (NORGP, %);
- fraction soil organic N available to mineralization (FMIN);
- NO_3^- content in soil solution (NO3, ppm or mg kg^{-1}); and
- NH_4^+ content in soil solution (NH4, ppm or mg kg^{-1}).

FG includes any soil particles greater than 2 mm such as stones, gravels, and coarse sands. FG and BD are necessary for calculation of soil mass and initial soil N.

Soil mass in the top layer (SOILM, g m^{-2}) is obtained from the thickness of the top layer (DEP1, mm), BD and FG.

$$SOILM = DEP1 \times BD \times (1 - FG) \times 1000 \quad (18.19)$$

The number 1000 is required for unit conversions.

Total organic soil N (NORG, in g N m^{-2}) can be found from SOILM and soil organic N percentage (NORGP, in %).

$$NORG = (NORGP \times 0.01) \times SOILM \quad (18.20)$$

Not all the soil organic N is subject to mineralization. Potentially mineralizable soil N comprises a small and variable portion of total soil N. This fraction is called FMIN and is an input for the submodel. FMIN depends on factors such as the amount and lability of organic N and microbial population of the soil. Stanford and Smith (1972) reported this fraction as 0.15 to 0.28 in Alfisols, 0.05 to 0.23 in Aridisols, 0.10 to 0.29 in Entisols, 0.10 to 0.26 in Mollisols, and 0.11 to 0.41 in Ultisols. A fraction of 0.15 has been used successfully in simulation exercises in different countries including the USA, Australia, India, and Israel (Sinclair and Amir, 1992; Sinclair and Muchow, 1995; Sinclair *et al.*, 1997).

Similarly, Jamieson *et al.* (1998) had to change net mineralization rate in their model from 0.12 kg N ha^{-1} day^{-1} under UK conditions to 0.42 kg N ha^{-1} day^{-1} for New Zealand conditions. They ascribed this higher rate to the greater lability of organic N in New Zealand soils due to inclusion of short-term pasture leys in New Zealand cropping rotations.

Some models replace the empiricism of FMIN in the current submodel with a large set of coefficients, which are generally empirical and require model calibration. Campbell *et al.* (1995) presented a method to estimate potentially mineralizable soil N. Using the method, FMIN can be estimated from N mineralization during the first 2 weeks of aerobic incubation at 35°C.

Using FMIN, potentially mineralizable soil N (MNORG, g N m^{-2}) can be found as:

$$MNORG = NORG \times FMIN \quad (18.21)$$

Another input for the submodel is NSOL at the beginning of the simulation, which is calculated from NO_3^- and NH_4^+ concentrations at sowing time. However, the amount of N in these soluble forms is usually low and little error results from general estimates of these two concentrations. These two variables are usually reported as mg per kg soil or ppm. The amounts of NO_3^- and NH_4^+ (i.e. ANO3 and ANH4 in g N m^{-2}) are obtained as:

$$ANO3 = NO3 \times (14 / 62) \times 0.000001 \times SOILM$$
$$ANH4 = NH4 \times (14 / 18) \times 0.000001 \times SOILM \quad (18.22)$$

where NO3 and NH4, both in mg kg^{-1}, represent concentration of NO_3^- and NH_4^+ in the top layer and 14, 62, and 18 are the molecular weights of N, NO_3^-, and NH_4^+, respectively. The number 0.000001 is used to convert the units from mg kg^{-1} to g N m^{-2}.

Initial soil N, NSOL in g N m^{-2}, is the sum of ANO3 and ANH4:

$$NSOL = ANO3 + ANH4 \quad (18.23)$$

Initial soil N concentration, NCON, is then obtained as:

$$NCON = NSOL / (WAT1 \times 1000) \quad (18.24)$$

Table 18.2. Soil carbon and total N in some default soils (Gijsman *et al.*, 2003).

Soil	Depth	Soil carbon (%)	Soil N (%)
Silty clay	Deep	1.60	0.158
	Medium	1.61	0.158
	Shallow	1.61	0.158
Silty loam	Deep	1.07	0.110
	Medium	1.07	0.110
	Shallow	1.07	0.110
Sandy loam	Deep	0.64	0.066
	Medium	0.64	0.066
	Shallow	0.64	0.066
Sand	Deep	0.27	0.026
	Medium	0.27	0.026
	Shallow	0.27	0.026

Below is an example to illustrate how these calculations are done in the submodel. For this example, it is supposed that top layer thickness is 600 mm, FG is equal to 0, BD is 1.35 g cm^{-3}, organic N is 0.08%, FMIN is 0.15, and initial concentrations of NO_3^- and NH_4^+ are 15 and 1.8 mg kg^{-1}, respectively.

SOILM = 600 × 1.35 × 1000 = 810,000 g soil m^{-2} or 8,100,000 kg ha^{-1}
NORG = 0.08 × 0.01 × 810,000 = 648 g N m^{-2} or 6480 kg ha^{-1}
MNORG = 648 × 0.15 = 97.2 g N m^{-2} or 972 kg ha^{-1}
ANO3 = 15 × (14/62) × 0.000001 × 810,000 = 2.74 g N m^{-2} or 27.4 kg ha^{-1}
ANH4 = 1.8 × (14/18) × 0.000001 × 810,000 = 1.13 g N m^{-2} or 11.3 kg ha^{-1}
NSOL = 2.74 + 1.13 = 3.88 g N m^{-2} or 38.8 kg N ha^{-1}

Table 18.2 includes estimates of soil carbon and N for some default soils. Other characteristics of the soils can be found in Table 13.1 (Chapter 13) and Table 14.2 (Chapter 14).

Programming

A program was written in VBA for the soil N submodel (Box 18.2). This submodel will be used in the next chapter to develop a model for N-limited conditions.

Box 18.2. Soil N submodel program. For names of variables refer to the text or Appendix III.

```
SoilN:
'---------------------------------------------- Parameters and initials
  If iniSNB = 0 Then
    FG = Sheet7.[b19] / 100          'Fraction soil > 2mm
    BD1 = Sheet7.[b20]               'Soil bulk density (g.cm-3)
```

Continued

Box 18.2. Continued.

```
    NORGP = Sheet7.[b21]          'Organic N (%)
    FMIN = Sheet7.[b22]           'Frac. Org N avail. for mineralization
    NO3 = Sheet7.[b23]            'ppm = mg.kg-1
    NH4 = Sheet7.[b24]            'ppm = mg.kg-1

    SOILM = DEP1 * BD1 * (1 – FG) * 1000      'g.m-2
    NORG = NORGP * 0.01 * SOILM               'g.m-2
    MNORG = NORG * FMIN                       'gN.m-2 org. N avail. for miner.

    NO3 = NO3 * (14 / 62) * 0.000001 * SOILM     'from ppm to gN.m-2
    NH4 = NH4 * (14 / 18) * 0.000001 * SOILM     'from ppm to gN.m-2

    NSOL = (NO3 + NH4)                        'gN.m-2
    NCON = NSOL / (WAT1 * 1000)               'gN.g-1 H2O

    INSOL = NSOL:       CNFERT = 0:     CNVOL = 0:
    CNLEACH = 0:        CNMIN = 0:      CNDNIT = 0:      iniSNB = 1:
  End If
'------------------------------ N net mineralization
  TMPS = TMP
  If TMPS > 35 Then TMPS = 35
  KNMIN = 24 * Exp(17.753 – 6350.5 / (TMPS + 273)) / 168
  KN = 1 – Exp(–KNMIN)
  If FTSW1 < 0.9 Then RN = 1.111 * FTSW1
  If FTSW1 >= 0.9 Then RN = 10 – 10 * FTSW1
  If RN < 0 Then RN = 0
  NMIN = MNORG * RN * KN
  NMIN = NMIN * (0.0002 – NCON) / 0.0002         'threshold = 200 mgN.L-1
  If NMIN < 0 Then NMIN = 0
  MNORG = MNORG – NMIN

  CNMIN = CNMIN + NMIN
'------------------------------ N application & volatilization
  NFERT = 0:    NVOL = 0:
  For N = 1 To FN
    If DAP = DAPNF(N) Then
      NFERT = NFERTI(N)                     'gN.m-2
      VOLF = VOLFI(N) / 100
      NVOL = VOLF * NFERT                   'gN.m-2
    End If
  Next N
  CNFERT = CNFERT + NFERT
  CNVOL = CNVOL + NVOL
'------------------------------ N downward movement
  NLEACH = NSOL * (DRAIN1 / (WAT1 + DRAIN1))  'gN.m-2
  If NCON <= 0.000001 Then NLEACH = 0          'threshold = 1 mgN.L-1

  CNLEACH = CNLEACH + NLEACH
```

Continued

Box 18.2. Continued.

```
'------------------------------ N denitrification
  NDNIT = 0
  If FTSW1 > 1 Then
    XNCON = NCON
    If XNCON > 0.0004 Then XNCON = 0.0004        'threshold = 400 mgN.L-1
    KDNIT = 6 * Exp(0.07735 * TMPS – 6.593)
    NDNIT = XNCON * (1 – Exp(–KDNIT))            'gN.g-1 H2O
    NDNIT = NDNIT * WAT1 * 1000                  'gN.m-2
  End If

  CNDNIT = CNDNIT + NDNIT

'------------------------------ Frac. top layer with roots
  FROOT1 = DEPORT / DEP1
  If FROOT1 > 1 Then FROOT1 = 1

'------------------------------ Updating
  NSOL = NSOL + NMIN + NFERT – NVOL – NLEACH – NDNIT – NUP
  NCON = NSOL / (WAT1 * 1000)

  SNAVL = (NCON – 0.000001) * ATSW1 * 1000 * FROOT1 'threshold = 1 mgN.L-1
  If SNAVL < 0 Then SNAVL = 0
Return
```

Exercises

1. Find soil N inputs for major soils of your location. How different/similar are they? Why?

2. Calculate soil mass (SOILM), organic N mass (NORG), potential mineralizable organic N (MNORG), and amount of NO_3^- and NH_4^+, and initial NSOL for a soil with a top layer of 450 mm, BD of 1.42 g cm^{-3}, organic N of 1%, FMIN of 0.1, and initial concentrations of NO_3^- and NH_4^+ of 20 and 0.9 mg kg^{-1}, respectively.

3. Repeat the calculation of Exercise 2 assuming the same conditions but a top layer thickness of 600 mm. How does this new depth affect the calculations? Discuss it.

19 A Model for Nitrogen-limited Conditions

In Chapters 17 and 18, submodels were developed to simulate plant nitrogen (N) budget and soil N balance as well as a submodel to account for N effect on leaf area development and senescence. In this chapter, the submodels are integrated into the model of water-limited conditions described in Chapter 16. The resultant model can be used to simulate crop growth and yield as influenced by N limitation.

Model Structure

The N-limited model of this chapter includes all the submodels from potential production and water-limited models as described in Chapters 12 and 16. As before, the model is written in Visual Basic for Application (VBA) in Excel. The entire model will be a subroutine (macro) in Excel. The model program includes a main part and submodels. The submodels are called by the main part when necessary. The main part of the program is presented in Box 19.1. The model, however, has four new submodels described in Chapters 17 and 18. They are: plant N submodels for non-legumes and legumes, crop leaf area index (LAI) submodel with inclusion of N effects, and soil N submodel.

To facilitate the application of all the three models, i.e. potential-production, water-limited, and N-limited models, a new control variable for N is used. In Chapter 16, a control variable named *"water"* was used to combine both potential-production and water-limited models. Similar to that variable the new control variable named *"nitrogen"* is applied here. Thus, users do not have to run three separate files and models.

Control variable *"water"* could have one of the three values as:

0 For potential production situation so that soil water submodel is bypassed.
1 For water-limited conditions with irrigation application.
2 For water-limited conditions without irrigation application (rainfed).

Box 19.1. Main part of the program of the N-limited crop model. Submodels are called by the main part when necessary.

```
'--------------------------- Main program
  GoSub ManagInputs
  GoSub InitialsHeaders
  GoSub FindSowingDate
  Do Until MAT = 1
     GoSub Weather
     GoSub Phenology
     If nitrogen = 0 Then GoSub CropLAI
     If nitrogen = 1 Or nitrogen = 2 Then GoSub CropLAIN
     GoSub DMProduction
     GoSub DMDistribution
     If nitrogen = 1 Then GoSub LegumPlantN
     If nitrogen = 2 Then GoSub NonLegumPlantN
     If water = 1 Or water = 2 Then GoSub SoilWater
     If nitrogen = 2 Then GoSub SoilN
     GoSub DailyPrintOut
  Loop
  GoSub SummaryPrintOut
  Exit Sub
'--------------------------- End of main program
```

The new control variable, *"nitrogen"*, can also have one of these values:

0 For potential production situation so that N submodels are bypassed.
1 For simulating a legume crop under N-limited conditions.
2 For simulating a non-legume crop under N-limited conditions.

To simulate N-limited conditions, "water" needs to be equal to 1 or 2 because simulating soil water balance is needed for N-limited conditions.

The flow diagram of the N-limited model and the sequence of events are similar to those described for potential production and water-limited models (see Fig. 12.2). The only difference is that in the N-limited model, N related submodels are called to simulate plant and soil N balance, provided that *"nitrogen"* has a value of 1 or 2 (Box 19.1).

Source codes of the N-limited model can be found in Box 19.2. Parameter estimates as used in this program belong to a wheat cultivar (cv. Tajan). The complete list of variables can be found in Appendix III.

Structure of Excel File Containing the Model

The structure of the Excel file that contains the N-limited model is the same as the water-limited model (Chapter 16) and similar to the potential production model (Chapter 12). The file includes several sheets and a module (model program macro). The program uses Excel's sheets for inputs and outputs as indicated in Fig. 12.3 in Chapter 12.

Box 19.2. Program of the model for simulating crop development, growth, and yield under N-limited conditions as well as potential production and water-limited conditions. The model can be requested from the authors or can be downloaded from the book's website (https://sites.google.com/site/CropModeling).

```
Sub nlm()
'--------------------------- A simple crop model for N-limited conditions

  ReDim DAPNF(10), NFERTI(10), VOLFI(10)
'--------------------------- Main program
  GoSub ManagInputs
  GoSub InitialsHeaders
  GoSub FindSowingDate
  Do Until MAT = 1
     GoSub Weather
     GoSub Phenology
     If nitrogen = 0 Then GoSub CropLAI
     If nitrogen = 1 Or nitrogen = 2 Then GoSub CropLAIN
     GoSub DMProduction
     GoSub DMDistribution
     If nitrogen = 1 Then GoSub LegumPlantN
     If nitrogen = 2 Then GoSub NonLegumPlantN
     If water = 1 Or water = 2 Then GoSub SoilWater
     If nitrogen = 2 Then GoSub SoilN
     GoSub DailyPrintOut
  Loop
  GoSub SummaryPrintOut
  Exit Sub
'--------------------------- End of main program

ManagInputs:
   pyear = Sheet1.[b7]
   pdoy = Sheet1.[b8]
   PDEN = Sheet1.[b9]

   water = Sheet1.[b11]
   VPDF = Sheet1.[b12]
   IRGLVL = Sheet1.[b13]
   nitrogen = Sheet1.[b17]

   If nitrogen = 2 Then
     FN = Sheet1.[b18]                              'Number of N application
     For N = 1 To FN
         DAPNF(N) = Sheet1.Range("B" & N + 21)     'Time of N application
         NFERTI(N) = Sheet1.Range("C" & N + 21)    'N amount (gN.m-2)
         VOLFI(N) = Sheet1.Range("D" & N + 21)     'Fraction volatilization
     Next N
   End If
Return

Weather:
     Row = Row + 1
```

Continued

Box 19.2. Continued.

```
    Yr = Sheet2.Range("A" & Row)
    DOY = Sheet2.Range("B" & Row)
    SRAD = Sheet2.Range("C" & Row)
    TMAX = Sheet2.Range("D" & Row)
    TMIN = Sheet2.Range("E" & Row)
    RAIN = Sheet2.Range("F" & Row)
    TMP = (TMAX + TMIN) / 2
Return

Phenology:
'--------------------------- Parameters and Initials
If iniPheno = 0 Then
  TBD = Sheet5.[b7]
  TP1D = Sheet5.[b8]
  TP2D = Sheet5.[b9]
  TCD = Sheet5.[b10]
  tuSOWEMR = Sheet5.[b11]
  tuEMRTLM = Sheet5.[b12]
  tuTLMBSG = Sheet5.[b13]
  tuBSGTSG = Sheet5.[b14]
  tuTSGMAT = Sheet5.[b15]

  tuEMR = tuSOWEMR
  tuTLM = tuEMR + tuEMRTLM
  tuBSG = tuTLM + tuTLMBSG
  tuTSG = tuBSG + tuBSGTSG
  tuMAT = tuTSG + tuTSGMAT

  DAP = 0:        CTU = 0:        WSFD = 1:
  iniPheno = 1
End If

'--------------------------- Temperature unit calculation
  If TMP <= TBD Or TMP >= TCD Then
    tempfun = 0
  ElseIf TMP > TBD And TMP < TP1D Then
    tempfun = (TMP – TBD) / (TP1D – TBD)
  ElseIf TMP > TP2D And TMP < TCD Then
    tempfun = (TCD – TMP) / (TCD – TP2D)
  ElseIf TMP >= TP1D And TMP <= TP2D Then
    tempfun = 1
  End If

  DTU = (TP1D – TBD) * tempfun
  If CTU > tuEMR Then DTU = DTU * WSFD
  CTU = CTU + DTU
  DAP = DAP + 1

  If CTU < tuEMR Then DTEMR = DAP + 1 'Saving days to EMR
  If CTU < tuTLM Then DTTLM = DAP + 1 'Saving days to TLM
  If CTU < tuBSG Then DTBSG = DAP + 1 'Saving days to BSG
```

Continued

Box 19.2. Continued.

```
  If CTU < tuTSG Then DTTSG = DAP + 1 'Saving days to TSG
  If CTU < tuMAT Then DTMAT = DAP + 1 'Saving days to MAT

  If CTU > tuMAT Then MAT = 1
Return

CropLAI:
'--------------------------- LAI initials and pars
  If iniLAI = 0 Then
    PHYL = Sheet5.[b17]
    PLACON = Sheet5.[b18]
    PLAPOW = Sheet5.[b19]
    SLA = Sheet5.[b20]

    MSNN = 1:   PLA2 = 0:   PLA1 = 0:   LAI = 0:
    MXLAI = 0:  WSFL = 1:   iniLAI = 1
  End If
'---------------------------Yesterday LAI to intercept PAR today
  LAI = LAI + GLAI – DLAI
  If LAI < 0 Then LAI = 0
  If LAI > MXLAI Then MXLAI = LAI          'Saving maximum LAI
'--------------------------- Daily increase and decrease in LAI today
  If CTU <= tuEMR Then
    GLAI = 0:    DLAI = 0
  ElseIf CTU > tuEMR And CTU <= tuTLM Then
    INODE = DTU / PHYL
    MSNN = MSNN + INODE
    PLA2 = PLACON * MSNN ^ PLAPOW
    GLAI = ((PLA2 – PLA1) * PDEN / 10000) * WSFL
    PLA1 = PLA2
    DLAI = 0
  ElseIf CTU > tuTLM And CTU <= tuBSG Then
    GLAI = GLF * SLA
    BSGLAI = LAI          'Saving LAI at BSG
    DLAI = 0
  ElseIf CTU > tuBSG Then
    GLAI = 0
    DLAI = DTU / (tuMAT – tuBSG) * BSGLAI
  End If
Return

DMProduction:
'--------------------------- Parameters and Initials
  If iniDMP = 0 Then
    TBRUE = Sheet5.[b22]
    TP1RUE = Sheet5.[b23]
    TP2RUE = Sheet5.[b24]
    TCRUE = Sheet5.[b25]
    KPAR = Sheet5.[b26]
```

Continued

Box 19.2. Continued.

```
    IRUE = Sheet5.[b27]

    WSFG = 1:   iniDMP = 1:
  End If
'--------------------------- Adjustment of RUE
  If TMP <= TBRUE Or TMP >= TCRUE Then
    TCFRUE = 0
  ElseIf TMP > TBRUE And TMP < TP1RUE Then
    TCFRUE = (TMP – TBRUE) / (TP1RUE – TBRUE)
  ElseIf TMP > TP2RUE And TMP < TCRUE Then
    TCFRUE = (TCRUE – TMP) / (TCRUE – TP2RUE)
  ElseIf TMP >= TP1RUE And TMP <= TP2RUE Then
    TCFRUE = 1
  End If

  RUE = IRUE * TCFRUE * WSFG
'--------------------------- Daily dry matter production
  FINT = 1 – Exp(–KPAR * LAI)
  DDMP = SRAD * 0.48 * FINT * RUE
Return

DMDistribution:
'--------------------------- Parameters and Initials
  If iniDMD = 0 Then
    FLF1A = Sheet5.[b29]
    FLF1B = Sheet5.[b30]
    WTOPL = Sheet5.[b31]
    FLF2 = Sheet5.[b32]
    FRTRL = Sheet5.[b33]
    GCC = Sheet5.[b34]

    WLF = 0.5:   WST = 0.5:   WVEG = WLF + WST:
    WGRN = 0:   iniDMD = 1:
  End If
'--------------------------- Biomass partitioning and yield formation
  If CTU <= tuEMR Or CTU > tuTSG Then
    DDMP = 0:   GLF = 0:   GST = 0:   TRANSL = 0:   SGR = 0
  ElseIf CTU > tuEMR And CTU <= tuTLM Then
    If WTOP < WTOPL Then FLF1 = FLF1A Else FLF1 = FLF1B
    GLF = FLF1 * DDMP
    GST = DDMP – GLF
    SGR = 0
  ElseIf CTU > tuTLM And CTU <= tuBSG Then
    GLF = FLF2 * DDMP
    GST = DDMP – GLF
    SGR = 0
    BSGDM = WTOP                     'Saving WTOP at BSG
  ElseIf CTU > tuBSG And CTU <= tuTSG Then
```

Continued

Box 19.2. Continued.

```
    GLF = 0:
    GST = 0:
    TRLDM = BSGDM * FRTRL
    TRANSL = DTU / (tuTSG – tuBSG) * TRLDM
    SGR = (DDMP + TRANSL) * GCC
  End If

  WLF = WLF + GLF
  WST = WST + GST
  WGRN = WGRN + SGR
  WVEG = WVEG + DDMP – (SGR / GCC)
  WTOP = WVEG + WGRN
Return

SoilWater:
'---------------------------- Parameters and Initials
  If iniSW = 0 Then
      DEPORT = Sheet5.[B36]
      MEED = Sheet5.[B37]
      GRTDP = Sheet5.[B38]
      tuBRG = tuEMR
      tuTRG = tuBSG
      TEC = Sheet5.[B39]
      WSSG = Sheet5.[B40]
      WSSL = Sheet5.[B41]
      WSSD = Sheet5.[B42]

      SOLDEP = Sheet7.[b7]
      DEP1 = Sheet7.[b8]
      SALB = Sheet7.[b9]
      CN = Sheet7.[b10]
      DRAINF = Sheet7.[b11]
      SAT = Sheet7.[b12]
      DUL = Sheet7.[b13]
      EXTR = Sheet7.[b14]
      CLL = DUL – EXTR

      MAI1 = Sheet7.[b16]
      MAI = Sheet7.[b17]

      IPATSW = SOLDEP * EXTR * MAI

      ATSW = DEPORT * EXTR * MAI1
      TTSW = DEPORT * EXTR
      FTSW = ATSW / TTSW
      WSTORG = IPATSW – ATSW

      ATSW1 = DEP1 * EXTR * MAI1
      TTSW1 = DEP1 * EXTR
      FTSW1 = ATSW1 / TTSW1
```

Continued

Box 19.2. Continued.

```
        WLL1 = DEP1 * CLL
        WAT1 = WLL1 + ATSW1
        WSAT1 = DEP1 * SAT
        EOSMIN = 1.5:   WETWAT = 10:   KET = 0.5:   CALB = 0.23:
        DYSE = 1:       CTR = 0:       CE = 0:       CRAIN = 0:
        CRUNOF = 0:     CIRGW = 0:     IRGNO = 0:    iniSW = 1
    End If
'---------------------------Irrigation
    If water = 1 And FTSW <= IRGLVL And CTU < tuTSG Then
        IRGW = (TTSW - ATSW)
        IRGNO = IRGNO + 1
    Else
        IRGW = 0
    End If
    CIRGW = CIRGW + IRGW
'--------------------------- Drainage
    If ATSW1 <= TTSW1 Then
        DRAIN1 = 0
    ElseIf ATSW1 > TTSW1 Then
        DRAIN1 = (ATSW1 - TTSW1) * DRAINF
    End If
    If ATSW <= TTSW Then
        DRAIN = 0
    ElseIf ATSW > TTSW Then
        DRAIN = (ATSW - TTSW) * DRAINF
    End If
    WSTORG = WSTORG + DRAIN - EWAT
    If WSTORG < 0 Then WSTORG = 0
'-------------------------- Water exploitation by root growth
    GRTD = GRTDP    'mm per day
    If CTU < tuBRG Then GRTD = 0
    If CTU > tuTRG Then GRTD = 0
    If DDMP = 0 Then GRTD = 0
    If DEPORT >= SOLDEP Then GRTD = 0
    If DEPORT >= MEED Then GRTD = 0
    If WSTORG = 0 Then GRTD = 0
    DEPORT = DEPORT + GRTD
    EWAT = GRTD * EXTR
    If EWAT > WSTORG Then EWAT = WSTORG
'-------------------------- Runoff
    RUNOF = 0
    If water = 2 And RAIN > 0.01 Then
        S = 254 * (100 / CN - 1)
        SWER = 0.15 * ((WSAT1 - WAT1) / (WSAT1 - WLL1))
```

Continued

Box 19.2. Continued.

```
    If SWER < 0 Then SWER = 0
    If (RAIN – SWER * S) > 0 Then
      RUNOF = (RAIN – SWER * S) ^ 2 / (RAIN + (1 – SWER) * S)
    Else
      RUNOF = 0
    End If
  End If

  If (WAT1 – DRAIN1) > WSAT1 Then
    RUNOF = RUNOF + (WAT1 – DRAIN1 – WSAT1)
  End If

  CRAIN = CRAIN + RAIN
  CRUNOF = CRUNOF + RUNOF

'---------------------------- LAI for soil evaporation
  If CTU <= tuBSG Then ETLAI = LAI Else ETLAI = BSGLAI

'---------------------------- Potential ET
  TD = 0.6 * TMAX + 0.4 * TMIN
  ALBEDO = CALB * (1 – Exp(–KET * ETLAI)) + SALB * Exp(–KET * ETLAI)
  EEQ = SRAD * (0.004876 – 0.004374 * ALBEDO) * (TD + 29)
  PET = EEQ * 1.1
  If TMAX > 34 Then PET = EEQ * ((TMAX – 34) * 0.05 + 1.1)
  If TMAX < 5 Then PET = EEQ * 0.01 * Exp(0.18 * (TMAX + 20))

'---------------------------- Soil evaporation
  EOS = PET * Exp(–KET * ETLAI)
  If PET > EOSMIN And EOS < EOSMIN Then EOS = EOSMIN

  SEVP = EOS
  If (RAIN + IRGW) > WETWAT Then DYSE = 1
  If ATSW1 < 1 or DYSE > 1 Or FTSW < 0.5 Then
    SEVP = EOS * ((DYSE + 1) ^ 0.5 – DYSE ^ 0.5)
    DYSE = DYSE + 1
  End If

  CE = CE + SEVP

'---------------------------- Plant transpiration
  VPTMIN = 0.6108 * Exp(17.27 * TMIN / (TMIN + 237.3))
  VPTMAX = 0.6108 * Exp(17.27 * TMAX / (TMAX + 237.3))
  VPD = VPDF * (VPTMAX – VPTMIN)
  TR = DDMP * VPD / TEC           'VPD in kPa, TEC in Pa
  If TR < 0 Then TR = 0

  CTR = CTR + TR

  If DEPORT <= DEP1 Then
    TR1 = TR
  ElseIf DEPORT > DEP1 Then
    If FTSW1 > WSSG Then RT1 = 1 Else RT1 = FTSW1 / WSSG
    TR1 = TR * RT1
  End If
```

Continued

Box 19.2. Continued.

```
'--------------------------- Updating
  ATSW1 = ATSW1 + RAIN + IRGW – DRAIN1 – RUNOF – TR1 – SEVP
  If ATSW1 < 0 Then ATSW1 = 0
  FTSW1 = ATSW1 / TTSW1
  WAT1 = WLL1 + ATSW1

  ATSW = ATSW + RAIN + IRGW + EWAT – DRAIN – RUNOF – TR – SEVP
  If ATSW < 0 Then ATSW = 0
  TTSW = DEPORT * EXTR
  FTSW = ATSW / TTSW
'--------------------------- Water-stress-factors
  If nitrogen = 1 Then
    If FTSW > WSSN Then WSFN = 1 Else WSFN = FTSW / WSSN
  End If
  If FTSW > WSSL Then WSFL = 1 Else WSFL = FTSW / WSSL
  If FTSW > WSSG Then WSFG = 1 Else WSFG = FTSW / WSSG
  WSFD = (1 – WSFG) * WSSD + 1

  If WAT1 > (0.95 * WSAT1) Then
    WSFN = 0:   WSFG = 0:   WSFL = 0:   WSFD = 0
  End If
Return

CropLAIN:
'--------------------------- LAI initials and pars
  If iniLAI = 0 Then
    PHYL = Sheet5.[b17]
    PLACON = Sheet5.[b18]
    PLAPOW = Sheet5.[b19]
    SLA = Sheet5.[b20]

    MSNN = 1:   PLA2 = 0:   PLA1 = 0:   LAI = 0:
    MXLAI = 0:  WSFL = 1:   SLNG = 2:   iniLAI = 1:
  End If
'--------------------------- Yesterday LAI to intercept PAR today
  If GLAI > (INLF / SLNG) Then GLAI = (INLF / SLNG)
  LAI = LAI + GLAI – DLAI
  If LAI < 0 Then LAI = 0
  If LAI > MXLAI Then MXLAI = LAI          'Saving maximum LAI
'--------------------------- Daily increase and decrease in LAI today
  If CTU <= tuEMR Then
    GLAI = 0:
  ElseIf CTU > tuEMR And CTU <= tuTLM Then
    INODE = DTU / PHYL
    MSNN = MSNN + INODE
    PLA2 = PLACON * MSNN ^ PLAPOW
    GLAI = ((PLA2 – PLA1) * PDEN / 10000) * WSFL
    PLA1 = PLA2
```

Continued

Box 19.2. Continued.

```
  ElseIf CTU > tuTLM And CTU <= tuBSG Then
    GLAI = GLF * SLA
    BSGLAI = LAI                    'Saving LAI at BSG
  ElseIf CTU > tuBSG Then
    GLAI = 0
  End If
  DLAI = XNLF / (SLNG – SLNS)
Return

NonLegumPlantN:
  If iniPNB = 0 Then
    SLNG = Sheet5.[B44]
    SLNS = Sheet5.[B45]
    SNCG = Sheet5.[B46]
    SNCS = Sheet5.[B47]
    GNC = Sheet5.[B48]
    MXNUP = Sheet5.[B49]

    NST = WST * SNCG:   NLF = LAI * SLNG:
    CNUP = NST + NLF:   NGRN = 0:             iniPNB = 1:
  End If

  If CTU <= tuEMR Or CTU > tuTSG Then
    NUP = 0:     XNLF = 0:        XNST = 0:
    INLF = 0:    INST = 0:        INGRN = 0:
  ElseIf CTU > tuEMR And CTU < tuBSG Then
    INGRN = 0
    NSTDF = (WST * SNCG) – NST
    If NSTDF < 0 Then NSTDF = 0
    NUP = (GST * SNCG) + (GLAI * SLNG) '+ NSTDF          '<------ Inactive!
    If NUP > MXNUP Then NUP = MXNUP
    If NUP < 0 Then NUP = 0
    If DDMP = 0 Then NUP = 0
    If NUP > SNAVL Then NUP = SNAVL

    If NST <= (WST * SNCS) Then
        INST = WST * SNCS – NST
        XNST = 0
        If INST >= NUP Then
          INLF = 0
          XNLF = INST – NUP
        ElseIf INST < NUP Then
          INLF = GLAI * SLNG
          If INLF > (NUP – INST) Then INLF = NUP – INST
          INST = NUP – INLF
          XNLF = 0
        End If
    ElseIf NST > (WST * SNCS) Then
```

Continued

Box 19.2. Continued.

```
            INLF = GLAI * SLNG
            XNLF = 0
            If INLF >= NUP Then
              INST = 0
              XNST = INLF – NUP
              If XNST > (NST – WST * SNCS) Then XNST = NST – WST * SNCS
              INLF = NUP + XNST
            ElseIf INLF < NUP Then
              INST = NUP – INLF
              XNST = 0
          End If
        End If

      ElseIf CTU >= tuBSG And CTU <= tuTSG Then
          INGRN = SGR * GNC
          NUP = INGRN
          If DDMP <= (SGR / GCC) Then NUP = 0
          If DDMP = 0 Then NUP = 0
          If NUP > SNAVL Then NUP = SNAVL

          If NUP > (SGR * GNC) Then
            'N is excess of seed needs
            INLF = 0
            INST = NUP – SGR * GNC
            XNLF = 0
            XNST = 0
          ElseIf NUP <= (SGR * GNC) Then
            'Need to transfer N from vegetative tissue
            INLF = 0
            INST = 0
            XNLF = (SGR * GNC – NUP) * FXLF
            XNST = (SGR * GNC – NUP) * (1 – FXLF)
          End If
      End If

      NST = NST + INST – XNST
      NLF = NLF + INLF – XNLF
      NVEG = NLF + NST
      NGRN = NGRN + INGRN
      CNUP = CNUP + NUP
      TRLN = LAI * (SLNG – SLNS) + (NST – WST * SNCS)
      FXLF = LAI * (SLNG – SLNS) / (TRLN + 0.000000000001)
      If FXLF > 1 Then FXLF = 1
      If FXLF < 0 Then FXLF = 0
    Return

    LegumPlantN:
      If iniPNB = 0 Then
        SLNG = Sheet5.[B44]
```

Continued

Box 19.2. Continued.

```
SLNS = Sheet5.[B45]
SNCG = Sheet5.[B46]
SNCS = Sheet5.[B47]
GNC = Sheet5.[B48]
MXNUP = Sheet5.[B49]
tuBNF = Sheet5.[B51]
WSSN = Sheet5.[B52]
INSOL = Sheet5.[B53]

NST = WST * SNCG:   NLF = LAI * SLNG:   WSFN = 1:
CNUP = NST + NLF:   NGRN = 0:           iniPNB = 1:
End If

If CTU <= tuEMR Or CTU > tuTSG Then
    NUP = 0:   XNLF = 0:   XNST = 0:   INLF = 0:   INST = 0:   INGRN = 0:
ElseIf CTU > tuEMR And CTU < tuBSG Then
    INGRN = 0
    NSTDF = (WST * SNCG) – NST
    If NSTDF < 0 Then NSTDF = 0
    NUP = (GST * SNCG) + (GLAI * SLNG) '+ NSTDF        '<------ Inactive!
    If CTU < tuBNF And CNUP > INSOL Then NUP = 0
    If NUP > MXNUP Then NUP = MXNUP
    NFC = NFC * 3 / 4 + NUP / WVEG * (1 / 4) 'from Sinclair et al. 2003
    NUP = NUP * WSFN
    If NUP < 0 Then NUP = 0
    If FTSW > 1 Then NUP = 0
    If DDMP = 0 Then NUP = 0

    If NST <= (WST * SNCS) Then
        INST = WST * SNCS – NST
        XNST = 0
        If INST >= NUP Then
            INLF = 0
            XNLF = INST – NUP
        ElseIf INST < NUP Then
            INLF = GLAI * SLNG
            If INLF > (NUP – INST) Then INLF = NUP – INST
            INST = NUP – INLF
            XNLF = 0
        End If

  ElseIf NST > (WST * SNCS) Then
    INLF = GLAI * SLNG
    XNLF = 0
    If INLF >= NUP Then
        INST = 0
        XNST = INLF – NUP
        If XNST > (NST – WST * SNCS) Then XNST = (NST – WST * SNCS)
        INLF = NUP + XNST
```

Continued

Box 19.2. Continued.

```
        ElseIf INLF < NUP Then
                INST = NUP – INLF
                XNST = 0
        End If
    End If

ElseIf CTU >= tuBSG And CTU <= tuTSG Then
    INGRN = SGR * GNC
    NUP = INGRN
    PDNF = NFC * WVEG
    If PDNF > NUP Then PDNF = NUP
    DNF = PDNF * WSFN
    If DNF < 0 Then DNF = 0
    If DDMP <= (SGR / GCC) Then DNF = 0
    If DDMP = 0 Then DNF = 0
    NUP = DNF

    If NUP > (SGR * GNC) Then
        'N is excess of seed needs
        INLF = 0
        INST = NUP – SGR * GNC
        XNLF = 0
        XNST = 0
    ElseIf NUP <= (SGR * GNC) Then
        'Need to transfer N from vegetative tissue
        INLF = 0
        INST = 0
        XNLF = (SGR * GNC – NUP) * FXLF
        XNST = (SGR * GNC – NUP) * (1 – FXLF)
    End If
  End If

  NST = NST + INST – XNST
  NLF = NLF + INLF – XNLF
  NVEG = NLF + NST
  NGRN = NGRN + SGR * GNC
  CNUP = CNUP + NUP

  TRLN = LAI * (SLNG – SLNS) + (NST – WST * SNCS)
  FXLF = LAI * (SLNG – SLNS) / (TRLN + 0.000000000001)
  If FXLF > 1 Then FXLF = 1
  If FXLF < 0 Then FXLF = 0
Return

SoilN:
'---------------------------- Parameters and initials
  If iniSNB = 0 Then
    FG = Sheet7.[b19] / 100        'Fraction soil > 2mm
    BD1 = Sheet7.[b20]             'Soil bulk density (g.cm-3)
```

Continued

Box 19.2. Continued.

```
    NORGP = Sheet7.[b21]          'Organic N (%)
    FMIN = Sheet7.[b22]           'Frac. Org N avail. for mineralization
    NO3 = Sheet7.[b23]            'ppm = mg.kg-1
    NH4 = Sheet7.[b24]            'ppm = mg.kg-1

    SOILM = DEP1 * BD1 * (1 – FG) * 1000 'g.m-2
    NORG = NORGP * 0.01 * SOILM          'g.m-2
    MNORG = NORG * FMIN                  'gN.m-2 org. N avail. for miner.

    NO3 = NO3 * (14 / 62) * 0.000001 * SOILM      'from ppm to gN.m-2
    NH4 = NH4 * (14 / 18) * 0.000001 * SOILM      'from ppm to gN.m-2

    NSOL = (NO3 + NH4)                   'gN.m-2
    NCON = NSOL / (WAT1 * 1000)          'gN.g-1 H2O

    INSOL = NSOL:      CNFERT = 0:      CNVOL = 0:
    CNLEACH = 0:       CNMIN = 0:       CNDNIT = 0:      iniSNB = 1:
  End If

'--------------------------- N net mineralization
  TMPS = TMP
  If TMPS > 35 Then TMPS = 35
  KNMIN = 24 * Exp(17.753 – 6350.5 / (TMPS + 273)) / 168
  KN = 1 – Exp(–KNMIN)
  If FTSW1 < 0.9 Then RN = 1.111 * FTSW1
  If FTSW1 >= 0.9 Then RN = 10 – 10 * FTSW1
  If RN < 0 Then RN = 0
  NMIN = MNORG * RN * KN
  NMIN = NMIN * (0.0002 – NCON) / 0.0002          'threshold = 200 mgN.L-1
  If NMIN < 0 Then NMIN = 0
  MNORG = MNORG – NMIN

  CNMIN = CNMIN + NMIN

'--------------------------- N application & volatilization
  NFERT = 0:    NVOL = 0:
  For N = 1 To FN
      If DAP = DAPNF(N) Then
         NFERT = NFERTI(N)       'gN.m-2
         VOLF = VOLFI(N) / 100
         NVOL = VOLF * NFERT     'gN.m-2
      End If
  Next N

  CNFERT = CNFERT + NFERT
  CNVOL = CNVOL + NVOL

'--------------------------- N downward movement
  NLEACH = NSOL * (DRAIN1 / (WAT1 + DRAIN1)) 'gN.m-2
  If NCON <= 0.000001 Then NLEACH = 0        'threshold = 1 mgN.L-1

  CNLEACH = CNLEACH + NLEACH
```

Continued

Box 19.2. Continued.

```
'---------------------------- N denitrification
  NDNIT = 0
  If FTSW1 > 1 Then
    XNCON = NCON
    If XNCON > 0.0004 Then XNCON = 0.0004        'threshold = 400 mgN.L-1
    KDNIT = 6 * Exp(0.07735 * TMPS – 6.593)
    NDNIT = XNCON * (1 – Exp(–KDNIT))              'gN.g-1 H2O
    NDNIT = NDNIT * WAT1 * 1000                    'gN.m-2
  End If

  CNDNIT = CNDNIT + NDNIT

'--------------------------- Frac. top layer with roots
  FROOT1 = DEPORT / DEP1
  If FROOT1 > 1 Then FROOT1 = 1

'--------------------------- Updating
  NSOL = NSOL + NMIN + NFERT – NVOL – NLEACH – NDNIT – NUP
  NCON = NSOL / (WAT1 * 1000)

  SNAVL = (NCON – 0.000001) * ATSW1 * 1000 * FROOT1 'threshold = 1 mgN.L-1
  If SNAVL < 0 Then SNAVL = 0
Return

FindSowingDate:
  Row = 10
  Do
    Row = Row + 1
    Yr = Sheet2.Range("A" & Row)
    DOY = Sheet2.Range("B" & Row)
    SRAD = Sheet2.Range("C" & Row)
    TMAX = Sheet2.Range("D" & Row)
    TMIN = Sheet2.Range("E" & Row)
    RAIN = Sheet2.Range("F" & Row)
  Loop Until Yr = pyear And DOY = pdoy
Return

InitialsHeaders:
'--------------------------- Initials
  MAT = 0
  iniPheno = 0
  iniLAI = 0
  iniDMP = 0
  iniDMD = 0
  iniSW = 0
  iniPNB = 0
  iniSNB = 0
'--------------------------- Headers
  Sheet4.Cells(2, 1) = "Year"
  Sheet4.Cells(2, 2) = "DOY"
```

Continued

Box 19.2. Continued.

```
Sheet4.Cells(2, 3) = "DAP"
Sheet4.Cells(2, 4) = "TMP"
Sheet4.Cells(2, 5) = "DTU"
Sheet4.Cells(2, 6) = "CTU"
Sheet4.Cells(2, 7) = "MSNN"
Sheet4.Cells(2, 8) = "GLAI"
Sheet4.Cells(2, 9) = "DLAI"
Sheet4.Cells(2, 10) = "LAI"
Sheet4.Cells(2, 11) = "TCFRUE"
Sheet4.Cells(2, 12) = "FINT"
Sheet4.Cells(2, 13) = "DDMP"
Sheet4.Cells(2, 14) = "GLF"
Sheet4.Cells(2, 15) = "GST"
Sheet4.Cells(2, 16) = "SGR"
Sheet4.Cells(2, 17) = "WLF"
Sheet4.Cells(2, 18) = "WST"
Sheet4.Cells(2, 19) = "WVEG"
Sheet4.Cells(2, 20) = "WGRN"
Sheet4.Cells(2, 21) = "WTOP"
If water = 1 Or water = 2 Then
Sheet4.Cells(2, 22) = "DEPORT"
Sheet4.Cells(2, 23) = "RAIN"
Sheet4.Cells(2, 24) = "IRGW"
Sheet4.Cells(2, 25) = "RUNOF"
Sheet4.Cells(2, 26) = "PET"
Sheet4.Cells(2, 27) = "SEVP"
Sheet4.Cells(2, 28) = "TR"
Sheet4.Cells(2, 29) = "DRAIN1"
Sheet4.Cells(2, 30) = "ATSW"
Sheet4.Cells(2, 31) = "FTSW"
Sheet4.Cells(2, 32) = "CRAIN"
Sheet4.Cells(2, 33) = "CIRGW"
Sheet4.Cells(2, 34) = "IRGNO"
Sheet4.Cells(2, 35) = "CRUNOF"
Sheet4.Cells(2, 36) = "CE"
Sheet4.Cells(2, 37) = "CTR"
Sheet4.Cells(2, 38) = "WSTORG"
End If
If nitrogen = 2 Then
Sheet4.Cells(2, 39) = "NFERT"
Sheet4.Cells(2, 40) = "CNFERT"
Sheet4.Cells(2, 41) = "NVOL"
Sheet4.Cells(2, 42) = "CNVOL"
Sheet4.Cells(2, 43) = "NLEACH"
Sheet4.Cells(2, 44) = "CNLEACH"
Sheet4.Cells(2, 45) = "NMIN"
Sheet4.Cells(2, 46) = "CNMIN"
```

Continued

Box 19.2. Continued.

```
  Sheet4.Cells(2, 47) = "NDNIT"
  Sheet4.Cells(2, 48) = "CNDNIT"
  Sheet4.Cells(2, 49) = "SNAVL"
  Sheet4.Cells(2, 50) = "NSOL"
  End If
  If nitrogen = 1 Or nitrogen = 2 Then
  Sheet4.Cells(2, 51) = "NUP"
  Sheet4.Cells(2, 52) = "NLF"
  Sheet4.Cells(2, 53) = "NST"
  Sheet4.Cells(2, 54) = "NVEG"
  Sheet4.Cells(2, 55) = "NGRN"
  Sheet4.Cells(2, 56) = "CNUP"
  End If
Return

SummaryPrintOut:
    Sheet1.[g8] = DTEMR
    Sheet1.[g9] = DTTLM
    Sheet1.[g10] = DTBSG
    Sheet1.[g11] = DTTSG
    Sheet1.[g12] = DTMAT
    Sheet1.[g15] = MXLAI
    Sheet1.[g16] = BSGLAI
    Sheet1.[g17] = BSGDM
    Sheet1.[G20] = WTOP
    Sheet1.[G21] = WGRN
    Sheet1.[G22] = WGRN / WTOP * 100

    If water = 1 Or water = 2 Then
      Sheet1.[G25] = IPATSW
      Sheet1.[G26] = CRAIN
      Sheet1.[G27] = CIRGW
      Sheet1.[G28] = IRGNO

      Sheet1.[G30] = ATSW
      Sheet1.[G31] = CRUNOF
      Sheet1.[G32] = CE
      Sheet1.[G33] = CTR
      Sheet1.[G34] = WSTORG

      Sheet1.[G36] = CE + CTR
      Sheet1.[G37] = CE / (CE + CTR)
    End If
    If nitrogen = 1 Or nitrogen = 2 Then
      Sheet1.[I8] = NLF
      Sheet1.[I9] = NST
      Sheet1.[I10] = (NLF + NST)
      Sheet1.[I11] = NGRN
      Sheet1.[I12] = CNUP
    End If
```

Continued

Box 19.2. Continued.

```
  If nitrogen = 2 Then
    Sheet1.[I15] = INSOL
    Sheet1.[I16] = CNFERT
    Sheet1.[I17] = CNMIN

    Sheet1.[I19] = NSOL
    Sheet1.[I20] = CNUP
    Sheet1.[I21] = CNVOL
    Sheet1.[I22] = CNLEACH
    Sheet1.[I23] = CNDNIT
  End If
Return

DailyPrintOut:
  Sheet4.Cells(DAP + 2, 1) = Yr
  Sheet4.Cells(DAP + 2, 2) = DOY
  Sheet4.Cells(DAP + 2, 3) = DAP
  Sheet4.Cells(DAP + 2, 4) = TMP
  Sheet4.Cells(DAP + 2, 5) = DTU
  Sheet4.Cells(DAP + 2, 6) = CTU
  Sheet4.Cells(DAP + 2, 7) = MSNN
  Sheet4.Cells(DAP + 2, 8) = GLAI
  Sheet4.Cells(DAP + 2, 9) = DLAI
  Sheet4.Cells(DAP + 2, 10) = LAI
  Sheet4.Cells(DAP + 2, 11) = TCFRUE
  Sheet4.Cells(DAP + 2, 12) = FINT
  Sheet4.Cells(DAP + 2, 13) = DDMP
  Sheet4.Cells(DAP + 2, 14) = GLF
  Sheet4.Cells(DAP + 2, 15) = GST
  Sheet4.Cells(DAP + 2, 16) = SGR
  Sheet4.Cells(DAP + 2, 17) = WLF
  Sheet4.Cells(DAP + 2, 18) = WST
  Sheet4.Cells(DAP + 2, 19) = WVEG
  Sheet4.Cells(DAP + 2, 20) = WGRN
  Sheet4.Cells(DAP + 2, 21) = WTOP
  If water = 1 Or water = 2 Then
  Sheet4.Cells(DAP + 2, 22) = DEPORT
  Sheet4.Cells(DAP + 2, 23) = RAIN
  Sheet4.Cells(DAP + 2, 24) = IRGW
  Sheet4.Cells(DAP + 2, 25) = RUNOF
  Sheet4.Cells(DAP + 2, 26) = PET
  Sheet4.Cells(DAP + 2, 27) = SEVP
  Sheet4.Cells(DAP + 2, 28) = TR
  Sheet4.Cells(DAP + 2, 29) = DRAIN1
  Sheet4.Cells(DAP + 2, 30) = ATSW
  Sheet4.Cells(DAP + 2, 31) = FTSW
  Sheet4.Cells(DAP + 2, 32) = CRAIN
  Sheet4.Cells(DAP + 2, 33) = CIRGW
```

Continued

Box 19.2. Continued.

```
    Sheet4.Cells(DAP + 2, 34) = IRGNO
    Sheet4.Cells(DAP + 2, 35) = CRUNOF
    Sheet4.Cells(DAP + 2, 36) = CE
    Sheet4.Cells(DAP + 2, 37) = CTR
    Sheet4.Cells(DAP + 2, 38) = WSTORG
    End If
    If nitrogen = 2 Then
    Sheet4.Cells(DAP + 2, 39) = NFERT
    Sheet4.Cells(DAP + 2, 40) = CNFERT
    Sheet4.Cells(DAP + 2, 41) = NVOL
    Sheet4.Cells(DAP + 2, 42) = CNVOL
    Sheet4.Cells(DAP + 2, 43) = NLEACH
    Sheet4.Cells(DAP + 2, 44) = CNLEACH
    Sheet4.Cells(DAP + 2, 45) = NMIN
    Sheet4.Cells(DAP + 2, 46) = CNMIN
    Sheet4.Cells(DAP + 2, 47) = NDNIT
    Sheet4.Cells(DAP + 2, 48) = CNDNIT
    Sheet4.Cells(DAP + 2, 49) = SNAVL
    Sheet4.Cells(DAP + 2, 50) = NSOL
    End If

    If nitrogen = 1 Or nitrogen = 2 Then
    Sheet4.Cells(DAP + 2, 51) = NUP
    Sheet4.Cells(DAP + 2, 52) = NLF
    Sheet4.Cells(DAP + 2, 53) = NST
    Sheet4.Cells(DAP + 2, 54) = NVEG
    Sheet4.Cells(DAP + 2, 55) = NGRN
    Sheet4.Cells(DAP + 2, 56) = CNUP
    End If
Return

End Sub '-------------------------------------------------------------------
```

The sheets remain unchanged from the water-limited model, except that more management and soil inputs are required in the "Run" and "Soil" sheets, respectively. In addition, "Outputs" and "Figures" sheets will have more data and figures as a result of N simulation.

In the "Run" sheet, the additional management inputs listed below are needed:

- *"nitrogen"*;
- the number of N (fertilizer) applications; and
- time (as days after sowing), amount (g N m^{-2}), and volatilization factor for each N application. Up to 10 N applications can be defined.

"Run" also includes important new outputs regarding N balance in plant and soil. The new summary outputs to this sheet are:

- accumulated N in leaves;
- accumulated N in stems;

- accumulated N in vegetative tissues;
- accumulated N in grains;
- total N accumulation by the crop;
- soil N at sowing time;
- cumulative N applied as fertilizer;
- cumulative N mineralization during growing season;
- soil N at maturity;
- cumulative N lost in volatilization;
- cumulative N lost through leaching; and
- cumulative N lost through denitrification.

Sheets "Outputs" and "Figures" are expanded to include plant and soil N outputs and their related figures. Similar to former models, the content of "Outputs" sheet can be used for further analyses and more figures might be included in "Figures" sheet.

Appendix I provides a practical guide for model troubleshooting, which can be used if model predictions do not follow field observations.

Sample Runs of the Model

With two examples it is indicated how the model can be used. In the first example, the model will be used to study the effect of different N fertilizer rates applied at sowing time on wheat yield in Gorgan, northeast Iran. The sowing date is 16 December 2005 and the weather data of growing season 2005/06 is used. The sowing density is 300 plants m^{-2}.

Six fertilizer amounts of 0, 5, 10, 15, 20, and 25 g N m^{-2} are applied. The fertilizer is urea and the method of application is incorporated into the soil at sowing time. A volatilization factor of 5% is used. The fertilizer treatments are examined for both irrigated and rainfed conditions. Under irrigated conditions, irrigation is applied automatically whenever fraction transpirable soil water (FTSW) in the root zone falls below 0.5.

Soil characteristics were: soil depth 1200 mm, top layer thickness 600 mm, soil albedo 0.13, curve number 79, drainage factor 0.5, saturation water content 0.36 mm mm^{-1}, drain upper limit 0.264 mm mm^{-1}, extractable soil water 0.13 mm mm^{-1}, initial soil moisture is 90% of field capacity, soil coarse fraction 0%, bulk density 1.35 g cm^{-3}, organic N 0.08%, fraction potential mineralizable soil N 0.1, NO_3^- concentration 15 mg kg^{-1}, and NH_4^+ concentration 1.8 mg kg^{-1}.

Under irrigated conditions, simulated grain yield without N application was 321 g m^{-2} or 3210 kg ha^{-1} (Fig. 19.1a). Total N loss through volatilization, denitrification and leaching from top layer was 3.3 g N m^{-2} (Fig. 19.1b). This N loss is probably overestimated because, as mentioned in Chapter 18, some of leached N below the top layer will be taken up by the crop, which is not considered by the model. By N application up to 15 g N m^{-2}, grain yield increased linearly, from 390 g m^{-2} at 5 g N m^{-2} to 505 g m^{-2} at 15 g N m^{-2} (Fig. 19.1a). From 15 to 20 g N m^{-2}, grain yield increased (35 g m^{-2}) but the amount of increase was lower. From 20 to 25 g N m^{-2}, grain yield did not increase and remained constant, i.e. 540 to 543 g m^{-2}.

Under rainfed conditions, simulated grain yield was 296 g m^{-2} without N application (Fig. 19.1a). But total N loss was considerably lower, 1.8 g N m^{-2}, due to lower losses via denitrification and leaching (Fig. 19.1b). Grain yield increased by N application up to 10 g N m^{-2} and there was no increase as a result of higher amounts of N application. Grain yield at 10 g N m^{-2} was 394 g m^{-2} and 408 g m^{-2} at N amounts of 25 g N m^{-2}.

While grain yield response to N amount was curvilinear with a diminishing return at higher N application, N loss increased linearly with greater N application under both irrigated and rainfed conditions (Fig. 19.1b). However, the loss was higher under irrigated conditions as irrigation caused more losses through denitrification and leaching.

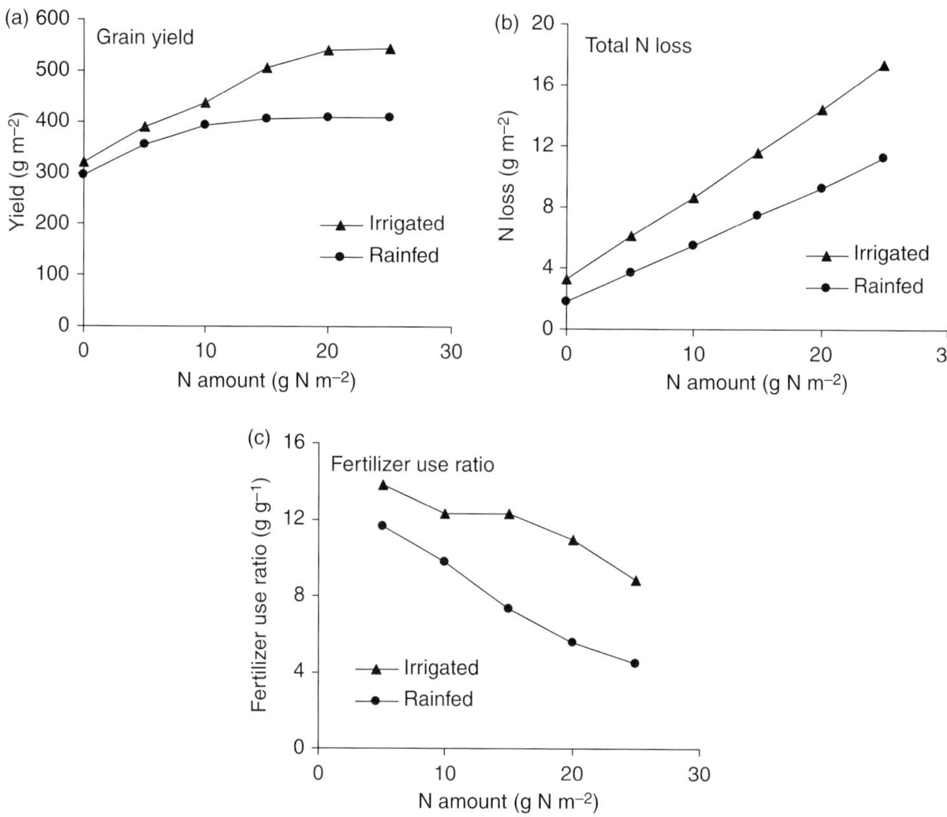

Fig. 19.1. Simulated responses to the amount of N fertilizer at sowing time for wheat crop in Gorgan during growing season of 2005/06 under irrigated and rainfed conditions: (a) grain yield, (b) total N losses via volatilization, denitrification, and leaching (from top 600 mm layer), and (c) fertilizer use efficiency as ratio of extra yield (fertilized yield minus non-fertilized control yield) to amount of N fertilizer.

Another important factor to assess the best N application is to achieve a high ratio of yield increase (yield of fertilized treatment minus yield under non-fertilized conditions) to amount of applied N. Figure 19.1c indicates simulated fertilizer use ratio (FUR, g yield per g applied N). For irrigated conditions, FUR was about 14 g g^{-1} for the 5 g N m^{-2} treatment but declined to 9 g g^{-1} for the 25 g N m^{-2} treatment. For rainfed conditions, the corresponding FURs were 12 and 4.5 g g^{-1} in 5 and 25 g N m^{-2} treatments, respectively.

The above analysis illustrates how the model can be used to study N limitation effects. For a practical application the model must be parameterized for the cultivar under consideration, its robustness must be determined and if the model is robust, it can be used to explore N limitation effects. In such analysis, a number of years should be simulated instead of the single year used here. Further, in the example, N amount at sowing was evaluated. A more comprehensive analysis should also include top-dressing treatments, i.e. split application of N.

In the second example, the best time for an N application of 5 g N m^{-2} is evaluated. The analysis is done for irrigated conditions. All soil and management inputs are identical to the first example. Five application times of 1, 30, 60, 90, and 120 days after sowing (DAP) are considered. In the model, any N application at sowing time must be defined for DAP of 1. Thus, DAP equal to 1 means N application at sowing time.

Grain yield with one N application of 5 g N m^{-2} at sowing time was 390 g m^{-2} (Fig. 19.2a). Delayed application after sowing up to 90 DAP resulted in increased yield. Grain yields of 424, 455, and 470 g m^{-2} were simulated with one N application of 5 g N m^{-2} at 30, 60, and 90 DAP, respectively. A DAP of 90 days is about the time of appearance of flag leaf tip. Delayed N application also resulted in less N losses (Fig. 19.2b). Total N loss was 3.5 g N m^{-2} for application time of 90 DAP. The highest N-use ratio was also simulated for 90 DAP treatment, i.e. about 30 g g^{-1} (Fig. 19.2c). The N-use ratio for application at sowing was about 14 g g^{-1}.

Application time of 120 DAP, 4 days before beginning seed growth, resulted in the lowest grain yield, a relatively high N loss, and the least FUR (Fig. 19.2). The reason is that N uptake is ceased at the beginning of seed growth, so there is not enough time for the crop to uptake N.

Exercises

1. Try to parameterize the N-limited model of this chapter for your crops/cultivars. Crop parameter estimates and soil inputs presented in previous chapters can be used as default values, but precise estimates of required temperature units are necessary.
2. Use the model of Exercise 1 to optimize crops/cultivars' yields at your location using different N application amounts and scenarios. A similar method as used in this chapter can be applied. Discuss your results and find physiological/agronomic justification for the results.

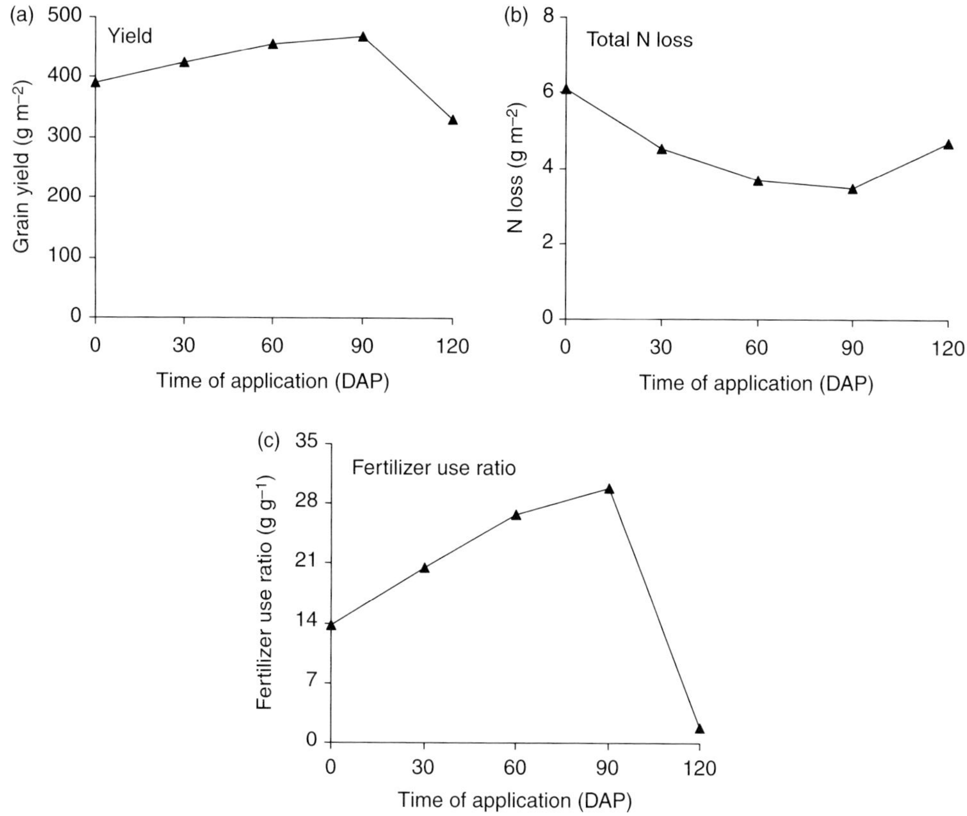

Fig. 19.2. Simulated responses to the time (as days after sowing) of one N fertilizer application with 5 g N m^{-2} in wheat in Gorgan during growing season of 2005/06 under irrigated conditions: (a) grain yield, (b) total N losses via volatilization, denitrification, and leaching (from top 600 mm layer), and (c) fertilizer use efficiency as ratio of extra yield (fertilized yield minus non-fertilized control yield) to the amount of N fertilizer.

3. Prepare a list of applications for the N-limited model.

4. Use the model of Exercise 1 to simulate crop yield response to N under current and future climates. Find physiological/agronomic reasons for the simulation results.

5. Try to use the model of Exercise 1 for other applications you have listed in Exercise 3.

Appendix I: A Practical Guide for Model Troubleshooting

Potential Production Model

On many occasions the results of model simulations do not appear to match observations. First, it should be remembered that there is variability in the observations. What is the quality of the observations, and are the simulation results within a reasonable error limit of the observations? Second, as discussed previously, the model is not reality and cannot be expected to provide results that match identically with observations. Hence, there is a need to establish criteria before beginning the simulations so that the acceptability of the simulation results can be judged. It is likely that agreement between simulation and observation results should not be expected to be any better than within 10 or 15%.

However, once simulation results are examined it may be clear that there is a major failure in the model to reproduce acceptable comparisons with observations. How do you approach the model results to attempt to isolate the model component that may be the source of the failure?

For most situations, isolating the problem may be achieved more quickly if there is a systematic approach to examining the model. In this section, a practical guide for such a systematic examination of the model is provided.

Many times, much of the failure of the model simulations to produce acceptable results can be traced to one of three general problems. These potential problems need to be checked one-by-one and in sequence:

1. Environmental data need to be checked. Environmental data include weather data, irrigation records, and soil input data in the case of water- and nitrogen-limited models. Confirm that weather data that are used are representative of the season and experimental site being simulated. Using weather data from a distant weather station and applying estimated solar radiation especially from temperature data can be sources of error in model predictions.

2. Errors in measurement and recording of observations may exist. Experimental data need to be examined closely for possible errors. Simple problems such as failure to completely dry plant samples at the proper temperature and for sufficient duration may result in incomplete drying and an overestimate of crop dry matter.
3. Error in the input parameters used in the simulation can obviously result in divergence between simulation and observation. A step-by-step procedure is explained below to identify difficulties in the input parameters. This procedure is usually most effective if it is performed in the sequence outlined below.

(i) Do predicted days to phenological stages fit observed data?
If predicted days from one phenological stage to another is shorter or longer than observed days, this is primarily due to lower or greater temperature unit requirements for that stage. Cardinal temperatures are conservative and reliable estimates can be found from references, so these parameters are not usually a major source of error. In addition to examining the input temperature unit duration of phenological stages, an important possibility is that photoperiod must be included in the simulation of phenological development. The inclusion of photoperiod is discussed in Chapter 7.

(ii) Are predicted node numbers on main stem close to those observed?
Underestimation or overestimation of the input phyllochron is the main reason for failure in predictions of main-stem node/leaf numbers. One possibility is that phyllochron cannot be left as a constant but needs to vary with crop development as indicated for wheat in Fig. 9.7. In this example, the model may predict a greater number of main-stem nodes/leaves during early vegetative growth and lower number of main-stem nodes/leaves during later vegetative growth. In such cases, two values of phyllochron should be used in the model with proper change in model codes.

(iii) Does predicted LAI during the main phase of leaf area development match observed LAI?
Matching LAI simulation results with observations can be one of the most challenging outputs of a model. Much of the reason is because of the high variability of measured leaf area among plants. Fortunately, models are not too sensitive to leaf area achieved in the latter part of the growing season when higher levels of LAI intercept most of the solar radiation even if LAI varies. However, earlier in the season discrepancies in LAI can indicate a major problem in the ability to simulate a crop. In the leaf area development presented in Chapter 9, LAI development was simulated using a power function (Eqns 9.3 and 9.4). Underestimation of LAI is an indication of underestimation of the exponent (i.e. PLAPOW) in the equations. Inversely, using a higher value for the exponent results in overestimation of LAI. One possibility is that PLAPOW may be sensitive to plant density, and simulations can be improved by including this effect.

(iv) Does predicted LAI during the period from termination of leaf growth on main stem (TLM) to beginning of seed growth (BSG) match observed LAI?
LAI during this stage is predicted using daily increase in leaf dry matter (GLF) and specific leaf area (SLA). Higher than observed values of LAI during the stage is sensitive to both of these variables and to the leaf dry matter partitioning coefficient during this phase (FLF2).

(v) Is predicted total crop mass close to observed data?
If prediction of LAI index is reasonable, any divergence in prediction of total (aboveground) crop mass can be ascribed to error in estimation of radiation use efficiency (RUE), extinction coefficient (KPAR), or both. However, both of these terms are conservative and potential values for each of these are not likely to vary a great deal. However, KPAR can be influenced by canopy architecture, especially horizontal leaf distribution, so the value of KPAR should be reviewed first. KPAR may be dependent on crop LAI (Chapter 10), thus it could be necessary to vary KPAR as a function of LAI.

(vi) Is predicted ratio of leaf-to-stem dry matter in accordance with observed ratio at termination of leaf development on the main stem (TLM)?
Sometimes, crop mass predictions during the period before TLM are acceptable but the leaf-to-stem dry matter ratio at TLM is not correct. This can result due to misappropriate estimation of leaf partitioning coefficients during the main phase of leaf area development before TLM, i.e. FLF1A and FLF1B. Crop mass at the turning point from FLF1A to FLF1B is fairly constant. Thus, estimation of FLF1A and FLF1B should be improved. Using higher values for the coefficients results in overestimation of leaf dry weight and leaf to stem dry matter ratio.

(vii) Is the predicted ratio of leaf-to-stem dry matter in accordance with the observed ratio at the beginning of seed growth (BSG)?
If leaf-to-stem dry matter at BSG is not in reasonable accordance with the observed ratio, improper estimate of FLF2 is likely the reason. Overestimating FLF2 results in overestimating the ratio and vice versa. However, if LAI predictions from TLM to BSG are acceptable but leaf-to-stem dry matter ratio at BSG is not, this indicates that FLF2 has been wrongly estimated and a biased estimation of SLA has masked the problem. Therefore, both FLF2 and SLA need to be checked again.

(viii) Is predicted crop yield appropriate?
Based on the procedure used to calculate crop yield, improper estimation of crop yield is due to improper estimates of fraction translocation of dry matter to the grains (FRTRL) and grain conversion coefficient (GCC) or both, provided that other model predictions mentioned in previous steps are correct. Lower values used for FRTRL and GCC lead to crop yield underestimation. Overestimations of the parameters result in crop yield overestimation. GCC is conservative and reliable estimates can be found from literature or biochemical analysis of grain and vegetative tissues. However, FRTRL is less conservative and even variable depending on crop, cultivar, and the location as discussed in Chapter 11. Defining FRTRL as a function of crop mass at BSG is a better option if the required data are available.

Water-limited Model

Water is often the critical environmental variable limiting yield, and hence has the greatest impact on crop yield. The prime reasons for a failure of a water-limited model to match simulations are the components accounting for the soil water balance and the crop response to soil water. Again, there is a step-by-step procedure as presented below to help identify the source of the model failure.

1. Environmental data including weather data and soil inputs need to be checked. Soil inputs, particularly, are often difficult to measure and may not be accurately determined for the location being simulated.
2. Results for water-deficit experiments are commonly quite variable due to the inherent variability that exists as a result of soil conditions and application of irrigation. The experimental data need to be examined closely to understand the sources of experimental variability. What is the reliability of these data?
3. Comparison of the ability to simulate the well-watered treatment versus water-deficit treatments helps to focus the problem on the water components of the model. If model predictions are acceptable under potential production conditions but unacceptable under water-limited conditions, then the following step-by-step procedure should be used for model troubleshooting.

(i) Do predicted days to phenological stages fit observed data?
Water limitation may accelerate or retard crop phenological development. If phenological predictions do not match observed dates, acceleration or retardation of development rate due to water deficit should be included in the model. More details can be found in Chapter 15. In some crops like chickpea or wheat, there may be a drought-accelerated development rate except for the sowing–emergence period. On the other hand, in some other crops like sorghum and soybean only grain-filling duration is shortened by water deficit. And, some crops respond to water deficit by retardation of development rate during specific phases.

(ii) Does predicted LAI match observed LAI?
FTSW threshold in response of leaf area development to water deficit (WSSL) is used in the model to adjust leaf area development in response to water deficit. A higher value of WSSL results in more severe retardation of leaf area development under drought. The inverse is also true. Underestimation (a lower value) of WSSL leads to greater increments in LAI under water-limited conditions.

(iii) Is predicted total crop mass close to observed data?
Similar to leaf area, FTSW threshold in response of dry matter production (transpiration) to water deficit (WSSG) is used in the model to correct daily rate of dry matter production for water-limited conditions. Therefore, faster or lower daily rates of dry matter production might be predicted due to incorrectly estimating WSSG. Three other crop parameters affect dry matter production under water-limited conditions by controlling rate of water loss in transpiration and determining water supply to the crop. They are transpiration efficiency coefficient (TEC), daily rate of increase in root depth (GRTD), and maximum effective extraction depth (MEED). Among the three parameters, TEC is conservative but the other two are variable depending on plant genetics and the nature of the soil profile. Inaccurate input information for GRTD and MEED are prime targets for an inability to simulate observed

results. An overestimation of GRTD results in more water available to the crop early in the growing season, or conversely a low GRTD results in more water available from the soil later in the season. High estimates of MEED, however, result in overestimation of total crop mass.

(iv) Is predicted crop yield acceptable?
The reasons for under- or overestimation of crop yield discussed above are also valid for water-limited conditions.

Nitrogen-limited Model

Generally, if model predictions are reasonable under potential production and water-limited conditions, they will be acceptable under N-limited conditions. The reason is that the N-limited model is based on simple parameters that are estimated from simple N measurements in plant tissues. However, if model predictions do not match crop observations under N-limited conditions, the model should be examined using the following steps:

1. Make sure that environmental data including weather data and, particularly, the soil nitrogen inputs are reasonable and representative of the experimental site.
2. Check model predictions under potential production and water-limited conditions to ensure that simulations for the case without nitrogen limitation are reasonable. If the model predictions are acceptable under potential production and water-limited conditions but unacceptable under N-limited conditions, the following step-by-step procedure should be used for model troubleshooting. It should be noted that phenological development and dry matter production and distribution between leaves and stems are assumed independent of plant N conditions (please refer to Chapter 17 for details). Therefore, these aspects are not checked below.

(i) Does predicted LAI match observed LAI?
Assuming that soil N predictions are correct, underestimation of LAI during vegetative growth (pre-seed growth period) may occur due to overestimation of specific leaf N in green leaves (SLNG) and vice versa. During senescence period (seed growth period), however, a higher estimate of specific leaf N in senesced leaves (SLNS) results in less mobilizable N per unit leaf area and higher simulated leaf senescence.

(ii) Is predicted N accumulation acceptable?
Having checked the estimation of SLNG and SLNS in (i), the problem should be related to N concentration in green (SNCG) and senesced (SNCS) stems. Over-prediction of N accumulation is related to overestimation of SNCG and/or SNCS and vice versa.

(iii) Is predicted crop yield underestimated or overestimated?
The reasons for under- or overestimation of crop yield discussed above for the potential production model are also valid here. However, grain N content (GNC) can have a major influence on yield formation under N-limited conditions. High estimates of GNC will result in low simulated grain yield, and vice versa.

Appendix II: Tables for Converting Date to Day of Year (DOY)

Non-leap years:

Day	Jan	Feb	March	April	May	June	July	Aug	Sept	Oct	Nov	Dec
1	1	32	60	91	121	152	182	213	244	274	305	335
2	2	33	61	92	122	153	183	214	245	275	306	336
3	3	34	62	93	123	154	184	215	246	276	307	337
4	4	35	63	94	124	155	185	216	247	277	308	338
5	5	36	64	95	125	156	186	217	248	278	309	339
6	6	37	65	96	126	157	187	218	249	279	310	340
7	7	38	66	97	127	158	188	219	250	280	311	341
8	8	39	67	98	128	159	189	220	251	281	312	342
9	9	40	68	99	129	160	190	221	252	282	313	343
10	10	41	69	100	130	161	191	222	253	283	314	344
11	11	42	70	101	131	162	192	223	254	284	315	345
12	12	43	71	102	132	163	193	224	255	285	316	346
13	13	44	72	103	133	164	194	225	256	286	317	347
14	14	45	73	104	134	165	195	226	257	287	318	348
15	15	46	74	105	135	166	196	227	258	288	319	349
16	16	47	75	106	136	167	197	228	259	289	320	350
17	17	48	76	107	137	168	198	229	260	290	321	351
18	18	49	77	108	138	169	199	230	261	291	322	352
19	19	50	78	109	139	170	200	231	262	292	323	353
20	20	51	79	110	140	171	201	232	263	293	324	354
21	21	52	80	111	141	172	202	233	264	294	325	355
22	22	53	81	112	142	173	203	234	265	295	326	356
23	23	54	82	113	143	174	204	235	266	296	327	357
24	24	55	83	114	144	175	205	236	267	297	328	358
25	25	56	84	115	145	176	206	237	268	298	329	359
26	26	57	85	116	146	177	207	238	269	299	330	360
27	27	58	86	117	147	178	208	239	270	300	331	361

Continued

Continued.

Day	Jan	Feb	March	April	May	June	July	Aug	Sept	Oct	Nov	Dec
28	28	59	87	118	148	179	209	240	271	301	332	362
29	29		88	119	149	180	210	241	272	302	333	363
30	30		89	120	150	181	211	242	273	303	334	364
31	31		90		151		212	243		304		365

Leap-years:

Day	Jan	Feb	March	April	May	June	July	Aug	Sept	Oct	Nov	Dec
1	1	32	61	92	122	153	183	214	245	275	306	336
2	2	33	62	93	123	154	184	215	246	276	307	337
3	3	34	63	94	124	155	185	216	247	277	308	338
4	4	35	64	95	125	156	186	217	248	278	309	339
5	5	36	65	96	126	157	187	218	249	279	310	340
6	6	37	66	97	127	158	188	219	250	280	311	341
7	7	38	67	98	128	159	189	220	251	281	312	342
8	8	39	68	99	129	160	190	221	252	282	313	343
9	9	40	69	100	130	161	191	222	253	283	314	344
10	10	41	70	101	131	162	192	223	254	284	315	345
11	11	42	71	102	132	163	193	224	255	285	316	346
12	12	43	72	103	133	164	194	225	256	286	317	347
13	13	44	73	104	134	165	195	226	257	287	318	348
14	14	45	74	105	135	166	196	227	258	288	319	349
15	15	46	75	106	136	167	197	228	259	289	320	350
16	16	47	76	107	137	168	198	229	260	290	321	351
17	17	48	77	108	138	169	199	230	261	291	322	352
18	18	49	78	109	139	170	200	231	262	292	323	353
19	19	50	79	110	140	171	201	232	263	293	324	354
20	20	51	80	111	141	172	202	233	264	294	325	355
21	21	52	81	112	142	173	203	234	265	295	326	356
22	22	53	82	113	143	174	204	235	266	296	327	357
23	23	54	83	114	144	175	205	236	267	297	328	358
24	24	55	84	115	145	176	206	237	268	298	329	359
25	25	56	85	116	146	177	207	238	269	299	330	360
26	26	57	86	117	147	178	208	239	270	300	331	361
27	27	58	87	118	148	179	209	240	271	301	332	362
28	28	59	88	119	149	180	210	241	272	302	333	363
29	29	60	89	120	150	181	211	242	273	303	334	364
30	30		90	121	151	182	212	243	274	304	335	365
31	31		91		152		213	244		305		366

Appendix III: List of Variables and Their Definitions

Variable	Unit	Definition
ALBEDO	–	Surface albedo
ALPHA	–	An intermediate variable in calculation of photoperiod
ATSW	mm	Actual transpirable soil water
ATSW1	mm	Actual transpirable soil water at top layer
bd	bd	Biological day per calendar day
BD1	$g\ cm^{-3}$	Bulk density of top layer soil
bdBRP	day	Biological day when response to photoperiod begins
bdBRV	day	Biological day when response to vernalization begins
bdTRP	day	Biological day when response to photoperiod terminates
bdTRV	day	Biological day when response to vernalization terminates
BSGDM	$g\ m^{-2}$	Crop above-ground dry matter at beginning seed growth
BSGLAI	$m^2\ m^{-2}$	Leaf area index at beginning seed growth
CALB	–	Crop albedo
CBD	day	Cumulative biological days
CE	mm	Cumulative soil evaporation
CEDSOC	–	An intermediate variable in calculation of photoperiod
CIRGW	mm	Cumulative irrigation water
CLL	$mm\ mm^{-1}$	Volumetric soil water content at crop lower limit
CN	–	Curve number
CNDNIT	$g\ N\ m^{-2}$	Accumulated nitrogen denitrification
CNFERT	$g\ N\ m^{-2}$	Accumulated nitrogen fertilizer applied
CNLEACH	$g\ N\ m^{-2}$	Accumulated nitrogen leached
CNMIN	$g\ N\ m^{-2}$	Accumulated nitrogen mineralization
CNUP	$g\ N\ m^{-2}$	Accumulated nitrogen in above-ground crop organs
CNVOL	$g\ N\ m^{-2}$	Accumulated nitrogen volatilization
cpp	h	Critical photoperiod
CRAIN	mm	Cumulative rainfall

Continued

Continued.

Variable	Unit	Definition
CRUNOF	mm	Cumulative run-off
CTR	mm	Cumulative transpiration
CTU	°C	Cumulative temperature unit
CUMVER	day	Cumulative vernalization days
DAP	–	Days after planting
DAPNF	–	Days after planting when N fertilizer is applied
DDMP	g m^{-2} day^{-1}	Daily dry matter production
DEC	–	An intermediate variable in calculation of photoperiod
DELT	mbar $°K^{-1}$	Slope of saturated vapor pressure versus temperature for daily temperature
DEP1	mm	Depth of top layer in calculation of water and nitrogen dynamics in soil
DEPORT	mm	Current depth of roots
DL	h	Day length
DLAI	m^2 m^{-2} day^{-1}	Daily decrease (death) in leaf area index
DNF	g N m^{-2} day^{-1}	Actual rate of biological nitrogen fixation
DOY	–	Day of year
DRAIN	mm	Amount of drainage water from soil profile
DRAIN1	mm	Amount of drainage water from top layer
DRAINF	–	Drainage factor
DTBRP	day	Days from sowing to beginning response to photoperiod
DTBSG	day	Days to beginning seed growth
DTEMR	day	Days to emergence
DTMAT	day	Days to maturity
DTTLM	day	Days to termination leaf growth on main stem
DTTRP	day	Days from sowing to termination response to photoperiod
DTTSG	day	Days to termination seed growth
DTU	°C	Daily temperature unit
DUL	mm mm^{-1}	Volumetric soil water content at drained upper limit
DYSE	day	Days since last rain and/or irrigation that has wetted soil top layer
EEQ	mm day^{-1}	Evaporation equivalent
EOS	mm day^{-1}	Potential soil evaporation
EOSMIN	mm day^{-1}	Minimum soil evaporation
ETLAI	m^2 m^{-2}	Leaf area index used in calculation of evapotranspiration
EWAT	mm	Amount of water available for crop use due to root growth
EXTR	mm mm^{-1}	Volumetric soil water content available for extraction by crop roots
FG	%	Fraction coarse material in soil
FINT	–	Fraction intercepted radiation (PAR)
FLF1A	g g^{-1}	Partitioning coefficient to leaves during main phase of leaf area development at lower levels of total crop mass
FLF1B	g g^{-1}	Partitioning coefficient to leaves during main phase of leaf area development at higher levels of total crop mass
FLF2	g g^{-1}	Partitioning coefficient to leaves from termination leaf growth on main stem to beginning seed growth

Continued

Continued.

Variable	Unit	Definition
FMIN	–	Fraction soil organic nitrogen available for mineralization
FN	#	Number of N fertilizer application
FROOT1	–	Fraction of soil top layer exploited by crop roots
FRTRL	$g\ g^{-1}$	Fraction crop mass at beginning seed growth which is translocatable to grains
FTSW	–	Fraction transpirable soil water
FTSW1	–	Fraction transpirable soil water at top layer
GCC	$g\ g^{-1}$	Grain conversion coefficient
GLAI	$m^2\ m^{-2}\ day^{-1}$	Daily increase (growth) in leaf area index
GLF	$g\ m^{-2}\ day^{-1}$	Daily increase (growth) in leaf weight
GNC	$g\ g^{-1}$	Grain nitrogen concentration
GRTD	$mm\ day^{-1}$	Daily increase (growth) in root depth
GRTDP	$mm\ day^{-1}$	Potential daily increase (growth) in root depth
GST	$g\ m^{-2}\ day^{-1}$	Daily increase (growth) in stem dry weight
INGRN	$g\ N\ m^{-2}\ day^{-1}$	Daily rate of nitrogen accumulation in grains
iniDMD	–	A variable that indicates if dry matter distribution submodel has been initialized (1) or not (0)
iniDMP	–	A variable that indicates if dry matter production submodel has been initialized (1) or not (0)
iniLAI	–	A variable that indicates if crop LAI submodel has been initialized (1) or not (0)
iniPheno	–	A variable that indicates if phenology submodel has been initialized (1) or not (0)
iniPNB	–	A variable that indicates if plant nitrogen budget submodel has been initialized (1) or not (0)
iniSNB	–	A variable that indicates if soil nitrogen balance submodel has been initialized (1) or not (0)
iniSW	–	A variable that indicates if soil water submodel has been initialized (1) or not (0)
INLF	$g\ N\ m^{-2}$	Daily rate of nitrogen accumulation in leaves
INODE	$\#\ day^{-1}$	Daily increase in main-stem node number
INSOL	$g\ N\ m^{-2}$	Initial soil nitrogen in soil solution
INST	$g\ N\ m^{-2}\ day^{-1}$	Daily rate of nitrogen accumulation in stems
IPATSW	mm	Actual transpirable soil water at sowing time
IRGLVL	–	FTSW when the crop needs to be irrigated; irrigation threshold level
IRGNO	#	Irrigation number
IRGW	mm	Irrigation water
IRUE	$g\ MJ^{-1}$	Radiation use efficiency under optimal growth conditions
KDNIT	day^{-1}	A coefficient indicating soil temperature effect on nitrogen denitrification
KET	–	Extinction coefficient for global solar radiation which is used in calculation of evapotranspiration
KN	–	A coefficient indicating soil temperature effect on nitrogen mineralization
KNMIN	day^{-1}	Exponent for calculation of nitrogen mineralization
KPAR	–	Extinction coefficient for photosynthetically active radiation (PAR)

Continued

Continued.

Variable	Unit	Definition
LAI	$m^2\ m^{-2}$	Leaf area index
LANDA	–	An intermediate variable in calculation of photoperiod
LAT	°	Latitude of location (negative for south latitudes)
MAI	–	Soil moisture availability index (0–1) in soil profile
MAI1	–	Soil moisture availability index (0–1) in top-layer
MAT	–	A variable that indicates crop maturity. It has a value of 0 before crop maturity and 1 at crop maturity
MEED	mm	Maximum effective depth of water extraction from soil by roots
MNORG	$g\ N\ m^{-2}$	Amount of soil organic nitrogen in top layer available for mineralization
MSNN	#	Main-stem node number
MXLAI	$m^2\ m^{-2}$	Maximum leaf area index
MXNUP	$g\ N\ m^{-2}\ day^{-1}$	Maximum uptake (fixation) rate of nitrogen
NCON	$g\ N\ g^{-1}$	Nitrogen concentration in soil solution
NDNIT	$g\ N\ m^{-2}\ day^{-1}$	Daily rate of nitrogen denitrification
NFC	$g\ N\ g^{-1}$	Coefficient of biological nitrogen fixation per unit vegetative mass
NFERT	$g\ N\ m^{-2}$	Amount of N fertilizer
NFERTI	$g\ N\ m^{-2}$	Amount of N fertilizer
NGRN	$g\ N\ m^{-2}$	Accumulated nitrogen in grains
NH4	$g\ m^{-2}$	Amount of nitrogen in NH4 form in soil top layer
nitrogen	–	A variable that specifies nitrogen simulation: 0 for non-N-limited conditions, 1 for legumes N-limited conditions, and 2 for non-legumes N-limited conditions
NLEACH	$g\ N\ m^{-2}\ day^{-1}$	Daily rate of nitrogen leached from the top layer
NLF	$g\ N\ m^{-2}$	Accumulated nitrogen in leaves
NMIN	$g\ N\ m^{-2}\ day^{-1}$	Daily rate of soil nitrogen mineralization
NO3	$g\ m^{-2}$	Amount of nitrogen in NO3 form in soil top layer
NORG	$g\ N\ m^{-2}$	Amount of soil organic nitrogen in top layer
NORGP	%	Percentage soil organic nitrogen
NSOL	$g\ N\ m^{-2}$	Total soil nitrogen in soil solution
NST	$g\ N\ m^{-2}$	Accumulated nitrogen in stems
NSTDF	$g\ N\ m^{-2}$	Amount of nitrogen required to bring stem nitrogen concentration to its target concentration
NUP	$g\ N\ m^{-2}\ day^{-1}$	Daily rate of nitrogen accumulation in crop above-ground organs
NVOL	$g\ N\ m^{-2}\ day^{-1}$	Daily rate of N volatilization
PDEN	$\#\ m^{-2}$	Plant density
PDNF	$g\ N\ m^{-2}\ day^{-1}$	Potential rate of biological nitrogen fixation
pdoy	–	Day of year (1–365) when crop is sown
PET	$mm\ day^{-1}$	Potential evapotranspiration
PHYL	°C per leaf	Phyllochron
Pi	–	π number (3.14)
PLA1	cm^2 per plant	Plant leaf area yesterday
PLA2	cm^2 per plant	Plant leaf area today
PLACON	–	A coefficient (constant) in power relationship between plant leaf area and main-stem node number

Continued

Continued.

Variable	Unit	Definition
PLAPOW	–	A coefficient (exponent) in power relationship between plant leaf area and main-stem node number
pp	h	Photoperiod duration
ppfun	–	Photoperiod function
ppres	–	Photoperiod response: 1=long day plant, 2=short day plant
ppsen	–	Photoperiod sensitivity coefficient
pyear	–	Year of sowing
RAIN	mm	Daily rainfall
RDN	–	A constant to convert degree to radian
RN	–	A coefficient indicating soil moisture effect on nitrogen mineralization
RUE	g MJ^{-1}	Radiation use efficiency adjusted for temperature and water stress
RUNOF	mm	Daily run-off
S	mm	Soil retention parameter in calculation of run-off
SABH	°	Sun angle below horizon
SALB	–	Soil albedo
SAT	mm mm^{-1}	Volumetric soil water content at saturation
SEVP	mm day^{-1}	Actual soil evaporation
SGR	g m^{-2} day^{-1}	Daily increase in seeds (grains) dry matter
SLA	m^2 g^{-1}	Specific leaf area
SLNG	g N m^{-2}	Specific leaf nitrogen in green leaves (target)
SLNS	g N m^{-2}	Specific leaf nitrogen in senesced leaves (minimum)
SNAVL	g N m^{-2}	Soil available nitrogen
SNCG	g N g^{-1}	Stem nitrogen concentration in green stems (target)
SNCS	g N g^{-1}	Stem nitrogen concentration in senesced stems (minimum)
SNOMLT	cm	Depth of melted snow
SNOW	cm	Snow depth
SOCRA	–	An intermediate variable in calculation of photoperiod
SOILM	g m^{-2}	Soil weight in top layer
SOLDEP	mm	Soil depth
SRAD	MJ m^{-2} day^{-1}	Solar radiation
SWER		A coefficient in run-off calculation that depends on soil water
TALSOC	–	An intermediate variable in calculation of photoperiod
TBD	°C	Base temperature for development
TBRUE	°C	Base temperature for dry matter production
TBVER	°C	Base temperature for vernalization
TCD	°C	Ceiling temperature for development
TCFRUE	°C	A correction factor of radiation use efficiency for daily temperature
TCRUE	°C	Ceiling temperature for dry matter production
TCVER	°C	Ceiling temperature for vernalization
TD	°C	Air temperature during the day
TEC	Pa	Transpiration efficiency coefficient
tempfun	–	Temperature function (0–1) for development
TMAX	°C	Maximum temperature
TMAXCR	°C	Crown maximum temperature

Continued

Continued.

Variable	Unit	Definition
TMIN	°C	Minimum temperature
TMINCR	°C	Crown minimum temperature
TMP	°C	Average daily temperature
TMPCR	°C	Crown mean temperature
TMPS	°C	Soil temperature
TP1D	°C	Lower optimum temperature for development
TP1RUE	°C	Lower optimum temperature for dry matter production
TP1VER	°C	Lower optimum temperature for vernalization
TP2D	°C	Upper optimum temperature for development
TP2RUE	°C	Upper optimum temperature for dry matter production
TP2VER	°C	Upper optimum temperature for vernalization
TR	mm	Daily transpiration
TR1	mm	Daily transpirational water that is uptaken from top layer
TRANSL	$g\ m^{-2}\ day^{-1}$	Daily rate of dry matter translocation from vegetative organs to grains
TRLDM	$g\ m^{-2}$	Total crop mass at beginning seed growth which is translocateable to grains
TTSW	mm	Total transpirable soil water
TTSW1	mm	Total transpirable soil water at top layer
tuBNF	°C	Temperature unit from sowing to beginning biological nitrogen fixation (in legume crops)
tuBRG	°C	Temperature unit from sowing to beginning root growth
tuBSG	°C	Temperature unit from sowing to beginning seed growth
tuBSGTSG	°C	Temperature unit from beginning seed growth to termination seed growth
tuEMR	°C	Temperature unit from sowing to emergence
tuEMRTLM	°C	Temperature unit from emergence to termination leaf growth on main stem
tuMAT	°C	Temperature unit from sowing to maturity
tuSOWEMR	°C	Temperature unit from sowing to emergence
tuTLM	°C	Temperature unit from sowing to termination leaf growth on main stem
tuTLMBSG	°C	Temperature unit from termination leaf growth on main stem to beginning seed growth
tuTRG	°C	Temperature unit from sowing to termination root growth
tuTSG	°C	Temperature unit from sowing to termination seed growth
tuTSGMAT	°C	Temperature unit from termination seed growth to maturity
VDSAT	day	Number of vernalization days which saturates vernalization response
VERDAY	$day\ day^{-1}$	Vernalization day per calendar day
verfun	–	Vernalization function
VOLF	$g\ g^{-1}$	Volatilization factor for a N application
VOLFI	$g\ g^{-1}$	Volatilization factor for a N application
VPD	kPa	Vapor pressure deficit
VPDF	–	A coefficient to calculate VPD; 0.65 for humid and subhumid climates and 0.75 for arid and semi-arid climates
VPTMAX	kPa	Air vapor pressure at maximum temperature

Continued

Continued.

Variable	Unit	Definition
VPTMIN	kPa	Air vapor pressure at minimum temperature
vsen	–	Vernalization sensitivity coefficient
vzres	–	Vernalization response: 1 if crop responds to vernalization. Otherwise it has a value of 0
WAT1	mm	Amount of soil water in top layer
water	–	A variable that specifies soil water simulation: 0 for potential production, 1 for irrigated water-limited conditions, and 2 for rainfed water-limited conditions
WETWAT	mm	Amount of rain and/or irrigation required to wet top layer and bring soil evaporation from stage II to I
WGRN	g m^{-2}	Accumulated grain dry matter
WLF	g m^{-2}	Accumulated leaf dry matter
WLL1	mm	Amount of soil water in top layer at lower limit
WSAT1	mm	Amount of soil water in top layer at saturation
WSFD	–	Water stress factor for development
WSFG	–	Water stress factor for growth (dry matter production)
WSFL	–	Water stress factor for leaf area development
WSSD	–	A coefficient that specifies acceleration or retardation in development in response to water deficit
WSSG	–	FTSW threshold when dry matter production starts to decline
WSSL	–	FTSW threshold when leaf area development starts to decline
WSSN	–	FTSW threshold when nitrogen fixation starts to decline
WST	g m^{-2}	Accumulated stem dry matter
WSTORG	mm	Amount of available soil water below crop root layer
WTOP	g m^{-2}	Accumulated crop (above-ground) dry matter
WTOPL	g m^{-2}	Total crop mass when leaf partitioning coefficient turns from FLF1A to FLF1B
WVEG	g m^{-2}	Accumulated vegetative (leaf + stem) dry matter
XNCON	g N g^{-1}	Nitrogen concentration in soil solution used in calculation of nitrogen denitrification
XNLF	g N m^{-2} day^{-1}	Daily rate of nitrogen mobilized from leaves
XNST	g N m^{-2} day^{-1}	Daily rate of nitrogen mobilized from stems
Yr	–	Year

References

Aase, J.K. and Siddoway, F.H. (1979) Crown depth soil temperatures and winter wheat survival. *Soil Science Society of America Journal* 43, 1229–1233.

Affholder, F., Scopel, E., Neto, J.M. and Capillon, A. (2003) Diagnosis of the productivity gap using a crop model: methodology and case study of small-scale maize production in central Brazil. *Agronomie* 23, 305–325.

Aggarwal, P.K. (1993) Agro-ecological zoning using crop simulation models: characterization of wheat environments of India. In: Penning de Vries, F.W.T., Teng, P.S. and Metselaar, K. (eds) *Systems Approaches for Agricultural Development*. Kluwer Academic Publishers, Dordrecht, the Netherlands, pp. 97–109.

Aggarwal, P.K., Kropff, M.J., Cassman, K.G. and ten Berg, H.F.M. (1997) Simulating genotypic strategies for increasing rice yield potential in irrigated, tropical environments. *Field Crops Research* 51, 5–17.

Aguera, F., Villalobos, F.J. and Orgaz, F. (1997) Evaluation of sunflower genotypes differing in early vigor using a simulation model. *European Journal of Agronomy* 7, 109–118.

Ahmadi, M. (2007) Predicting phenological development in wheat. MSc Thesis, Agronomy Group, Gorgan University of Agricultural Sciences and Natural Resources, Gorgan, Iran.

Albrizio, R. and Steduto, P. (2005) Resource use efficiency of field-grown sunflower, sorghum, wheat and chickpea. I. Radiation use efficiency. *Agricultural and Forest Meteorology* 116, 19–36.

Alcalde, J.A. and Larraín, M.F. (2006) Timing of photoperiod sensitivity in relation to floral initiation in contrasting genotypes of pea (*Pisum sativum* L.). *Field Crops Research* 96, 348–354.

Ali, I., Whisler, F., Iqbal, J., Jenkins, J.N. and Mckinion, J.M. (2004) Soil physical properties web database for GOSSYM and GLYCIM crop simulation models. *Agronomy Journal* 96, 1706–1710.

Allen, R.G., Pereira, L.S., Raes, D. and Smith, M. (1998) Crop evapotranspiration: guidelines for computing crop water requirements. *Irrigation and Drainage Paper 56*. UN-FAO, Rome, Italy.

Amir, J. and Sinclair, T.R. (1991) A model of water limitation on spring wheat growth and yield. *Field Crops Research* 29, 59–69.

Analla, M. (1998) Model validation through the linear regression fit to actual versus predicted values. *Agricultural Systems* 57, 115–119.

Anapalli, S.S., Ma, L., Nielsen, D.C., Vigil, M.F. and Ahuja, L.R. (2005) Simulating planting date effects on corn production using RZWQM and CERES-Maize models. *Agronomy Journal* 97, 58–71.

Andrade, F.H., Uhart, S.A. and Cirilo, A.G. (1993) Temperature affects radiation use efficiency in maize. *Field Crops Research* 32, 17–25.

Andrieu, B., Allirand, J.M. and Jaggard, K. (1997) Ground cover and leaf area index of maize and sugar beet crops. *Agronomie* 17, 315–321.

Arnold, J.G., Srinivasan, R., Muttiah, R.S. and Williams, J.R. (1998) Large area hydrologic modeling and assessment – Part I: model development. *Journal of the American Water Resources Association* 34, 73–89.

Asare, D.K., Sammis, T.W., Assadian, H. and Fowler, J.F. (1992) Evaluating three cotton simulation models under different irrigation regimes. *Agricultural Water Management* 22, 391–407.

Baker, D.E. and Ahern, C.R. (1989) Estimates of effective rooting depth for predicting available water capacity of Burdekin soils, Queensland. *Australian Journal of Soil Research* 27, 439–454.

Baloch, D.M., Karow, R.S., Marx, E., Kling, J.G. and Witt, M.D. (2003) Vernalization studies with Pacific Northwest wheat. *Agronomy Journal* 95, 1201–1208.

Bange, M.P., Hammer, G.L. and Rickert, K.G. (1998) Temperature and sowing date affect the linear increases of sunflower harvest index. *Agronomy Journal* 90, 324–328.

Bannayan, M. and Hoogenboom, G. (2008) Weather analogue: a tool for lead time prediction of daily weather data realizations based on a modified k-Nearest Neighbor approach. *Environmental Modelling and Software* 23, 703–713.

Bannayan, M., Crout, N.M.J. and Hoogenboom, G. (2003) Applying the CERES-Wheat model for real-time forecasting of winter wheat. *Agronomy Journal* 95, 114–125.

Bell, M.A. and Fischer, R.A. (1994) Using yield prediction models to assess yield gains: a case study for wheat. *Field Crops Research* 36, 161–166.

Benli, B., Pala, M., Stockle, C.O. and Oweis, T. (2007) Assessment of winter wheat production under early sowing with supplemental irrigation in a cold highland environment using CropSyst simulation model. *Agricultural Water Management* 93, 45–53.

Binder, J., Graeff, S., Link, J., Claupein, W., Liu, M., Dai, M. and Wang, P. (2008) Model-based approach to quantify production potentials of summer maize and spring maize in the North China Plain. *Agronomy Journal* 100, 862–873.

Bindi, M., Sinclair, T.R. and Harrison, J. (1999) Analysis of seed growth by linear increase in harvest index. *Crop Science* 39, 486–493.

Bindraban, P.S. (1999) Impact of canopy nitrogen profile in wheat on growth. *Field Crops Research* 63, 63–67.

Birch, C.J., Hammer, G.L. and Rickert, K.G. (1999) Dry matter accumulation and distribution in five cultivars of maize (*Zea mays*): relationships and procedures for use in crop modelling. *Australian Journal of Agricultural Research* 50, 513–527.

Bonhomme, R. (2000) Bases and limits to using 'degree.day' units. *European Journal of Agronomy* 13, 1–10.

Boote, K.J. and Pickering, P.K. (1994) Modeling photosynthesis of row crop canopies. *HortScience*, 29, 1423–1434.

Boote, K.J., Jones, J.W. and Pickering, N.B. (1996) Potential uses and limitations of crop models. *Agronomy Journal* 88, 704–710.

Boote, K.J., Jones, J.W., Hoogenboom, G. and Pickering, N.B. (1998) The CROPGRO model for grain legumes. In: Tsuji, G.Y., Hoogenboom, G. and Thornton, P.K. (eds) *Understanding Options for Agricultural Production*. Kluwer Academic Publishers, Dordrecht, the Netherlands, pp. 99–128.

Bostick, W.M., Koo, J., Walen, V.K., Jones, J.W. and Hoogenboom, G. (2004) A web-based data exchange system for crop model applications. *Agronomy Journal* 96, 853–856.

Bouman, B.A.M., Van Keulen, H., Van Laar, H.H. and Rabbinge, R. (1996) The 'School of de Wit' crop growth simulation models: a pedigree and historical overview. *Agricultural Systems* 56, 171–198.

Bouman, B.A.M., Kropff, M.J., Tuong, T.P., Wopereis, M.C.S., ten Berge, H.F.M. and Van Laar, H.H. (2001) *ORYZA2000: Modeling Lowland Rice*. International Rice Research Institute/ Wageningen University and Research Centre, Los Baños (the Philippines)/Wageningen.

Bradley, R.G. and Crout, N.M.J. (1994) *PARCH-User Guide*. Tropical Crops Research Unit, University of Nottingham, Sutton Bonington, UK.

Brisson, N., Gate, P., Gouache, D., Charmet, G., Oury, F.X. and Huard, F. (2010) Why are wheat yields stagnating in Europe? A comprehensive data analysis for France. *Field Crops Research* 119, 201–212.

Bristow, K.L. and Campbell, G.S. (1984) On the relationship between incoming solar radiation and daily maximum and minimum temperature. *Agricultural and Forest Meteorology* 31, 159–166.

Brown, H.E., Moot, D.J. and Teixeira, E.I. (2005) The components of lucerne (*Medicago sativa*) leaf area index respond to temperature and photoperiod in a temperate environment. *European Journal of Agronomy* 23, 348–358.

Brown, H.E., Moot, D.J. and Teixeira, E.I. (2006) Radiation use efficiency and biomass partitioning of lucerne (*Medicago sativa*) in a temperate climate. *European Journal of Agronomy* 25, 319–327.

Brown, H.E., Moot, D., Fletcher, A.L. and Jamieson, P.D. (2009) A framework for quantifying water extraction and water stress responses of perennial lucerne. *Crop and Pasture Science* 60, 785–794.

Caldiz, D.O., Gaspari, F.J., Hverkort, A.J. and Struik, P.C. (2001) Agro-ecological zoning and potential yield of single or double cropping of potato in Argentina. *Agriculture and Forest Meteorology* 109, 311–320.

Calvino, P.A., Sadras, V.O. and Andrade, F.H. (2003) Quantification of environmental and management effects on the yield of late-sown soybean. *Field Crops Research* 83, 67–77.

Campbell, C.A., Jame, Y.W., Akinremi, O.O. and Cabrera, M.L. (1995) Adapting the potentially mineralizable N concept for the prediction of fertilizer N requirements. *Fertilizer Research* 42, 61–75.

Campbell, G.S. and Norman, J.M. (1998) *An Introduction to Environmental Biophysics*. Springer-Verlag, New York.

Carberry, P.S. and Abrecht, D.G. (1991) Tailoring crop models to the semiarid tropics. In: Muchow, R.C. and Bellamy, J.A. (eds) *Climatic Risk in Crop Production: Models and Management for the Semiarid Tropics and Subtropics*. CAB International, Wallingford, UK.

Carberry, P.S., Ranganathan, R., Reddy, L.J., Chauhan, Y.S. and Robertson, M.J. (2001) Predicting growth and development of pigeonpea: flowering response to photoperiod. *Field Crops Research* 69, 151–162.

Carberry, P.S., Hochman, Z., McCown, R.L., Dalgliesh, N.P., Foale, M.A., *et al.* (2002) The FARMSCAPE approach to decision support: farmers', advisers', researchers' monitoring, simulation, communication, and performance evaluation. *Agricultural Systems* 74, 179–220.

Chapman, S.C., Hammer, G.L. and Meinke, H. (1993) A sunflower simulation model: I. model development. *Agronomy Journal* 85, 725–735.

Chapman, S.C., Cooper, M. and Hammer, G.L. (2002) Using crop simulation to interpret broad adaptation and genotype by environment interaction effects for sorghum in water-limited environments. *Australian Journal of Agricultural Research* 53, 1–11.

Chavas, D.R., Izaurralde, R., Thomson, A.M. and Gao, X. (2009) Long-term climate change impacts on agricultural productivity in eastern China. *Agricultural and Forest Meteorology* 149, 1118–1128.

Cheeroo-Nayamuth, B.F. (1999) *Crop Modelling/Simulation: An Overview*. Food and Agricultural Research Council, Réduit, Mauritius.

Chenu, K., Cooper, M., Hammer, G.L., Mathews, K.L., Dreccer, M.F. and Chapman, S.C. (2011) Environment characterization as an aid to wheat improvement: interpreting genotype–environment interactions by modelling water-deficit patterns in North-Eastern Australia. *Journal of Experimental Botany* 62, 1743–1755.

Corwin, D.L., Waggoner, B.L. and Rhoades, J.D. (1991) A functional model of solute transport that accounts for bypass. *Journal of Environmental Quality* 20, 647–658.

Cowan, I.R. (1977) Stomatal behaviour and environment. *Advances in Botanical Research* 4, 117–228.

Dalgliesh, N.P. and Foale, M.A. (2005) *Soil Matters: Monitoring Soil Water and Nutrients in Dryland Farming*. Agricultural Production Systems Research Unit, Toowoomba, Australia.

Dardanelli, D.L., Ritchie, J.T., Calmon, M., Andriani, J.M. and Collino, D.J. (2004) An empirical model for root water uptake. *Field Crops Research* 87, 59–71.

Debaeke, P., Caussanel, J.P., Kiniry, J.R., Kafiz, B. and Mondragon, G. (1997) Modeling crop:weed interactions in wheat with ALMANAC. *Weed Research* 37, 325–341.

Delgado, J.A., Gagliardi, P., Shaffer, M.J., Cover, H. and Hesketh, E. (2010) New tools to assess nitrogen management for conservation of our biosphere. In: Delgado, J.A. and Follett, R.F. (eds) *Advances in Nitrogen Management for Water Quality*. Soil and Water Conservation Society, Ankeny, Iowa, pp. 373–409.

Devi, M.J., Sinclair, T.R., Vadez, V. and Krishnamurthy, L. (2009) Peanut genotypic variation in transpiration efficiency and decreased transpiration during progressive soil drying. *Field Crops Research* 114, 280–285.

de Wit, C.T. (1958) *Transpiration and Crop Yields*. Verslagen van Landbouwkundige Onderzoekingen, No. 64.6. Institute for Biological and Chemical Research on Field Crops and Herbage, Wageningen, the Netherlands.

de Wit, C.T. (1965) Photosynthesis of leaf canopies. *Agricultural Research Report No. 663*. Pudoc, Wageningen, the Netherlands.

de Wit, C.T. (1970) Dynamic concepts in biology. In: Setlik, I. (ed.) *Prediction and Measurement of Photosynthetic Activity*. Pudoc, Wageningen, the Netherlands, pp. 17–23.

de Wit, C.T. (1978) *Simulation of Assimilation, Respiration and Transpiration of Crops* (Simulation Monographs). Pudoc, Wageningen, the Netherlands.

Donatelli, M., Bellocchi, G. and Fontana, F. (2003) RadEst3.00: software to estimate daily radiation data from commonly available meteorological variables. *European Journal of Agronomy* 18, 363–367.

Donatelli, M., Carlini, L. and Bellocchi, G. (2006) A software component for estimating solar radiation. *Environmental Modelling and Software* 21, 411–416.

Doraiswamy, P.C., McCarty, G.W., Hunt, E.R., Yost, R.S., Doumbia, M. and Franzluebbers, A.J. (2007) Modeling soil carbon sequestration in agricultural lands of Mali. *Agricultural Systems* 94, 63–74.

Dueri, S., Calanca, P.L. and Fuhrer, J. (2007) Climate change affects farm nitrogen loss – a Swiss case study with a dynamic farm model. *Agricultural Systems* 93, 191–214.

Duncan, W.G., Loomis, R.S., Williams, W.A. and Hanau, R. (1967) A model for simulating photosynthesis in plant communities. *Hilgardia* 38, 181–205.

Egli, D.B. and Bruening, W. (1992) Planting date and soybean yield: evaluation of environmental effects with a crop simulation model: SOYGRO. *Agricultural and Forest Meteorology* 62, 19–29.

Ellis, R.H., Collinson, S.T., Hudson, D. and Patefield, W.M. (1992) The analysis of reciprocal transfer experiments to estimate the durations of the photoperiod-sensitive and photoperiod-insensitive phases of plant development: an example in soya bean. *Annals of Botany* 70, 87–92.

Ellis, R.H., Lawn, R.J., Summerfield, R.J., Qi, A., Roberts, E.H., *et al.* (1994) Towards the reliable prediction of time to flowering in six annual crops. V: Chickpea. *Experimental Agriculture* 30, 271–282.

Fehr, W.R. and Caviness, C.E. (1977) *Stages of Soybean Development. Special Report No. 80.* Iowa State University Press, Ames, Iowa.

Fick, G.W., Loomis, R.S. and Williams, W.A. (1978) Sugar beet. In: Evans, L.T. (ed.) *Crop Physiology*. Cambridge University Press, Cambridge, UK, pp. 310–354.

Fila, G., Bellocchi, G., Acutis M. and Donatelli, M. (2003) IRENE: a software to evaluate model performance. *European Journal of Agronomy* 18, 369–372.

Flenet, F., Kiniry, J.R., Board, J.E., Westgate, M.E. and Reicosky, D.C. (1996) Row spacing effects on light extinction coefficients of corn, sorghum, soybean, and sunflower. *Agronomy Journal* 88, 185–190.

Fletcher, A.L. and Jamieson, P.D. (2006) Causes of variation in the rate of increase of wheat harvest index. *Field Crops Research* 113, 268–273.

Franzluebbers, A.J. (2010) Depth distribution of soil organic carbon as a signature of soil quality. 19th World Congress of Soil Science, Soil Solutions for a Changing World 1–6 August 2010, Brisbane, Australia. Published on DVD.

Gardner, F.P., Pearce, R.B. and Mitchell, R.L. (1985) *Physiology of Crop Plants*. Iowa State University Press, Ames, Iowa.

Gastal, F. and Lemaire, G. (2002) Nitrogen uptake and distribution in crops: an agronomical and ecophysiological perspective. *Journal of Experimental Botany* 53, 789–799.

Gauch H.G., Jr., Hwang, J.T.G. and Fick, G.W. (2003) Model evaluation by comparison of model-based predictions and measured values. *Agronomy Journal* 95, 1442–1446.

Gershenfeld, N.A. (1998) *The Nature of Mathematical Modeling*. Cambridge University Press, Cambridge, UK.

Ghaderi, F.A., Soltani, A. and Sadeghipour, H.R. (2008) Cardinal temperatures of germination in medicinal pumpkin (*Cucurbita pepo* convar. *Pepo* var *styriaca*), Borago (*Borago officinalis* L.) and black cumin (*Nigella sativa* L.). *Asian Journal of Plant Sciences* 7, 574–578.

Gholipoor, M., Sinclair, T.R. and Prasad, P.V. (2012) Genotypic variation within sorghum in transpiration response to drying soil. *Field Crops Research* (In review).

Gijsman, A.J., Jagtap, S.S. and Jones, J.W. (2003) Wading through a swamp of complete confusion: how to choose a method for estimating soil water retention parameters for crop models. *European Journal of Agronomy* 18, 77–106.

Gijsman, A.J., Thornton, P.K. and Hoogenboom, G. (2007) Using the WISE database to parameterize soil inputs for crop simulation models. *Computers and Electronics in Agriculture* 56, 85–100.

Gilmore E.C., Jr., and Rogers, J.S. (1958) Heat units as a method of measuring maturity in corn. *Agronomy Journal* 50, 611–615.

Godwin, D.C. and Jones, C.A. (1991) Nitrogen dynamics in soil-plant systems. In: Hanks, J. and Ritchie, J.T. (eds) *Modeling Plant and Soil Systems. Agronomy Monograph 31.* American Society of Agronomy, Crop Science Society of America, and Soil Science Society of America, Madison, Wisconsin, pp. 287–321.

Goudriaan, J. (1977) *Crop Micrometeorology: A Simulation Study*. Simulation Monographs. Pudoc, Wageningen, the Netherlands.

Goudriaan, J. (1996) Predicting crop yields under global change. In: Walker, B.H. and Steffen, W. (eds) *Global Change and Terrestrial Ecosystems*. Cambridge University Press, Cambridge, UK.

Goudriaan, J. and Monteith, J.L. (1990) A mathematical function for crop growth based on light interception and leaf area expansion. *Annals of Botany* 66, 695–701.

Goudriaan, J. and van Laar, H.H. (1994) *Modelling Potential Crop Growth Processes*. Kluwer Academic Press, Dordrecht, the Netherlands.

Graves, A.R., Hess, T., Matthews, R.B., Stephens, W. and Mason, T. (2002) Crop simulation models as tools in education. *Journal of Natural Resources and Life Sciences Education* 31, 48–54.

Grimm, S.S., Jones, J.W., Boote, K.J. and Hesketh, J.D. (1993) Parameter estimation for predicting flowering date of soybean cultivars. *Crop Science* 33, 137–144.

Grimm, S.S., Jones, J.W., Boote, K.J. and Hesketh, J.D. (1994) Modeling the occurrence of reproductive stages after flowering for four soybean cultivars. *Agronomy Journal* 86, 31–38.

Grindlay, D.J.C. (1997) Towards an explanation of crop nitrogen demand based on the optimisation of leaf nitrogen per unit leaf area. *Journal of Agricultural Science (Cambridge)* 128, 377–396.

Haefner, J.W. (2005) *Modeling Biological Systems: Principles and Application*. Springer, New York.

Hammer, G.L. and Broad, I.J. (2003) Genotype and environmental effects on dynamics of harvest index during grain filling in sorghum. *Agronomy Journal* 95, 199–206.

Hammer, G.L. and Muchow, R.C. (1994) Assessing climatic risk to sorghum production in water-limited subtropical environments: I. Development and testing of a simulation model. *Field Crops Research* 36, 221–234.

Hammer, G.L. and Wright, G.C. (1994) A theoretical analysis of nitrogen and radiation effects on radiation use efficiency in peanut. *Australian Journal of Agricultural Research* 45, 575–589.

Hammer, G.L., Carberry, P.S. and Muchow, R.C. (1993) Modeling genotype and environmental control of leaf area dynamics in grain sorghum. I. Whole plant level. *Field Crops Research* 33, 293–310.

Hammer, G.L., Sinclair, T.R., Boote, K.J., Wright, G.C., Meinke, H. and Bell, M.J. (1995) A peanut simulation model: I. Model development and testing. *Agronomy Journal* 87, 1085–1093.

Hammer, G.L., Kropff, M.J., Sinclair, T.R. and Porter, J.R. (2003) Future contributions of crop modelling from heuristics and supporting decision making to understanding genetic regulation and aiding crop improvement. *European Journal of Agronomy* 18, 15–31.

Hammer, G.L., van Oosterom, E., McLean, G., Chapman, S.C., Broad, I., *et al.* (2010) Adapting APSIM to model the physiology and genetics of complex adaptive traits in field crops. *Journal of Experimental Botany* 61, 2185–2202.

Hansen, S., Jensen, H.E., Nielsen, N.E. and Svendsen, H. (1991) Simulation of nitrogen dynamics and biomass production using the Danish simulation model DAISY. *Fertilizer Research* 27, 245–259.

Hargreaves, G.H. and Samani, Z.A. (1982) Estimating potential evapotranspiration. *Journal of Irrigation and Drainage Engineering, ASCE* 108 (IR3), 223–230.

Haskett, J.D., Pachepsky, Y.A. and Acock, B. (2000) Effect of climate and atmospheric change on soybean water stress: a study of Iowa. *Ecological Modelling* 135, 265–277.

Haun, J.R. (1973) Visual quantification of wheat development. *Agronomy Journal* 65, 116–119.

Hay, R.K.M. and Porter, J.R. (2006) *The Physiology of Crop Yield*. Wiley-Blackwell, Oxford, UK.

Heng, L.K., Asseng, S., Mejahed, K. and Rusan, M. (2007) Optimizing wheat productivity in two rain-fed environments of the West Asia–North Africa region using a simulation model. *European Journal of Agronomy* 26, 121–129.

Hochman, Z., Dalgliesh, N.P. and Bell, K.L. (2001) Contributions of soil and crop factors to plant available soil water capacity of annual crops on Black and Grey Vertosols. *Australian Journal of Agricultural Research* 52, 955–961.

Hodges, T. (1991) *Predicting Crop Phenology*. CRC Press, Boca Raton, Florida.

Hodges, T., Botner, D., Sakamoto, C. and Haug, J.H. (1987) Using the CERES-Maize model to estimate production of the US Cornbelt. *Agricultural and Forest Meteorology* 40, 293–303.

Hofstra, G., Hesketh, J.D. and Myhre, D.L. (1977) A plastochron model for soybean leaf and stem growth. *Canadian Journal of Plant Science* 57, 167–175.

Holzworth, D. and Hammer, G.L. (1996) *DEVEL: a crop development modeling tool*. Agricultural Production Systems Research Unit, Toowoomba, Australia.

Hunt, L.A., Reynolds, M.P., Sayre, K.D., Rajaram, S., White, J.W. and Yan, W. (2003) Crop modeling and identification of stable coefficients that may reflect significant groups of genes. *Agronomy Journal* 95, 20–31.

IPCC (2007) Summary for policymakers. Climate change 2007: the physical science basis. In: Solomon, S., Qin, D., Manning, M., Chen, Z., Marquis, M., Averyt, K.B., Tignor, M. and

Miller, H.L. (eds) *Contribution of Working Group I to the Fourth Assessment Report of the Intergovernmental Panel on Climate Change*. Cambridge University Press, Cambridge, UK.

Jame, Y.W. and Cutforth, H.W. (2004) Simulating the effect of temperature and sowing depth on germination and emergence of spring wheat. *Agricultural and Forest Meteorology* 124, 207–218.

Jamieson, P.D. and Semenov, M.A. (2000) Modeling nitrogen uptake and redistribution in wheat. *Field Crops Research* 68, 21–29.

Jamieson, P.D., Brooking, I.R., Porter, J.R. and Wilson, D.R. (1995) Prediction of leaf appearance in wheat: a question of temperature. *Field Crops Research* 41, 35–44.

Jamieson, P.D., Semenov, M.A., Brooking, I.R. and Francis, G.S. (1998) Sirius: a mechanistic model of wheat response to environmental variation. *European Journal of Agronomy* 8, 161–179.

Jara, J. and Stockle, C.O. (1999) Simulation of water uptake in maize using different levels of process detail. *Agronomy Journal* 91, 256–265.

Jones, C.A. and Kiniry, J.R. (1986) *CERES-Maize: A Simulation Model of Maize Growth and Development*. Texas A&M University Press, College Station, Texas.

Jones, J.W., Hoogenboom, G., Porter, C.H., Boote, K.J., Batchelor, W.D., *et al.* (2003) The DSSAT cropping system model. *European Journal of Agronomy* 18, 235–265.

Jordan, W.R., Dugas, W.A. and Shouse, P.J. (1983) Strategies for crop improvement for drought-prone regions. *Agricultural Water Management* 7, 281–299.

Kalra, N., Chakraborty, D., Kumar, P.R., Jolly, M. and Sharma, P.K. (2007) An approach to bridging yield gaps, combining response to water and other resource inputs for wheat in northern India, using research trials and farmers' fields data. *Agricultural Water Management* 93, 54–64.

Keating, B.A., Asseng, S., Brown, S.D., Carberry, P.S., Chapman, S., *et al.* (2003) An overview of APSIM, a model designed for farming systems simulation. *European Journal of Agronomy* 18, 267–288.

Keisling, T.C. (1982) Calculation of the length of day. *Agronomy Journal* 74, 758–759.

Kemanian, A., Stockle, C.O., Huggins, D.R. and Viega, L.M. (2007) A simple method to estimate harvest index in grain crops. *Field Crops Research* 103, 208–216.

Kichey, T., Hirel, B., Heumez, E., Dubois, F. and Le Gouis, J. (2007) In winter wheat (*Triticum aestivum* L.), post-anthesis nitrogen uptake and remobilisation to the grain correlates with agronomic traits and nitrogen physiological markers. *Field Crops Research* 102, 22–32.

Kiniry, J.R., Williams, J.R., Gassman, P.W. and Debaeke, P. (1992) A general, process-oriented model for two competing plant species. *Transactions of the American Society of Agricultural Engineers* 35, 801–810.

Kirby, E.J.M., Spink, J.H., Frost, D.L., Sylvester-Bradley, R., Scott, R.K., *et al.* (1999) A study of wheat development in the field: analysis by phases. *European Journal of Agronomy* 11, 63–82.

Ko, J., Piccinni, G., Guo, W. and Steglich, E. (2009) Parameterization of EPIC crop model for simulation of cotton growth in South Texas. *Journal of Agricultural Sciences (Cambridge)* 147, 169–178.

Kobayashi, K. and Salam, M.U. (2000) Comparing simulated and measured values using mean squared deviation and its components. *Agronomy Journal* 92, 345–352.

Koocheki, A., Nassiri, M., Soltani, A., Sharifi, H. and Ghorbani, R. (2006) Effects of climate change on growth criteria and yield of sunflower and chickpea crops in Iran. *Climate Research* 30, 247–253.

Kropff, M.J. and van Laar, H.H. (1993) *Modelling Crop–Weed Interactions*. CAB International, Wallingford, UK.

Large, E.G. (1954) Growth stages in cereals: illustration of the Feeke's scale. *Plant Pathology* 3, 128–129.

Lawn, R.J. and James, A.T. (2011) Application of physiological understanding in soybean improvement. I. Understanding phenological constraints to adaptation and yield potential. *Crop and Pasture Science* 62, 1–11.

Lecoeur, J. and Sinclair, T.R. (1996) Field pea transpiration and leaf growth in response to soil water deficits. *Crop Science* 36, 331–335.
Lecoeur, J. and Sinclair, T.R. (2001) Harvest index increase during seed growth of field pea. *European Journal of Agronomy* 14, 173–180.
Lee, J.H. and Heuvelink, E. (2003) Simulation of leaf area development based on dry matter partitioning and specific leaf area for cut chrysanthemum. *Annals of Botany* 91, 319–327.
Lemaire, G., van Oosterom, E., Sheehy, J., Jeuffroy, M.H., Massignam, A. and Rossato, L. (2007) Is crop N demand more closely related to dry matter accumulation or leaf area expansion during vegetative growth? *Field Crops Research* 100, 91–106.
Li, C., Frolking, S. and Frolking, T.A. (1992) A model of nitrous oxide evolution from soil driven by rainfall events. I. Model structure and sensitivity. *Journal of Geophysical Research* 97, 9759–9776.
Li, C., Farahbakhshazad, N., Jaynes, D.B., Dinnes, D.L., Salas, W. and McLaughlin, D. (2006) Modeling nitrate leaching with a biogeochemical model modified based on observations in a row-crop field in Iowa. *Ecological Modelling* 196, 116–130.
Lindquist, J.L., Arkebauer, T.J., Walters, D.T., Cassman, K.G. and Dobermann, A. (2005) Maize radiation use efficiency under optimal growth conditions. *Agronomy Journal* 97, 72–78.
Liu, J., Williams, J.R., Wang, X. and Yang, H. (2009) Using MODAWEC to generate daily weather data for the EPIC model. *Environmental Modelling and Software* 24, 655–664.
Lizaso, J.I., Batchelor, W.D., Boote, K.J. and Westgate, M.E. (2005) Development of a leaf-level canopy assimilation model for CERES-maize. *Agronomy Journal* 97, 722–733.
Lobell, D.B. and Ortiz-Monasterio, J.I. (2006) Evaluating strategies for improved water use in spring wheat with CERES. *Agricultural Water Management* 84, 249–258.
Loomis, R.S. and Connor, D.J. (1992) *Crop Ecology: Productivity and Management in Agricultural Systems*. Cambridge University Press, Cambridge, UK.
Loomis, R.S. and Williams, W.A. (1963) Maximum crop productivity: an estimate. *Crop Science* 3, 67–72.
Ludlow, M.M. and Muchow, R.C. (1990) A critical evaluation of traits for improving crop yields in water limited environments. *Advances in Agronomy* 43, 107–153.
Ludwig, F. and Asseng, S. (2006) Climate change impacts on wheat production in a Mediterranean environment in Western Australia. *Agricultural Systems* 90, 159–179.
Lugato, E., Zuliani, M., Alberti, G., Vedoveb, G.D., Gioli, B., *et al.* (2010) Application of DNDC biogeochemistry model to estimate greenhouse gas emissions from Italian agricultural areas at high spatial resolution. *Agriculture, Ecosystems and Environment* 139, 546–556.
Ma, L. and Shaffer, M.J. (2001) A review of carbon and nitrogen processes in nine US soil nitrogen dynamics models. In: Shaffer, M.J., Ma, L. and Hansen, S. (eds) *Modeling Carbon and Nitrogen Dynamics for Soil Management*. CRC Press, Boca Raton, Florida, pp. 55–102.
Maddonni, G.A., Otegui, M.E. and Cirilo, A.G. (2001) Plant population density, row spacing and hybrid effects on maize canopy architecture and light attenuation. *Field Crops Research* 72, 183–193.
Major, D.J. and Kiniry, J.R. (1991) Predicting daylength effects on phenological processes. In: Hodges, T. (ed.) *Predicting Crop Phenology*. CRC Press, Boca Raton, Florida, pp. 15–28.
Manschadi, A.M., Christopher, J., deVoil, P. and Hammer, G.L. (2006) The role of root architectural traits in adaptation of wheat to water-limited environments. *Functional Plant Biology* 33, 823–837.
Mayer, D.G., Stuart, M.A. and Swain, A.J. (1994) Regression of real-world data on model output: an appropriate overall test of validity. *Agricultural Systems* 45, 93–104.
Mayr, E. (1982) *The Growth of Biological Thought*. Harvard University Press, Cambridge, Massachusetts.
McGechan, M.B. and Wu, L. (2001) A review of carbon and nitrogen processes in European soil nitrogen dynamics models. In: Shaffer, M.J., Ma, L. and Hansen, S. (eds) *Modeling Carbon and Nitrogen Dynamics for Soil Management*. CRC Press, Boca Raton, Florida, pp. 103–172.

McKinion, J.M., Jenkins, J.N., Akins, D., Turner, S.B., Willers, J.L., Jallas, E. and Whisler, F.D. (2001) Analysis of a precision agriculture approach to cotton production. *Computers and Electronics in Agriculture* 32, 213–228.

McLaren, J.S. and Craigon, J. (1981) A computer assisted learning system for use in the teaching of crop production. *Agricultural Progress* 56, 69–83.

McMaster, G.S. and Wilhelm, W.W. (1997) Growing degree-days: one equation, two interpretations. *Agricultural and Forest Meteorology* 87, 291–300.

McMaster, G.S., White, J.W., Weiss, A., Baenziger, P.S., Wilhelm, W.W., *et al.* (2009) Simulating crop phenological responses to water deficits. In: Ahuja, L.R., Reddy, V.R., Anapalli, S.A. and Yu, Q. (eds) *Modeling the Response of Crops to Limited Water: Recent Advances in Understanding and Modeling Water Stress Effects on Plant Growth Processes*. American Society of Agronomy – Crop Science Society of America – Soil Science Society of America, Madison, Wisconsin, pp. 277–300.

Mearns, L.O., Rosenzweig, C. and Goldberg, R. (1996) The effect of changes in daily and interannual climatic variability on CERES-wheat: a sensitivity study. *Climate Change* 32, 257–292.

Meinke, H. and Hammer, G.L. (1995) A peanut simulation model. II: Assessing regional production potential. *Agronomy Journal* 87, 1093–1099.

Meisinger, J.J. and Randall, J.W. (1991) Estimating nitrogen budgets for soil-crop systems. In: Follett, R.F., Keeney, D.R. and Cruse, R.M. (eds) *Managing Nitrogen for Groundwater Quality*. Soil Science Society of America, Madison, Wisconsin, pp. 85–124.

Mirdavardoost, P. (2009) Quantifying vernalization response in some Iranian wheat cultivars. MSc Thesis, Agronomy Group, Gorgan University of Agricultural Sciences and Natural Resources, Gorgan, Iran.

Monsi, M. and Saeki, T. (1953) Über den Lichtfaktor in den Pflanzengesellschaften und seine Bedeutung für die Stoffproduktion. *Japanese Journal of Botany* 14, 22–52.

Monteith, J.L. (1977) Climate and the efficiency of crop production in Britain. *Philosophical Transactions of the Royal Society B* 281, 277–294.

Monteith, J.L. (1986) How do crops manipulate supply and demand? *Philosophical Transactions of the Royal Society A* 316, 245–259.

Monteith, J.L. (1990) Conservative behavior in the response of crops to water and light. In: Rabbinge, R. (ed.) *Theoretical Production Ecology: Hindsight and Perspectives*. Pudoc, Wageningen, the Netherlands, pp. 3–14.

Monteith, J.L. (1996) The quest for balance in crop modeling. *Agronomy Journal* 88, 695–697.

Monteith, J.L. and Unsworth, H.M. (2007) *Principles of Environmental Physics*, 3rd edn. Academic Press.

Monteith, J.L., Huda, A.K.S. and Midya, D. (1989) RECAP: a resource capture model for sorghum and pearl millet. In: Virmani, S.M., Tandon, H.L.S. and Alagarswamy, G. (eds) *Modeling the Growth and Development of Sorghum and Pearl Millet*. ICRISAT Research Bulletin 12, Patancheru, India.

Montgomery, D.C., Peck, E.A. and Vining, G.G. (2001) *Introduction to Linear Regression Analysis*. Wiley, New York.

Monzon, J.P., Sadras, V.O., Abbate, P.A. and Caviglia, O.P. (2007) Modeling management strategies for wheat–soybean double crops in the south-eastern Pampas. *Field Crops Research* 101, 44–52.

Moot, D.J., Jamieson, P.D., Henderson, A.L., Ford, M.A. and Porter, J.R. (1996) Rate of change in harvest index during grain filling of wheat. *Journal of Agricultural Sciences (Cambridge)* 126, 387–395.

Muchow, R.C. (1988) Effect of nitrogen supply on the comparative productivity of maize and sorghum in a semi-arid tropical environment. III. Grain yield and nitrogen accumulation. *Field Crops Research* 18, 31–43.

Muchow, R.C. and Carberry, P.S. (1993) Designing improved plant types for the semiarid tropics: agronomist's viewpoints. In: Penning de Vries, F.W.T., Teng, P.S. and Metselaar, K. (eds)

Systems Approaches for Agricultural Development. Kluwer Academic Publishers, Dordrecht, the Netherlands, pp. 37–61.

Muchow, R.C. and Sinclair, T.R. (1991) Water deficit effects on maize yields modeled under current and "greenhouse" climates. *Agronomy Journal* 83, 1052–1059.

Muchow, R.C., Hammer, G.L. and Vanderlip, R.L. (1994) Assessing climatic risk to sorghum production in water-limited subtropical environments. II: Effects of planting date, soil water at planting, and cultivar phenology. *Field Crops Research* 36, 235–246.

Muchow, R.C., Cooper, M. and Hammer, G.L. (1996) Characterizing environmental challenges using models. In: Cooper, M. and Hammer, G.L. (eds) *Plant Adaptation and Crop Improvement*. CAB International, Wallingford, UK, pp. 349–364.

Nemecek, T. and Kägi, T. (2007) Life Cycle Inventories of Swiss and European Agricultural Production Systems. Final report ecoinvent V2.0 No. 15a, Agroscope Reckenholz Taenikon Research Station ART. Swiss Centre for Life Cycle Inventories, St-Gallen, Switzerland.

Ney, B., Dore, T. and Sagan, M. (1997) Grain legumes. In: Lemaire, G. (ed.) *Diagnosis on the Nitrogen Status in Crops*. Springer-Verlag, Heidelberg, pp. 107–117.

Ng, N. and Loomis, R.S. (1984) *Simulation of Growth and Yield of the Potato Crop*. Simulation monographs. Pudoc, Wageningen, the Netherlands.

Nikoobin, M. (2009) Quantifying vernalization response in some Iranian canola cultivars. MSc Thesis, Agronomy Group, Gorgan University of Agricultural Sciences and Natural Resources, Gorgan, Iran.

O'Leary, G.J. and Connor, D.J. (1996) A simulation model of the wheat crop in response to water and nitrogen supply: I. Model construction. *Agricultural Systems* 52, 1–29.

O'Leary, G.J. and Connor, D.J. (1998) A simulation study of wheat crop response to water supply, nitrogen nutrition, stubble retention and tillage. *Australian Journal of Agricultural Research* 49, 11–19.

Olesen, J.E., Berntsen, J., Hansen, E.M., Petersen, B.M. and Petersen, J. (2002) Crop nitrogen demand and canopy area expansion in winter wheat during vegetative growth. *European Journal of Agronomy* 16, 279–294.

Palanisamy, S., Penning de Vries, F.T.W., Mohandass, S., Thiyagarajany, T.M. and Karim, A.A. (1993) Simulation in pre-testing of rice genotype in Tamil Nadu. In: Penning de Vries, F.W.T., Teng, P.S. and Metselaar, K. (eds) *Systems Approaches for Agricultural Development*. Kluwer Academic Publishers, Dordrecht, the Netherlands, pp. 63–75.

Passioura, J.B. (1996) Simulation models: science, snake oil, education, or engineering. *Agronomy Journal* 88, 690–694.

Paz, J.O., Batchelor, W.D., Tylka, G.L. and Hartzler, R.G. (2001a) A modeling approach to quantifying the effects of spatial soybean yield limiting factors. *Transactions of the American Society of Agricultural Engineers* 44 (5), 1329–1334.

Paz, J.O., Batchelor, W.D. and Tylka, G.L. (2001b) Estimating potential economic return for variable rate management in soybeans. *Transactions of the American Society of Agricultural Engineers* 44 (5), 1335–1341.

Paz, J.O., Batchelor, W.D. and Pedersen, P. (2004) WebGro: a web-based soybean management decision support system. *Agronomy Journal* 96, 1771–1779.

Pease, C.M. and Bull, J.J. (1992) Is science logical? *BioScience* 42, 293–298.

Pengelly, B.C., Muchow, R.C. and Blamey, F.P.C. (1999) Predicting leaf area development in response to temperature in three tropical annual forage legumes. *Australian Journal of Agricultural Research* 50, 253–259.

Penning de Vries, F.W.T. (1975) Use of assimilates in higher plants. In: Cooper, J.P. (ed.) *Photosynthesis and Productivity in Different Environments*. Cambridge University Press, Cambridge, UK, pp. 459–480.

Penning de Vries, F.W.T., Brunsting, A.H.M. and Van Laar, H.H. (1974) Products, requirements and efficiency of biosynthesis: a quantitative approach. *Journal of Theoretical Biology* 45, 339–377.

Penning de Vries, F.W.T., Jansen, D.M., ten Berge, H.F.M. and Bakema, A. (1989) *Simulation of Ecophysiological Processes of Several Annual Crops*. Simulation Monographs 29, Pudoc, Wageningen, the Netherlands and International Rice Research Institute, Manila, the Philippines.

Peralta, J.M. and Stockle, C.O. (2001) Dynamics of nitrate leaching under irrigated potato rotation in Washington State: a long-term simulation study. *Agriculture, Ecosystems and Environment* 88, 23–34.

Piper, E.L., Boote, K.J., Jones, J.W. and Grimm, S.S. (1996) Comparison of two phenology models for predicting flowering and maturity date of soybean. *Crop Science* 36, 1606–1614.

Porter, J.R. and Gawith, M. (1999) Temperatures and the growth and development of wheat: a review. *European Journal of Agronomy* 10, 23–36.

Porter, J.R., Jamieson, P.D. and Wilson, D.R. (1993) Comparison of the wheat simulation models AFRCWHEAT2, CERES-wheat and SWHEAT for non-limiting conditions of crop growth. *Field Crops Research* 33, 131–157.

Priestley, C.H.B. and Taylor, R.J. (1972) On the assessment of surface heat flux and evaporation using large-scale parameters. *Monthly Weather Review* 100, 81–92.

Purcell, L.C. (2000) Soybean canopy coverage and light interception measurements using digital imagery. *Crop Science* 40, 834–837.

Purcell, L.C. (2003) Comparison of thermal units derived from daily and hourly temperatures. *Crop Science* 43, 1874–1879.

Ranganathan, R., Chauhan, Y.S., Flower, D.J., Robertson, M.J., Sanetra, C. and Silim, S.N. (2001) Predicting growth and development of pigeonpea: leaf area development. *Field Crops Research* 69, 163–172.

Ratliff, L.F., Ritchie, J.T. and Cassel, D.K. (1983) Field-measured limits of soil water availability as related to laboratory-measured properties. *Soil Science Society of America Journal* 47, 770–775.

Ray, J.D. and Sinclair, T.R. (1997) Stomatal closure of maize hybrids in response to drying soil. *Crop Science* 37, 803–807.

Reddy, K.R. (1976) Nitrification-denitrification reactions in flooded soil. Ph.D Thesis, Louisiana State University, Baton Rouge, Louisiana.

Reddy, K.R., Kakani, V.G., McKinion, J.M. and Baker, D.N. (2002) Application of cotton simulation model GOSSYM for crop management, economics, and policy decisions. In: Ahuja, L.R., Ma, L. and Howell, TA. (eds) *Agricultural Systems Models: In Field Research and Technology Transfer*. Lewis Publishers, Boca Raton, Florida, pp. 33–54.

Reyenga, P.J., Howden, S.M., Meinke, H. and Hall, W.B. (2001) Global change impacts on wheat production along an environmental gradient in South Australia. *Environment International* 27, 195–200.

Reynolds, J.F. and Acock, B. (1985) Predicting the response of plants to increasing carbon dioxide: a critique of plant growth models. *Ecological Modelling* 29, 107–129.

Rietveld, M.R. (1978) A new method for estimating the regression coefficients in the formula relating solar radiation to sunshine. *Agricultural and Forest Meteorology* 19, 243–252.

Rinaldi, M. (2004) Water availability at sowing and nitrogen management of durum wheat: a seasonal analysis with the CERES-wheat model. *Field Crops Research* 89, 27–37.

Ritchie, J.T. (1981) Water dynamics in the soil-plant-atmosphere system. *Plant and Soil* 58, 81–96.

Ritchie, J.T. (1991) Wheat phasic development. In: Hanks, R.J. and Ritchie, J.T. (eds) *Modeling Plant and Soil Systems*. Agronomy Monograph No. 31, American Society of Agronomy – Crop Science Society of America – Soil Science Society of America, Madison, Wisconsin, pp. 31–54.

Ritchie, J.T. (1998) Soil water balance and plant water stress. In: Tsuji, G.Y., Hoogenboom, G. and Thornton, P.K. (eds) *Understanding Options for Agricultural Production*. Kluwer Academic Publishers, Dordrecht, the Netherlands pp. 41–54.

Ritchie, J.T. and NeSmith, D.S. (1991) Temperature and crop development. In: Hanks, R.J. and Ritchie, J.T. (eds) *Modeling Plant and Soil Systems*. Agronomy Monograph No. 31, American Society of Agronomy – Crop Science Society of America – Soil Science Society of America, Madison, Wisconsin, pp. 5–29.

Ritchie, J.T., Gerakis, A. and Suleiman, A. (1999) Simple model to estimate field-measured soil water limits. *Transactions of the American Society of Agricultural Engineers* 42, 1609–1614.

Ritchie, S.W., Hanway, J.I. and Benson, G.O. (1992) *How a Corn Plant Develops*. Cooperative Extension Service Special Report No. 48. Iowa State University Press, Ames, Iowa.

Roberts, E.H. and Summerfield, R.J. (1987) Measurement and prediction of flowering in annual crops. In: Atherton, J.G. (ed.) *Manipulation of Flowering*. Butterworth-Heinemann, London, pp. 17–50.

Robertson, M.J. and Fukai, S. (1994) Comparison of water extraction models for grain sorghum under continuous soil drying. *Field Crops Research* 36, 145–160.

Robertson, M.J., Carberry, P.S., Huth, N.I., Turpin, J.E., Probert, M.E., *et al.* (2002a) Simulation of growth and development of diverse legume species in APSIM. *Australian Journal of Agricultural Research* 53, 429–446.

Robertson, M.J., Watkinson, A.R., Kirkegaard, J.A., Holland, J.F., Potter, T.D., Burton, W., *et al.* (2002b) Environmental and genotypic control of time to flowering in canola and Indian mustard. *Australian Journal of Agricultural Science* 53, 793–809.

Robertson, M.J., Sakala, W., Benson, T. and Shamudzarira, Z. (2005) Simulating response of maize to previous velvet bean (*Mucuna pruriens*) crop and nitrogen fertilizer in Malawi. *Field Crops Research* 91, 91–105.

Sadras, V.O. and Milroy, S.P. (1996) Soil-water thresholds for the responses of leaf expansion and gas exchange: a review. *Field Crops Research* 47, 253–266.

Salado-Navarro, L.R. and Sinclair, T.R. (2009) Crop rotations in Argentina: analysis of water balance and yield using crop models. *Agricultural Systems* 102, 11–16.

Saseendran, S.A., Ahuja, L.R., Ma, L., Timlin, D.T., Stockle, C.O., *et al.* (2008) Current water deficit stress simulations in selected agricultural system simulation models. In: Ahuja, L.R., Reddy, V.R., Anapalli, S.A. and Yu, Q. (eds) *Modeling the Response of Crops to Limited Water: Recent Advances in Understanding and Modeling Water Stress Effects on Plant Growth Processes*. American Society of Agronomy – Crop Science Society of America – Soil Science Society of America, Madison, Wisconsin, pp. 1–38.

Saxton, K.E. and Rawls, W.J. (2006) Soil water characteristic estimates by texture and organic matter for hydrologic solutions. *Soil Science Society of America Journal* 70, 1569–1578.

Saxton, K.E., Rawls, W.J., Romberger, J.S. and Papendick, R.I. (1986) Estimating generalized soil-water characteristics from texture. *Soil Science Society of America Journal* 50, 1031–1036.

Semenov, M.A. (2008) Simulation of weather extreme events by stochastic weather generator. *Climate Research* 35, 203–212.

Semenov, M.A. (2009) Impacts of climate change on wheat in England and Wales. *Journal of the Royal Society Interface* 6, 343–350.

Semenov, M.A. and Porter, J.P. (1995) Climatic variability and the modelling of crop yields. *Agriculture and Forest Meteorology* 73, 265–283.

Semenov, M.A. and Stratonovitch, P. (2010) The use of multi-model ensembles from global climate models for impact assessments of climate change. *Climate Research* 41, 1–14.

Serraj, R., Sinclair, T.R. and Purcell, L.C. (1999) Symbiotic N_2 fixation response to drought. *Journal of Experimental Botany* 50, 143–155.

Shibu, M.E., Leffelaar, P.A., van Keulen, H. and Aggarwal, P.K. (2010) LINTUL3, a simulation model for nitrogen-limited situations: application to rice. *European Journal of Agronomy* 32, 255–271.

Shiraiwa, T. and Sinclair, T.R. (1993) Distribution of nitrogen among leaves in soybean canopies. *Crop Science* 33, 804–808.

Silim, S.N. and Saxena, M.C. (1993) Adaptation of spring sown chickpea to the Mediterranean basin: I. Response to moisture supply. *Field Crops Research* 34, 121–136.

Sinclair, T.R. (1984) Leaf area development in field-grown soybeans. *Agronomy Journal* 76, 141–146.

Sinclair, T.R. (1986) Water and nitrogen limitations in soybean grain production: model development. *Field Crops Research* 15, 125–141.

Sinclair, T.R. (1991) Canopy carbon assimilation and crop radiation-use efficiency dependence on leaf nitrogen content. In: Boote, K.J. and Loomis, R.S. (eds) *Modeling Crop Photosynthesis – From Biochemistry to Canopy*. Crop Science Society of America Special Publication No. 19, Madison, Wisconsin, pp. 95–107.

Sinclair, T.R. (1994) Limits to crop yield? In: Boote, K.J., Bennett, J.M., Sinclair, T.R. and Paulsen, G.M. (eds) *Physiology and Determinations of Crop Yield*. American Society of Agronomy, Madison, Wisconsin, pp. 509–532.

Sinclair, T.R. (2000) Model analysis of plant traits leading prolonged crop survival during severe drought. *Field Crops Research 68*, 211–217.

Sinclair, T.R. (2004) Improved carbon and nitrogen assimilation for increased yield. In: Boerma, H.R. and Specht, J.E. (eds) *Soybeans: Improvement, Production, and Uses*. American Society of Agronomy, Madison, Wisconsin, pp. 537–568.

Sinclair, T.R. (2005) Theoretical analysis of soil and plant traits influencing daily plant water flux on drying soils. *Agronomy Journal* 97, 1148–1152.

Sinclair, T.R. (2006) A reminder of the limitations in using Beer's Law to estimate daily radiation interception by vegetation. *Crop Science* 46, 2342–2347.

Sinclair, T.R. (2010) Precipitation: the thousand-pound gorilla in crop response to climate change. In: Hillel, D. and Rosenzweig, C. (eds) *Handbook of Climate Change and Agroecosystems: Impacts, Adaptation, and Mitigation*. World Scientific Books, Singapore.

Sinclair, T.R. (2011) Challenges in breeding for yield increase for drought. *Trends in Plant Science* 16, 289–293.

Sinclair, T.R. and Amir, J. (1992) A model to assess nitrogen limitations on the growth and yield of spring wheat. *Field Crops Research* 30, 1–11.

Sinclair, T.R. and Amir, J. (1996) Model analysis of a straw mulch system for continuous wheat in an arid climate. *Field Crops Research* 47, 33–41.

Sinclair, T.R. and deWit, C.T. (1975) Photosynthate and nitrogen requirements for seed production by various crops. *Science* 189, 565–567.

Sinclair, T.R. and de Wit, C.T. (1976) Analysis of the carbon and nitrogen limitations to soybean yield. *Agronomy Journal* 68, 319–324.

Sinclair, T.R. and Horie, T. (1989) Leaf nitrogen, photosynthesis, and crop radiation use efficiency: a review. *Crop Science* 29, 90–98.

Sinclair, T.R. and Jamieson, P.D. (2006) Grain number, wheat yield, and bottling beer: an analysis. *Field Crops Research* 98, 60–67.

Sinclair, T.R. and Ludlow, M.M. (1985) Who taught plants thermodynamics? The unfulfilled potential of plant water potential. *Australian Journal of Plant Physiology* 12, 213–217.

Sinclair, T.R. and Ludlow, M.M. (1986) Influence of soil water supply on the plant water balance of four tropical grain legumes. *Australian Journal of Plant Physiology* 13, 329–341.

Sinclair, T.R. and Muchow, R.C. (1995) Effect of nitrogen supply on maize yield: I. Modeling physiological responses. *Agronomy Journal* 87, 632–641.

Sinclair, T.R. and Muchow, R.C. (1999) Radiation use efficiency. *Advances in Agronomy* 65, 215–265.

Sinclair, T.R. and Muchow, R.C. (2001) System analysis of plant traits to increase grain yield on limited water supplies. *Agronomy Journal* 93, 263–270.

Sinclair, T.R. and Rawlins, S.L. (1993) Inter-seasonal variation in soybean and maize yields under global environment change. *Agronomy Journal* 85, 406–409.

Sinclair, T.R. and Seligman, N. (1996) Crop modeling: from infancy to maturity. *Agronomy Journal* 88, 698–704.
Sinclair, T.R. and Seligman, N. (2000) Criteria for publishing papers on crop modeling. *Field Crops Research* 68, 165–172.
Sinclair, T.R. and Sheehy, J.E. (1999) Erect leaves and photosynthesis in rice. *Science* 283, 1456–1457.
Sinclair, T.R. and Shiraiwa, T. (1993) Soybean radiation-use efficiency as influenced by nonuniform specific leaf nitrogen distribution and diffuse radiation. *Crop Science* 33, 808–812.
Sinclair, T.R. and Weiss, A. (2010) *Principles of Ecology in Plant Production*. CAB International, Wallingford, UK.
Sinclair, T.R., Muchow, R.C., Ludlow, M.M., Leach, G.J., Lawn, R.J. and Foale, M.A. (1987) Field and model analysis of water deficits on carbon and nitrogen accumulation by soybean, cowpea and black gram. *Field Crops Research* 17, 121–140.
Sinclair, T.R., Kitani, S., Hinson, K., Bruniard, J. and Horie, T. (1991) Soybean flowering date: linear and logistic models based on temperature and photoperiod. *Crop Science* 31, 786–790.
Sinclair, T.R., Muchow, R.C. and Monteith, J.L. (1997) Model analysis of sorghum response to nitrogen in subtropical and tropical environments. *Agronomy Journal* 89, 201–207.
Sinclair, T.R., Farias, J.R., Neumaier, N. and Nepomuceno, A.L. (2003) Modeling nitrogen accumulation and use by soybean. *Field Crops Research* 81, 149–158.
Sinclair, T.R., Gilbert, R.A., Perdomo, R.E., Shine, J.M., Jr., Powell, G. and Montes, G. (2004) Sugarcane leaf area development under field conditions in Florida, USA. *Field Crops Research* 88, 171–178.
Sinclair, T.R., Hammer, G.L. and van Oosterom, E.J. (2005a) Potential yield and water-use efficiency benefits in sorghum from limited maximum transpiration rate. *Functional Plant Biology* 32, 945–952.
Sinclair, T.R., Neumaier, N., Farias, J.R.B. and Nepomuceno, A.L. (2005b) Comparison of vegetative development in soybean cultivars for low-latitude environments. *Field Crops Research* 92, 53–59.
Sinclair, T.R., Messina, C.D., Beatty, A. and Samples, M. (2010) Assessment across the United States of the benefits of altered soybean drought traits. *Agronomy Journal* 102, 475–482.
Soil Survey Staff (2011) US general soil map (STATSGO2), Natural Resources Conservation Service, United States Department of Agriculture. Available at: http://soildatamart.nrcs.usda.gov (accessed 21 March 2011).
Soltani, A. and Galeshi, S. (2002) Importance of rapid canopy closure for wheat production in a temperate sub-humid environment: experimentaion and simulation. *Field Crops Research* 77, 17–30.
Soltani, A. and Hoogenboom, G. (2003a) A statistical comparison of stochastic weather generators WGEN and SIMMETEO. *Climate Research* 24, 215–230.
Soltani, A. and Hoogenboom, G. (2003b) Minimum data requirement for parameter estimation of stochastic weather generators. *Climate Research* 25, 109–119.
Soltani, A. and Hoogenboom, G. (2007) Assessing crop management options with crop simulation models based on generated weather data. *Field Crops Research* 103, 198–207.
Soltani, A. and Sinclair, T.R. (2011) A simple model for chickpea development, growth and yield. *Field Crops Research* 124, 252–260.
Soltani, A., Ghassemi-Golezani, K., Khooie, F.R. and Moghaddam, M. (1999) A simple model for chickpea growth and yield. *Field Crops Research* 62, 213–224.
Soltani, A., Khooie, F.R., Ghassemi-Golezani, K. and Moghaddam, M. (2000) Thresholds for chickpea leaf expansion and transpiration response to soil water deficit. *Field Crops Research* 68, 205–210.
Soltani, A., Khooie, F.R., Ghassemi-Golezani, K. and Moghaddam, M. (2001a) A simulation study of chickpea crop response to limited irrigation in a semiarid environment. *Agricultural Water Management* 49, 225–237.

Soltani, A., Zeinali, E. and Galeshi, S. (2001b) Simulating geophysical fluid dynamics laboratory predicted climate change impacts on rice cropping in Iran. *Journal of Agricultural Science Technology (Tehran)* 3, 81–90.

Soltani, A., Galeshi, S. and Attarbashi, M.R. (2004a) Comparison of two methods for estimating parameters of harvest index increase during seed growth. *Field Crops Research* 89, 369–378.

Soltani, A., Meinke, H. and de Voil, P. (2004b) Assessing linear interpolation to generate daily radiation and temperature data for use in crop simulations. *European Journal of Agronomy* 21, 133–148.

Soltani, A., Torabi, B. and Zarei, H. (2005) Modeling crop yield using a modified harvest index-based approach: application in chickpea. *Field Crops Research* 91, 273–285.

Soltani, A., Robertson, M.J., Torabi, B., Yousefi-Daz, M. and Sarparast, R. (2006a) Modeling seedling emergence in chickpea as influenced by temperature and sowing depth. *Agricultural and Forest Meteorology* 138, 156–167.

Soltani, A., Hammer, G.L., Torabi, B., Robertson, M.J. and Zeinali, E. (2006b) Modeling chickpea growth and development: phenological development. *Field Crops Research* 99, 1–13.

Soltani, A., Robertson, M.J., Mohammad-Nejad, Y. and Rahemi-Karizaki, A. (2006c) Modeling chickpea growth and development: leaf production and senescence. *Field Crops Research* 99, 14–23.

Soltani, A., Robertson, M.J., Rahemi-Karizaki, A., Poorreza, J. and Zarei, H. (2006d) Modeling biomass accumulation and partitioning in chickpea (*Cicer arietinum* L.). *Journal of Agronomy and Crop Science* 192, 379–389.

Soltani, A., Robertson, M.J. and Manschadi, A.M. (2006e) Modeling chickpea growth and development: nitrogen accumulation and use. *Field Crops Research* 99, 24–34.

Soltani, A., Gholipoor, M. and Ghassemi-Golezani, K. (2007) Analysis of temperature and atmospheric CO_2 effects on radiation use efficiency in chickpea (*Cicer arietinum* L.). *Journal of Plant Sciences* 2 (1), 89–95.

Soltani, A., Gholipoor, M. and Torabi, B. (2009) Comparison of thermal time calculated using different time steps for use in crop simulation. 5th International Scientific Conference of Iran and Russia on Agricultural Development Problems, 8–9 October, Saint-Petersburg.

Speath, S.C. and Sinclair, T.R. (1985) Linear increase in soybean harvest index during seed-filling. *Agronomy Journal* 77, 207–211.

Spitters, C.J.T. (1990) Crop growth models: their usefulness and limitations. *Acta Horticulturae* 267, 349–362.

Spitters, C.J.T. and Schapendonk, A.H.C.M. (1990) Evaluation of breeding strategies for drought tolerance in potato by means of crop growth simulation. *Plant and Soil* 123, 193–203.

Stanford, G. and Smith, S.J. (1972) Nitrogen mineralization potentials of soils. *Soil Science Society of America Proceedings* 36, 465–472.

Stanford, G., Dzienia, S. and Vander-Pol, R.A. (1975) Effects of temperature on denitrification rate in soils. *Soil Science Society of America Proceedings* 39, 867–870.

Stapper, M. (1984) *SIMTAG: a simulation model of wheat genotypes*. ICARDA: Allepo, Syria, and University of New England, Armidale, Australia.

Stockle, C.O. and Debaeke, P. (1997) Modeling crop nitrogen requirement: a critical analysis. *European Journal of Agronomy* 7, 161–169.

Stockle, C.O., Martin, S. and Campbell, G.S. (1994) CropSyst, a cropping systems model: water/nitrogen budgets and crop yield. *Agricultural Systems* 46, 335–359.

Stockle, C.O., Donatelli, M. and Nelson, R. (2003) CropSyst, a cropping systems simulation model. *European Journal of Agronomy* 18, 289–307.

Streck, N.A., Weiss, A. and Baenziger, S.P. (2003a) A generalized vernalization response function for winter wheat. *Agronomy Journal* 95, 155–159.

Streck, N.A., Weiss, A., Xuec, Q. and Baenziger, P.S. (2003b) Improving predictions of developmental stages in winter wheat: a modified Wang and Engel model. *Agricultural and Forest Meteorology* 115, 139–150.

Summerfield, R.J. and Roberts, E.H. (1987) Effects of illuminance on flowering in long- and short-day grain legumes: a reappraisal and unifying model. In: Atherton, J.G. (ed.) *Manipulation of Flowering*. Butterworth-Heinemann, London, UK, pp. 203–223.

Sylvester-Bradley, R. and Makepeace, R.J. (1984) A code for stages of development in oilseed rape (*Brassica napus* L.). *Aspects of Applied Biology* 6, 399–419.

Taiz, L. and Zeiger, E. (2010) *Plant Physiology*, 5th edn. Sinauer Associates, Sunderland, Massachusetts.

Tanner, C.B. and Sinclair, T.R. (1983) Efficient water use in crop production: research or re-search? In: Taylor, H.M., Jordan, W.R. and Sinclair, T.R. (eds) *Limitations to Efficient Water Use in Crop Production*. American Society of Agronomy, Crop Science Society of America, and Soil Science Society of America, Madison, Wisconsin, pp. 1–27.

Teh, C.B.S. (2006) *Introduction to Mathematical Modeling of Crop Growth*. Brown Walker Press, Boca Raton, Florida.

Tilman, D. (1999) Global environmental impacts of agricultural expansion: the need for sustainable and efficient practices. *Proceedings of the National Academy of Sciences* 96, 5995–6000.

Tsuji, G.Y., Hoogenboom, G. and Thornton, P.K. (eds) (1998) *Understanding Options for Agricultural Production*. Kluwer Academic Publishers, Dordrecht, the Netherlands.

Turpin, J.E., Robertson, M.J., Hillcoat, N.S. and Herridge, D.F. (2002) Fababean (*Vicia faba*) in Australia's northern grains belt: canopy development, biomass, and nitrogen accumulation and partitioning. *Australian Journal of Agricultural Research* 53, 227–237.

Van Delden, A., Kropff, M.J. and Haverkort, A.J. (2001) Modeling temperature- and radiation-driven leaf area expansion in the contrasting crops potato and wheat. *Field Crops Research* 72, 119–142.

Van Ittersum, M.K., Leffelaar, P.A., van Keulen, H., Kropff, M.J., Bastiaans, L. and Goudriaan, J. (2003) On approaches and applications of the Wageningen crop models. *European Journal of Agronomy* 18, 201–234.

Van Keulen, H. (1975) *Simulation of Water Use and Herbage Growth in Arid Regions*. Simulation Monographs. Pudoc, Wageningen, the Netherlands.

van Oosterom, E.J., Borrell, A.K., Chapman, S.C., Broad, I.J. and Hammer, G.L. (2010) Functional dynamics of the nitrogen balance of sorghum. I. N demand of vegetative plant parts. *Field Crops Research* 115, 29–38.

Viets, F.G. Jr (1965) The plant's need for and use of nitrogen. In: Bartholomew, W.V. and Clark, F.E. (eds) *Soil Nitrogen*. American Society of Agronomy, Madison, Wisconsin, pp. 503–549.

Villalobos, F.J. and Ritchie, J.T. (1992) The effects of temperature on leaf emergence rates of sunflower genotypes. *Field Crops Research* 29, 37–46.

Villalobos, F.J., Hall, A.J., Ritchie, J.T. and Orgaz, F. (1996) OILCROP-SUN: a development, growth and yield model of the sunflower crop. *Agronomy Journal* 88, 403–415.

Vinocur, M.G. and Ritchie, J.T. (2001) Maize leaf development biases caused by air-apex temperature differences. *Agronomy Journal* 93, 767–772.

Wahbi, A. and Sinclair, T.R. (2005) Simulation analysis of relative yield advantage of barley and wheat in an eastern Mediterranean climate. *Field Crops Research* 91, 287–296.

Wallach, D. (2006) Evaluating crop models. In: Wallach, D., Makowski, D. and Jones J.W. (eds) *Working with Dynamic Crop Models: Evaluating, Analyzing, Parameterization and Applications*. Elsevier, Amsterdam, the Netherlands.

Wang, E.L. and Engel, T. (1998) Simulation of phenological development on wheat crops. *Agricultural Systems* 58, 1–24.

Wang, X., Gassman, P.W., Williams, J.R., Potter, S. and Kemanian, A.R. (2008) Modeling the impacts of soil management practices on runoff, sediment yield, maize productivity, and soil organic carbon using APEX. *Soil and Tillage Research* 101, 78–88.

Watts, D.G. and Hanks, R.J. (1978) A soil-water-nitrogen model for irrigated corn on sandy soils. *Soil Science Society of America Journal* 42, 492–499.

Wessolek, W. and Asseng, S. (2006) Trade-off between wheat yield and drainage under current and climate change conditions in northeast Germany. *European Journal of Agronomy* 24, 333–342.

Whisler, F.D., Acock, B., Baker, D.N., Fye, R.F., Hodges, H.F., *et al.* (1986) Crop simulation models in agronomic systems. *Advances in Agronomy* 40, 141–207.

White, J.W. and Hoogenboom, G. (2003) Gene-based approaches to crop simulation: past experiences and future opportunities. *Agronomy Journal* 95, 52–64.

Wilhelm, W.W. and McMaster, G.S. (1995) Importance of the phyllochron in studying development and growth in grasses. *Crop Science* 35, 1–3.

Wilkerson, G.G., Jones, J.W., Boote, K.J., Ingram, K.T. and Mishoe, J.W. (1983) Modeling soybean growth for crop management. *Transactions of the American Society of Agricultural Engineers* 26, 63–73.

Williams, J.R. (1991) Runoff and soil erosion. In: Hanks, R.J. and Ritchie, J.T. (eds) *Modeling Plant and Soil Systems*. Agronomy Monograph No. 31, American Society of Agronomy – Crop Science Society of America – Soil Science Society of America, Madison, Wisconsin, pp. 439–456.

Williams, J.R., Jones, C.A. and Dyke, P.T. (1984) A modeling approach to determining the relationship between erosion and soil productivity. *Transactions of the American Society of Agricultural Engineers* 27, 129–144.

Williams, J.R., Jones, C.A., Kiniry, J.A. and Spanel, D.A. (1989) The EPIC crop growth model. *Transactions of the American Society of Agricultural Engineers* 32, 497–510.

Wilson, D.R., Muchow, R.C. and Murgatroyd, C.J. (1995) Model analysis of temperature and solar radiation limitations to maize potential productivity in a cool climate. *Field Crops Research* 43, 1–18.

Wu, D., Yu, Q., Lu, C. and Hengsdijk, H. (2006) Quantifying production potentials of winter wheat in the North China Plain. *European Journal of Agronomy* 24, 226–235.

Wullschleger, S.D., Lynch, J.P. and Berntson, G.M. (1994) Modeling the belowground response of plants and soil biota to edaphic and climate change: what can we expect to gain? *Plant and Soil* 165, 149–160.

Xie, Y., Kiniry, J.R. and Williams, J.R. (2003) The ALMANAC model's sensitivity to input variables. *Agricultural Systems* 78, 1–16.

Yin, X. and van Laar, H.H. (2005) *Crop Systems Dynamics: an ecophysiological simulation model of genotype-by-environment interactions*. Wageningen Academic Publishers, Wageningen, the Netherlands.

Yin, X., Kropff, M.J., Aggarwal, P.K., Peng, S. and Horie, T. (1997) Optimal preflowering phenology of irrigated rice for high yield potential in three Asian environments: a simulation study. *Field Crops Research* 51, 19–27.

Yin, X., Lantinga, E.A., Schapendonk, A.H.C.M. and Zhong, X. (2003a) Some quantitative relationships between leaf area index and canopy nitrogen content and distribution. *Annals of Botany* 91, 893–903.

Yin, X., Stam, P., Kropff, M.J. and Schapendonk, A.H.C.M. (2003b) Crop modelling, QTL mapping, and their complementary role in plant breeding. *Agronomy Journal* 95, 90–98.

Zadoks, J.C., Chang, T.T. and Konzak, C.F. (1974) A decimal code for the growth stages of cereals. *Weed Research* 14, 415–421.

Index

Page numbers in **bold** indicate figures and tables.